T0135627

Thomas Ruedas

Convection and Melting Processes in a Mantle Plume under a Spreading Ridge, with Application to the Iceland Plume

Bibliografische Information Der Deutschen Bibliothek

Die Deutsche Bibliothek verzeichnet diese Publikation in der Deutschen
Nationalbibliografie; detaillierte bibliografische Daten sind im Internet über
http://dnb.ddb.de abrufbar.

ISBN 3-8325-0627-6

Logos Verlag Berlin
Comeniushof, Gubener Str. 47,
10243 Berlin
Tel.: +49 030 42 85 10 90
Fax: +49 030 42 85 10 92
INTERNET: http://www.logos-verlag.de

Convection and Melting Processes in a Mantle Plume under a Spreading Ridge, with Application to the Iceland Plume

Dissertation
zur Erlangung des Doktorgrades
der Naturwissenschaften

vorgelegt beim Fachbereich 11 (Geowissenschaften/Geographie)
der Johann Wolfgang Goethe-Universität
in Frankfurt am Main

von
Thomas Ruedas
aus Hamburg

Frankfurt 2004
(DF1)

vom Fachbereich 11 (Geowissenschaften/Geographie) der
Johann Wolfgang Goethe-Universität als Dissertation angenommen.

Dekan: Prof. Dr. Ulrich Schmidt

Gutachter: Prof. Dr. Harro Schmeling

Prof. Dr. Wolfgang Jacoby

Datum der Disputation: 14. Mai 2004

To my parents

SIGVRDI STEPHANII TERRARVM HYPERBOREARVM DELINEATIO ANNO 1570
after a copy from "Gronlandia Antiqua" by Þormóður Torfason (1706)

ABSTRACT

Mid-oceanic ridges and plumes are geodynamic key features of the earth which play prominent, though different, roles in igneous petrogenesis and crust formation, and they have therefore received much attention over the past decades. Usually, they do not exist at the same place: mid-oceanic ridges evolve from rifts and are most fundamentally distinguished by their individual spreading velocities, which controls several characteristics of their structure and melting style, but as a whole, they share many features worldwide, in particular their rather uniform crustal thickness of 6–7 km; in contrast, plumes are related to hotspots of the lithosphere, but their depth of origin is still controversial and possibly not confined to a single, unique depth, and the style of melt generation in their upper part strongly depends on their own size and the lithospheric framework they encounter when approaching the surface. A very particular situation arises, when a ridge and a plume merge, because the plume is located under very thin lithosphere in this case, which allows it to rise to very shallow depths and to unfold the capacity for melt production provided by its high temperature to maximum extent, without being oppressed and weakened by a thick lithospheric lid; the melt-producing potential of the ridge and of the plume are superimposed here. This constellation is currently realized in Iceland, which was, and still is, entirely created as a volcanic island resulting from the boost of volcanic activity since the convergence and coalescence of the Iceland plume and northern mid-Atlantic ridge southeast of Greenland some 20 Ma ago. This study focusses on the idealized constellation of a ridge-centered plume similar to present-day Iceland, and strives to contribute to the insight gained from numerical models of mantle convection and melt dynamics developped by several workers during the last decade.

The first chapter develops the mathematical background of the convection of melting mantle additionally undergoing solid-state phase transitions and outlines the influence of some basic material properties as discussed in the literature. It also gives an account of previous work on the ascent of mantle plumes and their interaction with the lithosphere.

In the second chapter, several important physical properties of mantle melts are presented and discussed, and their effect on physical and chemical properties of the bulk mantle as well as physical and compositional aspects of melt generation are summarized in order to give an impression of the extraordinary complexity of the process and with respect to the models of this study and the simplifications made therein; in this context, parameterizations of some particularly important properties and variables for use in the numerical models are given special emphasis. The mathematical fundamentals of melt percolation through the partially molten rock are summarized, and a number of fast melt transport models is reviewed. A formulation for melt flow in a medium with macroscopically anisotropic permeability and an outline of its integration into a convection model is developped. Finally, crust formation and transport are shortly described, and their mathematical formulation as implemented in the numerical models is presented along with a discussion of some analytical solutions.

Chapter III describes and discusses the numerical methods applied for solving the convection, melt segregation, and crust transport equations introduced in chapters I and II; in particular, this includes a series of resolution and accuracy tests and the comparison of two different methods for solving the advection problem, namely upwind and semi-Lagrangian advection. As a result of this comparison, it has been decided to use the semi-Lagrangian method for almost all of the models in this study to take advantage of its greater efficiency and its potential to minimize numerical diffusion. Furthermore, the treatment of rheology and phase transitions and the influence of water as implemented in the modelling program are described. A final section is dedicated to the inclusion or derivation of thermoelastic properties of the mantle from numerical models.

In chapter IV, an extensive literature review and synopsis of the current state of research on the general structure of the Icelandic crust and mantle is given, and an attempt is made to synthesize a comprehensive and conclusive picture of the geodynamical situation and to highlight unresolved issues. The view resulting from this survey serves for the design and evaluation of the models of Iceland generated in this study.

In the fifth chapter, the results of the numerical models produced in this study are presented in two groups; they are discussed in the context of the previous chapters in chapter VI. In the first group, several series of models of increasing complexity attempt to highlight particular general effects of solid-phase transitions, variable thermal expansivity α, different rheological models, water and dehydration, melting enthalpy, the use of different solidus and melting parameterizations, and the density and viscosity of melt on convection and/or melt generation and segregation in the setting of a ridge-centered plume, but not yet always with close consideration of Iceland. The models on phase transitions show that the olivine transitions have a significant impact on the upwelling of a plume in the deep upper mantle, whereas the importance of the transitions of the Al-bearing phases in the shallow upper mantle rather lies in their effect on the geotherm within the melting region and on melt chemistry. The series with different thermal expansivity models – constant, depth-dependent, depth/temperature-dependent, and with varying compositions – highlights the significant variability of α in the upper mantle and gives an idea of its influence on convection dynamics; the models suggest that a simplified compositional model can capture essential, though not all, effects resulting from p–T-dependent thermal expansivity. The models on the effects of different rheologies and the role of water emphasize the substantial change in the style of melting and of mantle convection in the melting zone induced by the presence and subsequent loss of water; the deeper onset of melting influences the amount of melt produced and the spatial distribution of melt production and eruption, and the stiffening of the mantle in the melting plume greatly reduces active upwelling and the productivity of the plume. However, all melting models depend on experimental data and how they are parameterized for use in computations, and the models on melting entropy and the comparisons of different parameterizations reveal that the existing uncertainties in the data and simplifications or assumptions inherent in the parameterizations can result in sometimes substantial differences in certain results of convection/melting models; this underscores the need of more precise experimental foundations and of a good knowledge of mantle petrology and chemistry. The short model series on melt properties indicates that it is worthwhile to consider variations of both density and viscosity of the melt in segregation models, because both lead to noticeable deviations of melt dynamics from the behaviour of melt with constant properties. – The second group of models consists of several model sets which strive to reproduce several features of the Iceland plume in the upper mantle reasonably well to be compared with independent observations as those summarized in chapter IV; in the four main series, the plume's excess temperature, the threshold porosity for the extraction of melt, the radius of the plume, the basal influx velocity of plume material into the numerical model box, the solidus and melting parameterizations, and the water content and associated rheology are varied. The existence of ambiguities between several variables is paid special attention: it turns out that there is a trade-off *e.g.* between temperature, extraction threshold, and solidus and melting parameterizations in terms of crust production and, partly, also in terms of melt production and the extent of the melting zone, which means that there is considerable freedom in several dimensions of parameter space, and a multitude of different parameter combinations can produce quite similar results. Considering all results, it is suggested that, in a mantle with a typical water content and a potential temperature of *ca.* 1400 °C and under the assumption of a melt extraction threshold of 1 %, a plume with an elevated water content and an excess temperature of some 200 K (in 200 km depth) exists beneath Iceland, and that this plume has a volume flux of at least 4, but probably between 5 and 6 km^3/a, given that the radius of its stem is 100–150 km. For two selected models, one dry and one hydrous, the anisotropic permeability pattern was determined with the method developped in chapter II; especially in the dry model, but also in parts of the hydrous one, the orientation of tensile dikes as resulting from the stress field would lead to defocussing of melt in off-axis parts of the melting zone, but in vertical ascent near the axis or in slight focussing in near-surface regions dominated by passive upwelling. Furthermore, assessments of some seismological, electromagnetic, and geochemical observables have been derived from the temperature, porosity, and composition fields resulting from some of the convection/melt models. In the geophysical observables, there are strong differences between water-free and water-bearing models, which suggest that in the latter, the deeper parts of the anomalies, which lie below the melting region, are stronger than in the former; on the other hand, the calculation of the seismic anomaly with respect to a reference representative for mantle beneath normal 20 Ma-old oceanic lithosphere reveals a strong decrease of the low-velocity anomaly within parts of the melting zone of the plume head.

The five appendices contain some supplementary mathematical analyses of issues related to model design, present own parameterizations of solidus and liquidus functions from literature data, provide some more detailed information on a number of material properties referred to in the main text, including tabulations of several datasets used mostly in chapter V, and present a tentative attempt to predict thermoelastic properties in a model mantle on the basis of the algorithm outlined in chapter III.

Mittelozeanische Rücken und Plumes sind geodynamische Schlüsselmerkmale der Erde, die herausragende, wenn auch verschiedene Rollen bei der magmatischen Gesteinsbildung und der Erzeugung der Kruste spielen, und sie haben daher im Lauf der vergangenen Jahrzehnte viel Aufmerksamkeit erhalten. Üblicherweise kommen sie nicht an den gleichen Stellen vor: mittelozeanische Rücken entwickeln sich aus Rifts und sind am grundlegendsten durch ihre individuellen Spreizungsgeschwindigkeiten, die zahlreiche Charakteristika ihrer Struktur und des Stils ihrer Schmelzbildung steuert, unterscheidbar, aber insgesamt haben sie weltweit viele Eigenschaften gemein, insbesondere die ziemlich gleichmäßige Mächtigkeit der von ihnen erzeugten Kruste von 6–7 km. Im Gegensatz dazu stehen Plumes in Beziehung zu Hotspots der Lithosphäre, aber ihre Ursprungstiefe ist noch immer umstritten und möglicherweise nicht auf eine einzige, eindeutige Tiefe beschränkt, und der Charakter der Schmelzbildung in ihrem oberen Teil hängt stark von ihrer Größe und den lithosphärischen Rahmenbedingungen ab, auf die sie treffen, wenn sie sich der Oberfläche nähern; sie unterscheiden sich auch in chemisch-petrologischer Hinsicht vom normalen oberen Mantel.

Eine besondere Situation liegt vor, wenn ein Rücken und ein Plume sich vereinigen, weil der Plume in diesem Fall unter sehr dünner Lithosphäre liegt, was ihm erlaubt, bis in sehr geringe Tiefen aufzusteigen und seine durch seine hohe Temperatur gegebene Kapazität zur Schmelzproduktion in vollem Umfang zu entfalten, ohne durch eine dicke Lithosphären- schicht gehemmt und geschwächt zu werden; das Schmelzerzeugungspotential des Rückens und des Plumes überlagern sich hier, wobei allerdings die individuellen Produktivitäten nicht einfach addiert werden können. Diese Konstellation ist derzeit in einmaliger und exemplari- scher Form in Island realisiert, das vollständig als Vulkaninsel infolge des Schubs vulkanischer Aktivität seit der Konvergenz und Verschmelzung des Island-Plumes und des nördlichen Mit- telatlantischen Rückens südöstlich von Grönland vor etwa 20 Ma geschaffen wurde und immer noch wird. Die vorliegende Studie konzentriert sich auf die idealisierte Konstellation eines rückenzentrierten Plumes ähnlich dem heutigen Island und versucht, einen Beitrag zu den Erkenntnissen zu leisten, die im Lauf des letzten Jahrzehnts von einer Reihe von Arbeits- gruppen mit Hilfe numerischer Modelle von Mantelkonvektion und Schmelzdynamik erreicht wurden.

Das erste Kapitel entwickelt die mathematische Grundlage der Konvektion schmelzenden Mantels, der außerdem Festkörper-Phasenübergänge durchläuft, und umreißt den Einfluß einiger grundlegender Materialeigenschaften, wie sie in der Literatur diskutiert worden sind. Es liefert zudem eine Darstellung früherer Arbeiten über den Aufstieg von Mantelplumes und ihrer Wechselwirkung mit der Lithosphäre.

Im zweiten Kapitel wird eine Reihe wichtiger physikalischer Eigenschaften von Mantel- schmelzen vorgestellt und diskutiert, und ihre Wirkung auf physikalische und chemische Eigenschaften des Mantels insgesamt sowie physikalische und chemisch-petrologische Aspek- te der Schmelzerzeugung werden zusammengefaßt, um einen Eindruck der außerordentlichen Komplexität des Prozesses zu vermitteln, aber auch im Hinblick auf die Modelle dieser Studie und die in ihnen gemachten Vereinfachungen. In diesem Zusammenhang wird besonderes Au- genmerk auf Parametrisierungen einiger besonders wichtiger Eigenschaften und Variablen für die Verwendung in den numerischen Modellen gelegt; es werden dabei auch aus Literaturda- ten eigene Parametrisierungen für den Solidus von Lherzolith (KLB-1) und für die Schmelz- kurve von Forsterit ermittelt, die in folgenden Kapiteln für eigene Modelle verwendet werden. Außerdem werden die mathematischen und gesteinsphysikalischen Grundlagen der Schmelz- perkolation durch das partiell geschmolzene Gestein zusammengefaßt. Dabei werden zum einen die zahlreichen, zum Teil problematischen experimentellen Befunde zur mikroskopi-

schen Struktur partiell geschmolzenen Gesteins und ihre Bedeutung für die Geschwindigkeit
von Zweiphasenströmungen und die maximale Porosität der Mantelmatrix diskutiert, und
zum anderen wird eine Übersicht über eine Anzahl schneller Schmelztransportmodelle, die
in der Literatur vorgeschlagen wurden, gegeben. Da letztere Modelle zum Teil implizieren,
daß die Permeabilität des partiell geschmolzenen Mantels anisotrop ist, werden eine Formulierung für Schmelzströmung in einem Medium mit makroskopisch anisotroper Permeabilität
und ein Schema für ihre Integration in ein Konvektionsmodell entwickelt. Schließlich werden
kurz Krustenbildung und -transport an mittelozeanischen Rücken beschrieben, und es wird
ihre mathematische Formulierung, wie sie in den numerischen Modellen implementiert ist,
zusammen mit einer Diskussion einiger einfacher analytischer Lösungen vorgestellt.

Kapitel III beschreibt und diskutiert die numerischen Methoden, die zur Lösung der
Gleichungen für die Konvektion, die Schmelzsegregation und den Krustentransport verwendet werden; dies schließt insbesondere eine Reihe von Auflösungs- und Genauigkeitstests
und den Vergleich zweier verschiedener Verfahren zur Lösung des Advektionsproblems, das
Upwind-Verfahren und das Semi-Lagrange-Verfahren, ein. Veranlaßt durch das Ergebnis die-
ses Vergleichs wurde entschieden, für fast alle Modelle dieser Studie das Semi-Lagrange-
Verfahren zu verwenden, um dessen größere Effizienz und dessen Potential zur Minimierung
numerischer Diffusion auszunutzen. Desweiteren werden die Behandlung von Rheologie und
Festkörper-Phasenübergängen sowie der Einfluß von Wasser auf Viskosität und Solidustem-
peratur und die Abhängigkeit seiner Konzentration im Gestein vom Schmelzgrad, wie sie im
Modellierungsprogramm implementiert sind, beschrieben. Ein abschließender Abschnitt ist
der Einbeziehung thermoelastischer Eigenschaften des Mantels in numerische Modelle oder
ihrer Ableitung aus ihnen gewidmet.

In Kapitel IV wird eine umfangreiche Literaturübersicht des derzeitigen Standes der
Erforschung der allgemeinen Struktur der isländischen Kruste und des Mantels mit seismi-
schen/seismologischen, gravimetrischen, elektromagnetischen, geologischen und geochemischen Verfahren, Laborexperimenten sowie numerischen und analytischen Studien gegeben,
und es wird der Versuch unternommen, daraus ein umfassendes und schlüssiges Bild der geo-
dynamischen Situation zu gewinnen und zu bewerten sowie ungeklärte Fragen aufzuzeigen.
Die aus dieser Zusammenschau gewonnene Sicht dient dem Design und der Bewertung der
Modelle von Island, die in dieser Studie erzeugt wurden.

Im fünften Kapitel werden in zwei Gruppen die Ergebnisse der numerischen Modelle
vorgestellt, die in dieser Studie hergestellt wurden; sie werden in Kapitel VI im Zusammenhang der vorangegangenen Kapitel diskutiert. In der ersten Gruppe versuchen zahlreiche Serien von Modellen zunehmender Komplexität bestimmte allgemeine Effekte von
Mineralphasenübergängen, variablem thermischen Ausdehnungskoeffizienten α, verschiedenen rheologischen Modellen, Wasser und Dehydration, Schmelzenthalpie, der Anwendung
unterschiedlicher Solidus- und Schmelzparametrisierungen und der Dichte und Viskosität
von Schmelze auf die Konvektion und/oder Schmelzbildung und -segregation in der Konstellation eines rückenzentrierten Plumes hervorzuheben, aber noch nicht immer mit engem
Bezug zu Island. Die Modelle zu Phasenübergängen zeigen, daß die Olivin-Transformationen
einen gewichtigen Einfluß auf den Aufstrom eines Plumes im tiefen oberen Mantel haben,
während die Bedeutung der aluminiumhaltigen Phasen im flachen oberen Mantel eher in
ihrem Effekt auf die Geotherme innerhalb der Schmelzzone und auf die Schmelzchemie liegt.
Die Serie mit unterschiedlichen Modellen des thermischen Ausdehnungskoeffizienten – kon-
stantem, tiefenabhängigem, tiefen-/temperaturabhängigem und für variable Zusammenset-
zungen – zeigt die erhebliche Variabilität von α im oberen Mantel auf und vermittelt einen
Eindruck seines Einflusses auf die Konvektionsdynamik; die Modelle legen nahe, daß ein
vereinfachtes Modell der Zusammensetzung wesentliche, wenn auch nicht alle Effekte, die

aus p–T-abhängigem thermischen Ausdehnungskoeffizienten resultieren, wiedergeben kann. Die Modelle zu den Effekten verschiedener Rheologien und der Rolle des Wassers betonen die grundlegende Veränderung im Charakter des Schmelzens und der Mantelkonvektion in der Schmelzzone, die durch die Anwesenheit und den anschließenden Verlust von Wasser verursacht wird; das tiefere Einsetzen von Schmelzen beeinflußt die produzierte Menge von Schmelze und die räumliche Verteilung der Schmelzerzeugung und Eruption, und die Verfestigung des Mantels im schmelzenden Plume verringert drastisch das aktive Aufströmen und die Produktivität des Plumes. In jedem Fall hängen aber alle Schmelzmodelle von experimentellen Daten und deren Parametrisierung für die Verwendung in Rechnungen ab, und die Modelle zur Schmelzentropie und die Vergleiche verschiedener Parametrisierungen zeigen, daß die bestehenden Unsicherheiten in den Daten und die den Parametrisierungen innewohnenden Vereinfachungen oder Annahmen in zuweilen substantiellen Unterschieden in bestimmten Ergebnissen von Konvektions-/Schmelzmodellen resultieren können; dies unterstreicht den Bedarf an genaueren experimentellen Grundlagen und an einer guten Kenntnis der Zusammensetzung des Mantels. Die kurze Modellserie zu den Eigenschaften der Schmelze deutet darauf hin, daß es sich lohnt, Variationen sowohl der Dichte als auch der Viskosität der Schmelze in Segregationsmodellen zu berücksichtigen, weil beide zu merklichen Abweichungen der Schmelzdynamik von dem Verhalten einer Schmelze mit konstanten Eigenschaften führen.

Die zweite Gruppe von Modellen besteht aus mehreren Modellsätzen, die bestrebt sind, eine Reihe von Eigenschaften des Island-Plumes im oberen Mantel in einer Form zu reproduzieren, die hinreichend für einen Vergleich mit unabhängigen Beobachtungen wie den in Kapitel IV angegebenen ist; in den vier Hauptserien werden die Überschußtemperatur des Plumes, der Porositätsschwellwert für die Extraktion von Schmelze, der Plumeradius, die Einstromgeschwindigkeit von Plumematerial durch den Modellboden in die numerische Modellbox, die Solidus- und Schmelzparametrisierungen und der Wassergehalt und die damit verbundene Rheologie variiert. Dem Vorhandensein von Mehrdeutigkeiten zwischen mehreren Variablen wird dabei besondere Aufmerksamkeit gewidmet: es stellt sich heraus, daß z.B. die Werte für die potielle Temperatur und die Temperaturanomalie und für den Extraktionsschwellwert sowie die Solidus- und die Schmelzparametrisierung im Hinblick auf die Krustenproduktion und teilweise auch im Hinblick auf die Schmelzproduktion und die räumliche Ausdehnung der Schmelzzone in gewissem Rahmen frei wählbar sind, so daß erheblicher Spielraum in mehreren Dimensionen des Parameterraums besteht und eine Vielzahl verschiedener Parameterkombinationen recht ähnliche Resultate liefern kann. Modelle, in denen die Rolle von Wasser nicht berücksichtigt wurde, legen für den Island-Plume eher moderate Temperaturen von 130–140 K über der Temperatur des Hintergrundmantels nahe, da andernfalls eine zu mächtige Kruste über dem Plumezentrum erzeugt würde. Die Modelle, in denen die Effekte von Wasser einbezogen wurden, erfordern dagegen Plumetemperaturen, die einige Dutzend Kelvin höher liegen; da sie die Form der anomalen Kruste besser beschreiben und auch mit geochemischen Rahmenbedingungen besser vereinbar sind, werden sie als im allgemeinen realistischer eingestuft. Desweiteren legen die Modelle in Übereinstimmung mit einer früheren numerischen Untersuchung von Ito *et al.* (1999) die Vermutung nahe, daß frühere Abschätzungen des Volumenflusses des Plumes mit analytischen Modellen für Island zu niedrige Werte geliefert haben.

Unter Berücksichtigung aller Ergebnisse wird vorgeschlagen, daß in einem Mantel mit typischem Wassergehalt und einer potentiellen Temperatur von *ca.* 1400 °C und unter der Annahme eines Schmelzextraktionsschwellwerts von 1 % ein Plume mit erhöhtem Wassergehalt und einer Überschußtemperatur von etwa 200 K (in 200 km Tiefe) unter Island existiert und daß dieser Plume einen Volumenfluß von mindestens 4, wahrscheinlich aber zwischen 5 und 6 $^{km^3}/_a$ bei einem Radius des Plumestamms von 100–150 km hat. Die Aufstromgeschwindigkeit des Plumes nahe dem unteren Rand der Schmelzzone im Plumezentrum liegt bei 15–20 $^{cm}/_a$.

Für zwei ausgewählte Modelle, ein trockenes und ein wasserhaltiges, wurde das Muster der anisotropen Permeabilität mit der in Kapitel II entwickelten Methode bestimmt; besonders im trockenen Modell, aber auch in Teilen des wasserhaltigen, würde die Orientierung von dehnungsinduzierten Dikes, wie sie sich aus dem Spannungsfeld ergibt, zu einer Defokussierung von Schmelze in den achsenfernen Teilen der Schmelzzone, aber zu vertikalem Aufstieg nahe der Achse oder leichter Fokussierung in oberflächennahen Regionen, die von passivem Aufstrom dominiert werden, führen.

Darüber hinaus wurden Abschätzungen einiger seismologischer, elektromagnetischer und geochemischer Observablen von den räumlichen Verteilungen der Temperatur, Porosität und chemischen Zusammensetzung, die sich aus den Konvektions-/Schmelzmodellen ergeben hatten, vorgenommen. In den geophysikalischen Observablen treten starke Unterschiede zwischen wasserfreien und wasserhaltigen Modellen auf, die darauf hinweisen, daß in letzteren die tieferen Teile der Anomalien, die unterhalb der Schmelzzone liegen, stärker sind als in ersteren. Andererseits zeigt die Berechnung der seismischen Anomalie relativ zu einer Referenz, die für Mantel unter normaler 20 Ma alter Lithosphäre repräsentativ ist, eine starke Abnahme der Niedriggeschwindigkeitsanomalie in Teilen der Schmelzzone des Plumekopfes. Die starken Anomalien im ungeschmolzenen tiefen oberen Mantel, die für die wasserhaltigen Modelle vorausgesagt werden, sind größer als die gemeinhin mittels seismischer Tomographie beobachteten, können aber möglicherweise mit ihnen in Einklang gebracht werden, wenn neuere Befunde bestätigt werden, die besagen, daß strahlentomographische Abbildungen tendenziell Anomalien zu schwach oder unscharf darstellen.

Im Hinblick auf die Tatsache, daß der Island-Plume nicht während seiner ganzen Existenz unter dem Mittelatlantischen Rücken, sondern zunächst unter dem Grönländischen Kraton gelegen hat, also seitlich versetzt zum Rücken war, ist die Modellserie um zwei Modelle – eines ohne und eines mit Berücksichtigung der Effekte von Wasser – ergänzt worden, in denen die Schmelz- und Krustenbildung in nicht-rückenzentrierten, aber ortsfesten Plumes betrachtet wird. Die Modelle, die jedoch als vorläufige Ergebnisse betrachtet werden sollten, lassen vermuten, daß die Variationen der Dicke der thermischen Lithosphäre mit der Entfernung zum Spreizungszentrum eine merkliche Reduktion der Krustendicke und vor allem des Volumens der anomalen, plume-generierten Kruste bewirken.

Über das konkrete Beispiel Island hinaus können aus der Gesamtheit aller Modelle der Studie folgende allgemeine Schlußfolgerungen gezogen werden:

– Während thermischer Auftrieb sicher der wichtigste Antrieb für Plumes ist, haben mehrere mit der chemischen Zusammensetzung verbundene Beiträge zum Auftrieb – sowohl durch Schmelzprozesse bedingte als auch durch Festkörper-Phasenübergänge bedingte – lokal einige Bedeutung und sollten in Modellen, die die Simulation von realen Situationen anstreben, berücksichtigt werden. Die Phasenübergänge von aluminiumhaltigen Phasen sind allerdings hauptsächlich für die Geotherme und im Hinblick auf die Chemie der Schmelzen wichtig, aber nicht so sehr für den Auftrieb.

– Neben Druck und Temperatur hat der Wassergehalt und seine Änderung durch Schmelzprozesse einen grundlegenden Einfluß auf die Dynamik von Plumes und den Stil und Umfang der Schmelzbildung im allgemeinen; während noch immer mehr experimentelle Information benötigt wird, um die zahlreichen Einzelheiten der Wirkung von Wasser weiter aufzuklären, kann das Ergebnis, daß der Viskositätsanstieg des Mantels durch

Dehydration und die wasserabhängige Verschiebung des Solidus in Schmelzmodelle einbezogen werden sollten, als robust betrachtet werden.

– Die Studie bestätigt die Bedeutung von Wasser für eine korrekte Interpretation von geophysikalischen Beobachtungen wie seismologischen oder elektromagnetischen Darstellungen des Mantels: die Berücksichtigung von Wasser kann die Form und Stärke von Anomalien grundlegend ändern, und es wäre inkonsistent, Wasser in einem dynamischen Modell einzubeziehen, es aber in Modellen der seismischen Geschwindigkeit oder der elektrischen Leitfähigkeit, die aus ihm abgeleitet werden, zu vernachlässigen.

– Schmelz- und Krustenbildung hängen stark vom Solidus und der Form der Schmelzfunktion ab; verschiedene in der Literatur gefundene oder aus Literaturdaten abgeleitete Parametrisierungen führen zu erheblichen Unterschieden in den Ergebnissen, obwohl sie alle für die Charakterisierung des Schmelzens von Peridotit konzipiert waren. Zusätzliche Unsicherheiten werden durch die Streuung der experimentellen Daten zur Schmelzentropie bzw. -enthalpie eingeführt. Da solche Probleme schon bei einem rein peridotitischen Mantel auftreten, läßt sich vorhersehen, daß das Vorhandensein einer zweiten, eklogitischen Komponente die Mehrdeutigkeit noch verstärken würde, wenn man nicht über eine gute Kenntnis der Zusammensetzung des Quellmaterials und seines Schmelzverhaltens verfügt.

– Unter Berücksichtigung aller genannten Unsicherheiten scheint der folgende Satz von Parametern für das Schmelzen in Mantel mit niedrigem Wassergehalt unter langsam spreizenden mittelozeanischen Rücken angemessen zu sein, wenn man eine Schmelzfunktion annimmt, die konvex von unten ist: eine potentielle Temperatur von ungefähr 1380 °C, ein Extraktionsschwellwert für den schnellen Entzug von Schmelze aus dem Mantel von einigen Zehntelprozent, wobei ein höherer Schwellwert mit einer höheren potentiellen Temperatur einhergeht, und ein Solidus, der einige Dutzend Kelvin unter dem ursprünglich von Hirschmann (2000) hergeleiteten liegt. Die Festlegung dieser Parameter beinhaltet, daß die Schmelzfunktion zwischen dem Solidus von Lherzolith und der Schmelzkurve von Forsterit „gestreckt" wurde, um das fraktioniertes Schmelzen, das in den der Parametrisierung des Schmelzens zugrundeliegenden Experimenten nicht realisiert werden kann, näherungsweise zu beschreiben. Ferner ist zu betonen, daß die angenommene petrologische Zusammensetzung eine rein peridotitische ist, d.h. es ist keine eklogitische Komponente einbezogen worden.

– Die Segregation von Schmelzen führt zu ihrer Umverteilung im partiell geschmolzenen Mantel, was auch geophysikalische Observable beeinflussen kann; um die damit zusammenhängenden Phänomene realistisch zu modellieren, sollten die Dichte und die Viskosität der Schmelze nicht als Konstanten, sondern eher als Funktionen von Druck, Temperatur und Zusammensetzung behandelt werden. Die Segregationsgeschwindigkeiten im Regime der porösen Zweiphasenströmung liegen in der Größenordnung von Zentimetern pro Jahr, bei niedrigen Viskositäten möglicherweise auch einigen Dezimetern pro Jahr. Im Hinblick auf die räumliche Verteilung der Krustenbildung ist es auch wünschenswert, die Umverteilung durch schnellen Fluß in Kanälen mit großräumiger, z.B. aus dem Spannungsfeld zu berechnender Ausrichtung in Betracht zu ziehen, wenngleich die hier angewendete instantane Extraktion aus dem Modell in vertikalen Querschnittsflächen senkrecht zum Rücken und Anlagerung am Spreizungszentrum als erste Näherung ausreichend ist.

Der Hauptteil der Arbeit wird durch fünf Anhänge ergänzt, in denen auf einige Seitenaspekte genauer eingegangen wird. Anhang B enthält einige ergänzende mathematische Analysen der Eigenschaften, die der Quellregion des Plumes in den numerischen Modellen zugewiesen werden, und der Massenbilanz. Die in Kapitel II vorgestellten eigenen Parametrisierungen von Solidus- und Liquidusfunktionen anhand von Literaturdaten und auch die Berechnung der Mineralzusammensetzung des schmelzenden Mantels werden in Anhang C ausführlicher diskutiert und hergeleitet. Appendix D bietet zusätzliche detailliertere Informationen zu einer Reihe von Materialeigenschaften, auf die im Haupttext Bezug genommen wird, einschließlich Auflistungen von verschiedenen Datensätzen, die hauptsächlich in Kapitel V benutzt werden. Der Exkurs über die thermoelastischen Eigenschaften von Mantelgestein und den Algorithmus zu ihrer Berechnung in Kapitel III wird in Anhang E noch einmal aufgegriffen und in einem vorläufigen Versuch, thermoelastische Eigenschaften eines Modellmantels vorauszusagen, zur Anwendung gebracht.

CONTENTS

CHAPTER I

SOME FUNDAMENTALS OF MANTLE CONVECTION AND PLUMES

*For the things of this world cannot
be made known without a knowledge
of mathematics.*

—ROGER BACON

There is common agreement about large-scale convection driven by thermal or compositional differences to take place in the earth's mantle. Although the mantle itself is not liquid in the common sense, the high temperatures and pressures in the earth's deep interior make the rock behave like a very viscous material which is able to flow at large timescales by processes summarized under the concept of creep. Thus, fluid dynamical concepts (*e.g.* Batchelor, 1967) can be used to describe the dynamics of the mantle and to implement algorithms for numerical modelling on mantle convection.

Two different main convection modes seem to exist in the mantle: a large-scale, roughly planar Rayleigh–Bénard convection mode, which accounts for the global flow pattern, and the more local convection of mantle plumes, which emerge from thermal (or chemical) boundary layer instabilities and seem to be fairly fixed in space, compared with lithospheric plate motion; their manifestation at the surface are so-called "hotspots" and associated features such as flood basalt provinces and geoid swells. It is still unclear whether whole-mantle or layered convection prevails in the mantle, particularly in the case of plumes, *i.e.* whether the plumes start from the core–mantle boundary (CMB) (Morgan, 1971) or from a shallower level as the 660 km discontinuity. For instance, geochemical analysis indicates that mid-oceanic ridge basalts (MORB) and plumes sample different reservoirs, of which the latter is more primordial, and seismic tomography suggests that subducted slabs also penetrated the lower mantle, but on the other hand the kinetics and thermodynamics of the phase change from γ-olivine to perovskite is expected to hinder the crossing of mantle currents through the 660 km discontinuity (Christensen and Yuen, 1985) (see section I.2.2). At least in the uppermost mantle, the large-scale flow is partly related to the world-wide volcanic system of the mid-ocean ridges, which are quantitatively the most important source of volcanic material with a production of $18^{km^3}/a$ of basaltic crust and the largest heat leak of the planet; the mantle source material is a relatively uniform peridotite depleted in trace elements. In contrast, plumes only produce $0.5^{km^3}/a$ at present, although the eruption rate of plume-related volcanic structures can show large temporal variations; the source material is a complex mixture of upper and lower mantle material and subducted crust. It seems that a large part of the plume material is not erupted, though, but ponded at the base of the lithosphere, which leads to the assumption that melt extraction is quite inefficient (Olson, 1990). – It should be mentioned that although the plume concept is accepted by most earth scientists, a minority rejects it (*e.g.* Anderson *et al.*, 1992), and different modifications and extensions have been proposed in the course of the last two decades; Courtillot *et al.* (2003) make an attempt to evaluate both models and propose a preliminary compromise. In this study, however, no reason is seen to doubt the existence of the plumes at least in the upper mantle, on which it concentrates.

In this chapter, the fluid-mechanical laws applicable to mantle convection are presented with regard to their implementation and use in the numerical models developped later; as this study deals with mid-ocean ridges and plumes, emphasis lies on phenomena and processes specific for these structures, whereas subduction zones will not be considered.

I.1 BASIC EQUATIONS OF MANTLE CONVECTION

Three basic laws are governing the motion of a fluid: the conservation laws for mass, momentum, and energy.

I.1.1 Conservation of mass

Conservation of mass can be expressed as[1]

$$\frac{1}{\varrho'}\left(\frac{\partial}{\partial t'} + \vec{v}' \cdot \nabla'\right)\varrho' + \nabla' \cdot \vec{v}' = 0 \qquad (\mathrm{I.1})$$

(*e.g.* Peyret and Taylor, 1985; Schubert *et al.*, 2001), where ϱ' is the density and $\vec{v}'$ is the velocity. Quite often in mantle convection, it is assumed that the material is incompressible ($\varrho' = const.$); after non-dimensionalization, eq. I.1 can then be simplified to:

$$\nabla \cdot \vec{v} = 0. \qquad (\mathrm{I.1a})$$

The assumption of incompressibility is also made in the convection models of this study. However, if melt is generated in, and extracted from, the mantle by the mechanisms discussed in section II.4, this conservation law is violated to some extent, because some mass in fact disappears from a certain volume, and it would be more consistent to include the corresponding decrease in volume in the compressibility term on the LHS of eq. I.1. On the other hand, the rather broad extent of the melting region and the fact that only low to moderate melting degrees are generally reached in a large part of it lets one expect the error introduced by this simplification to be minor (see Turcotte and Phipps Morgan, 1992).

I.1.2 Conservation of momentum: Navier–Stokes equation

Conservation of momentum in a flowing medium with bulk and shear viscosity η_b' and η' is expressed by the Navier–Stokes equation:

$$-\nabla'p' + \nabla' \cdot \left\{\left(\eta_b' - \frac{2}{3}\eta'\right)(\nabla' \cdot \vec{v}')\mathbf{1} + \eta'[\nabla'\vec{v}' + (\nabla'\vec{v}')^{\mathrm{T}}]\right\} + \vec{F}' = \varrho'\left(\frac{\partial}{\partial t'} + \vec{v}' \cdot \nabla'\right)\vec{v} \quad (\mathrm{I.2a})$$

(*e.g.* Peyret and Taylor, 1985; Schubert *et al.*, 2001). The LHS represents the pressure forces, viscous deviatoric stresses and some additional body force $\vec{F}$, and the RHS represents the inertial force; $\mathbf{1}$ is the unity tensor. The assumption of incompressibility eq. I.1a can be used to simplify eq. I.2a, and subsequent application of the non-dimensionalization rules for ϱ', η', $\vec{v}'$, p', (x', y', z') and t' in appendix A yields

$$-\nabla p + \nabla \cdot \left\{\eta[\nabla\vec{v} + (\nabla\vec{v})^{\mathrm{T}}]\right\} + \vec{F} = \frac{\varrho_0'\kappa'}{\eta_0'}\varrho\left(\frac{\partial}{\partial t} + \vec{v} \cdot \nabla\right)\vec{v} \qquad (\mathrm{I.2b})$$

---

[1]In most instances non-dimensional variables will be used; see appendix A for the non-dimensionalization rules. If dimensional variables are used in places where confusion might happen, they are primed. All variables used in this study are tabulated in the list of symbols on p. 265ff.

The inverse of the factor $\varrho_0'\kappa'/\eta_0'$ is called Prandtl number Pr and describes the ratio of the kinematic viscosity (momentum diffusivity) η_0'/ϱ_0' and the thermal diffusivity κ'. In the earth, it can be assumed that the inertial forces are negligible and $Pr \to \infty$, because the momentum diffusivity is more than twenty orders of magnitude greater than the thermal, so that the RHS vanishes:

$$-\nabla p + \nabla \cdot \left\{\eta[\nabla\vec{v} + (\nabla\vec{v})^\mathrm{T}]\right\} + \vec{F} = 0 \qquad (\mathrm{I.2c})$$

(*e.g.* Schubert *et al.*, 2001). Note that this simplification makes the Navier–Stokes equation time-independent and removes one non-linear term. This has important implications for solving the equation numerically (see section III.1).

Here and in the following sections, the body force term $\vec{F}$ will be used to introduce a number of driving forces. In mantle dynamics, the most common cause for the onset of convection is an instable thermal layering *e.g.* due to basal heating of the mantle at the CMB, so that the flow is driven by the buoyancy of hot material with reduced density; this condition must hold even if an incompressible medium has been assumed. Thus, density variations have to be allowed for at least at one point, namely, as a buoyancy term in the Navier–Stokes equation, while they can be neglected in the other places; this assumption is called the Boussinesq approximation. With these simplifications and using a reference density ϱ_0' and a vertical temperature difference $\Delta T'$ across the convecting system, eq. I.2c can be written as

$$-\nabla'\tilde{p}' + \nabla' \cdot \left\{\eta'[\nabla'\vec{v}' + (\nabla'\vec{v}')^\mathrm{T}]\right\} - \varrho_0' g'\alpha'(T' - T_0')\vec{e}_z = 0$$

$$\text{or} \qquad -\nabla\tilde{p} + \nabla \cdot \left\{\eta[\nabla\vec{v} + (\nabla\vec{v})^\mathrm{T}]\right\} - Ra\,T\vec{e}_z = 0 \qquad (\mathrm{I.2d})$$

(Schubert *et al.*, 2001), where $Ra = \varrho_0' g'\alpha'\Delta T' z_\mathrm{m}'^3/\eta'\kappa'$ is the Rayleigh number, which gives the ratio of advective and conductive timescales and is a measure of the vigour of thermal convection. Note that the hydrostatic pressure has been subtracted from p to yield $\tilde{p}$, and that a constant, vertically oriented g is assumed, which is reasonable for the upper mantle. Additional possible buoyancy terms are related to compositional heterogeneities and are discussed in more detail in section I.2.1.

The fluid dynamical equations impose certain conditions for (thermal) mantle plumes to form and ascend. The most fundamental prerequisite is the existence of a thermal boundary layer at the bottom of the convecting layer. It is well-established that the D″ at the CMB is such a boundary, and several known plumes seem to have their origin there, but uncertainties concerning the material properties of mantle rock in the transition zone also admit the possibility that they can be initiated at the 660 km discontinuity (spinel–perovskite boundary). In each case, Ra must have a value high enough to cause time-dependent behaviour; this is the case in the earth, where $Ra = 5 \cdot 10^5 \ldots 5 \cdot 10^7$. Instabilities will then develop into plumes in some tens or a few hundred million years, if the residence time of material in the boundary layer is greater than this value. Experiments and theoretical models show that the developping plume will form a growing head connected to the source region by a thinner conduit and that the head will ultimately reach a velocity at which it separates from the tail because mass flux from the source cannot keep pace (Whitehead and Luther, 1975); the flow in these tails is likely to be perturbed in the mantle, which will lead to the development of solitary waves in the diapir (Olson, 1990). – In principle, chemical buoyancy can lead to the formation of plumes as well, and it is possible that the D″ is also of interest in this respect. However, the importance of chemical convection is less well known.

I.1.3 Conservation of energy

The third law formulates the conservation of energy and describes how the temperature T' evolves in time:

$$\left(\frac{\partial}{\partial t'} + \vec{v}' \cdot \nabla'\right) T' - \frac{\alpha'}{\varrho' c_p'} T' \left(\frac{\partial}{\partial t'} + \vec{v}' \cdot \nabla'\right) p' =$$

$$\kappa' \nabla'^2 T + \frac{H'}{c_p'} + \frac{\eta_b' - \frac{2}{3}\eta'}{\varrho' c_p'} (\nabla' \cdot \vec{v}')^2 + \frac{\eta'}{2\varrho' c_p'} [\nabla' \vec{v}' + (\nabla' \vec{v}')^{\mathrm{T}}] \cdot [\nabla' \vec{v}' + (\nabla' \vec{v}')^{\mathrm{T}}] \quad (\text{I.3})$$

(*e.g.* Peyret and Taylor, 1985; Schubert *et al.*, 2001). The LHS consists of two terms describing the advection of the temperature field and temperature changes due to adiabatic (de)compression; on the RHS, the terms for conductive heat transport, internal heat production and two terms for viscous dissipation are present. Although this equation will not be solved in the above form in this study, the Boussinesq approximation will be extended by including adiabatic heating, which is a consequence of compressibility, and for consistency, dissipation also has to be kept, although with taking into account eq. I.1a. However, p' is assumed to be constant in time and set equal to the hydrostatic reference pressure; internal heat production will also be neglected, because for the time spans considered in this study, which are on the order of some ten million years, and with regard to the focus on the upper mantle, it is considered to be of minor importance. Setting $\nabla' \cdot \vec{v}' = 0$, $\vec{v}' \cdot \nabla' p' = v_z' \varrho' g'$ and using Ra and the dissipation number $Di = z_{\mathrm{m}}' g' \alpha' / c_p' = (\partial T' / \partial z')_{S'} z_{\mathrm{m}}' / T'$, which measures the adiabatic heating during ascent and the concomitant viscous dissipation, one can non-dimensionalize the temperature equation and arrives, after rearranging, at

$$\frac{\partial T}{\partial t} = \nabla^2 T - \vec{v} \cdot \nabla T + Di \left\{ \underline{T} v_z + \frac{\eta}{2Ra} [\nabla \vec{v} + (\nabla \vec{v})^{\mathrm{T}}] \cdot [\nabla \vec{v} + (\nabla \vec{v})^{\mathrm{T}}] \right\}; \quad (\text{I.3a})$$

note the use of the absolute temperature $\underline{T}$ with Di here. It is necessary for internal consistency that the dissipative term and the adiabatic temperature gradient appear together, because they scale both with Di (*e.g.* White and McKenzie, 1995). – For the mantle, the ratio of advective and conductive heat transport expressed by the Péclet number $Pe = v' z_{\mathrm{m}}'/\kappa'$ is large, *i.e.* $Pe \gg 1$.

However, with respect to possible changes in mantle composition or properties, this is quite a simple form which might have to be extended if special circumstances appear. In the present study, the change in latent heat $L_{\mathrm{m}} = \underline{T} \Delta S_{\mathrm{m}}$ due to melting or freezing is of particular interest; melting of a mass M causes a decrease, freezing an increase of T, which significantly influences the temperature distribution in the melting region, as already shown by Olson (1994). Describing this process by the melting/crystallization rate $(\partial/\partial t + \vec{v} \cdot \nabla) M := \Gamma$, eq. I.3a becomes

$$\frac{\partial T}{\partial t} = \nabla^2 T - \vec{v} \cdot \nabla T + Di \left\{ \underline{T} v_z + \frac{\eta}{2Ra} [\nabla \vec{v} + (\nabla \vec{v})^{\mathrm{T}}] \cdot [\nabla \vec{v} + (\nabla \vec{v})^{\mathrm{T}}] \right\} - \underline{T} \Delta S_{\mathrm{m}} \Gamma. \quad (\text{I.3b})$$

Often it is assumed that L_{m} is independent from p and T for simplicity.

I.1.4 Chemical composition

It is also frequently of interest to be able to distinguish chemically different reservoirs and their evolution in a model. One common method is the use of tracer particles, which are introduced into the model and whose path is then followed; this model, however, is computationally quite expensive, in particular with respect to memory requirements. An alternative approach is the use of a compositional field C (*e.g.* Farnetani and Richards, 1995) which is ruled by the advection–diffusion equation

$$\left(\frac{\partial}{\partial t} + \vec{v} \cdot \nabla\right) C = \frac{1}{Le}\nabla^2 C, \tag{I.4}$$

where the Lewis number $Le = \kappa/\kappa_C$ describes the ratio of thermal to chemical diffusivity; κ is introduced for scaling only, and in the mantle, $\kappa \gg \kappa_C$, so that transport of chemical heterogeneities essentially reduces to an advection problem. Possible applications of such a compositional field would be the possibility to distinguish a plume from normal mantle or to introduce an eclogitic component into the plume (see section I.2.1).

Melting of course also changes the chemical composition of the mantle by leaving it depleted in the more fusible components. This process may be represented by the melting degree f, which would follow the flow of the solid by advection:

$$\left(\frac{\partial}{\partial t} + \vec{v}_{\mathrm{s}} \cdot \nabla\right) f = \frac{\Gamma}{\varrho_{\mathrm{f}}} \tag{I.5}$$

(*e.g.* Schmeling, 2000; Ghods and Arkani-Hamed, 2000). On the long term, this results in the formation of a depleted layer in the uppermost mantle which spreads out from the melting region beneath the spreading center (see section I.3.2).

I.2   THE INFLUENCE OF MATERIAL PROPERTIES ON CONVECTION AND PLUME ASCENT

I.2.1   Density and petrology of upwelling melting mantle

As will be discussed in more detail in chapter II, upwelling mantle undergoes certain changes of its chemical and petrological composition due to solid state phase transitions and melting. Among other properties, the bulk density of the mantle rock is directly affected by these processes, and it is to be expected that the variations in density have a noticeable impact on the style of convection in the surroundings of an ascending mantle current.

One can introduce compositional buoyancy terms into eq. I.2d in the same way as the thermal buoyancy and define corresponding compositional Rayleigh numbers Rc similar to Ra. In the context of this study, compositional variations due to melting processes are of special interest, and two effects related to the generation and extraction of melt are obvious: the modification of mantle density due to the melting degree (depletion) f of the matrix[2] and the retention of melt in its pores. In a similar way phase transformations such as the olivine–spinel or the spinel–perovskite transition cause density changes (see section I.2.2).

By melting of mantle rock, the remaining matrix rock becomes depleted in the more fusible components. In spinel and garnet lherzolite, this causes a density reduction, because the denser Fe components of the Mg–Fe solid solutions dominating mantle mineralogy are removed to a relatively larger extent than the Mg endmembers, and the dense phases spinel resp. garnet are consumed during melting (*e.g.* Oxburgh and Parmentier, 1977)[3]; the density

---

[2]For completeness it is noted that the petrological changes due to depletion also affect the thermal expansivity α and thereby the thermal buoyancy introduced in eq. I.2d; however, this is a secondary effect.

[3]Oxburgh and Parmentier (1977) only consider garnet, although most of the melting under normal MOR occurs in the spinel stability field. Spinel, however, also has a higher density than the other phases (see table D.2), so that their reasoning might basically remain valid, although the fractions of spinel and garnet in the respective lherzolites are not the same.

reduction resulting from 20 % melting might be equivalent to the effect of thermal expansion due to a temperature increase of 500 K (Sotin and Parmentier, 1989), although models with a more detailed treatment of compositional properties suggest that this might be an overestimate (Niu and Batiza, 1991; Forsyth, 1992, also see section V.1.1). On the other hand, a residue of melting plagioclase lherzolite is probably denser than the fertile material due to the low density of plagioclase; however, this is probably of minor importance for convection dynamics, because primary plagioclase is unlikely to exist in significant amounts in the largely depleted uppermost oceanic mantle and because the plagioclase–spinel boundary lies at the very top of the convecting mantle or already in the lithosphere (Forsyth, 1992). He estimates the density change for 15 % depletion to be 35, -20, and -40 kg/m³ in the plagioclase, spinel, and garnet stability fields, respectively.

Let ϱ_0 and ϱ_{dp} be the densities of the unmolten rock and the depleted residue, respectively, and $\Delta\varrho_{dp,max}$ the density difference between unmolten and maximally depleted rock. Then ϱ_{dp} can be expressed as

$$\varrho_{dp} = \varrho_0 + \Delta\varrho_{dp}(f) := \varrho_0 + \Delta\varrho_{dp,max}f, \tag{I.6}$$

where $\Delta\varrho_{dp}(f) = \varrho_{dp} - \varrho_0$ is the density change due to the depletion f; note that negative values or increments for f are in principle possible and can be interpreted as enrichment/refertilization[4]. The second step in the above equation makes the assumption that the density changes linearly with f; however, varying melting proportions and phase exhaustion would have to be considered in a model which includes the petrological changes in detail; to make the approximation useful, one might have to use a $\Delta\varrho_{dp,max}$ which is not directly derived from a real rock composition. After non-dimensionalization of $\Delta\varrho_{dp}$, the depletion Rayleigh number $Rc_f = {}^{\Delta\varrho'_{dp,max}gz_m'^3}/{\eta'\kappa'}$ can be defined to introduce the depletion buoyancy term into eq. I.2d.

Melt usually has a density which is different from the matrix, so that it is separated from its region of generation (see section II.4) when the melt content (porosity) φ exceeds a threshold φ_{max} of a few percent or even less; however, a small fraction of the melt can be expected to remain in the pores of the source region (see section II.1.5) and will also modify the bulk density of the mantle, and even the melt fraction percolating through the porous matrix will contribute to the buoyancy. As Ghods and Arkani-Hamed (2002) showed, quite high melt contents might be reached in molten diapirs and make the melt retention term the most important of the buoyancy terms.

Let ϱ_f be the melt density. Then the density of the depleted matrix retaining a volume fraction φ of melt is

$$\varrho_\varphi = \varrho_0 + \Delta\varrho_{dp} + \Delta\varrho_\varphi = (1-\varphi)\varrho_{dp} + \varphi\varrho_f \tag{I.7}$$
$$\Rightarrow \Delta\varrho_\varphi = \varrho_\varphi - \varrho_{dp} = \varphi(\varrho_f - \varrho_{dp})$$
$$= (\varrho_f - \varrho_0 - \Delta\varrho_{dp})\varphi \approx (\varrho_f - \varrho_0)\varphi = \Delta\varrho_{f0}\varphi;$$

the last step is an approximation for small f and $\Delta\varrho_{dp,max}$, which should be valid in the earth's mantle. Similar to Rc_f, a melt Rayleigh number $Rc_\varphi = {}^{\Delta\varrho'_{f0}gz_m'^3}/{\eta'\kappa'}$ can be defined for the inclusion of melt retention buoyancy in eq. I.2d. It should be noted that it is not appropriate to assume incompressibility for melt as well, so a reference melt density ϱ_{f0} is used, and a factor $\varrho_f - \varrho_{dp} = \Delta\varrho_f = \Delta\varrho_f(p,T)$ (see section II.1.2 and appendix D.2) must be included in the buoyancy term; the effect of retained melt would be particularly significant

---

[4]The convention used throughout this work is that $f > 0$ means depletion and $f < 0$ means enrichment in fusible phases.

at low pressures (Niu and Batiza, 1991) if larger porosities are reached. Taking these effects together, the Stokes equation can finally be written in the following form:

$$-\nabla \tilde{p} + \nabla \cdot \left\{ \eta[\nabla \vec{v} + (\nabla \vec{v})^{\mathrm{T}}] \right\} + \left(-Ra\,T + Rc_f\,f + Rc_\varphi\,\frac{\Delta \varrho'_f}{\Delta \varrho'_{f0}}\varphi \right) \vec{e}_z = 0. \qquad (\mathrm{I.8})$$

Several authors (Sotin and Parmentier, 1989; Scott and Stevenson, 1989; Su and Buck, 1993; Cordery and Phipps Morgan, 1993; Barnouin-Jha *et al.*, 1997) have conducted series of numerical experiments on the effect of density variations on convection beneath ridges which show that density heterogeneities due to melting can have strong influence on the convection pattern, in particular at slow-spreading ridges, because the position of the thermal boundary layer exerts more control on the horizontal extent of the melting region, leading to steeper lateral density gradients (Cordery and Phipps Morgan, 1993). Scott and Stevenson (1989) included density changes due to phase transitions and, more significantly, due to melting and assumed that the segregation of melt is exclusively driven by its effective buoyancy; the latter assumption, although justified at least as a first approximation (*e.g.* Phipps Morgan, 1987; Cordery and Phipps Morgan, 1993), is a simplification which is dropped in this study (cf. section II.4.1). Su and Buck (1993) investigated the influence of density variations due to thermal expansion, depletion, and melt retention and also used melt buoyancy as the only cause for segregation. The additional upwelling velocity due to buoyancy effects close to the spreading center plane needs to be compensated by a local convection roll whose descending limb forces low-density residue downwards to depth levels after a certain time where less depleted material resides; the more buoyancy factors are included in the model, the more confined is the upwelling and melt generation region to the spreading center. This constellation counteracts the dynamics of the local convection and limits the lateral extent of the upwelling, with consequences for the width of the eruption zone (see section II.4.2). The resulting flow pattern is controlled by the large-scale plate-driven flow and the distribution of porosity and melting degree. Similar experiments of Barnouin-Jha *et al.* (1997) confirmed the appearance of small axis-parallel convection rolls, whose distance to the ridge decreases with falling spreading rates. They also demonstrated that at low spreading rate and intermediate to strong buoyancy, melt production along the axis is not coherent on a large scale, which leads to axis segmentation, while an increase in the spreading rate degrades this variability. – Scott and Stevenson (1989) and Su and Buck (1993) showed that at high spreading rates, buoyancy effects are of minor importance for the larger-scale convection pattern, whereas a transition from steady to episodic behaviour takes place if the spreading rate becomes very low; this breakdown happens because the residue does not circulate fast enough from the upwelling back to the conduit of the upward current.

A particular issue related to mantle composition is source heterogeneity in terms of petrology; it is shortly mentioned here for completeness, but will not be included in later models of this study. It has repeatedly been suggested that eclogitic fragments of very old subducted lithospheric slabs collect at the CMB or at the base of the upper mantle, warm up and are carried upward by a plume (*e.g.* Hofmann and White, 1982). As eclogite has a lower solidus and liquidus than peridotite for pressures prevailing in the uppermost 250–300 km, it will start to melt earlier than the peridotite and deliver substantial amounts of melt, which might erupt in a short time interval as typical for flood basalts and create thick basalt layers even if the plume has only a moderate excess temperature (Leitch and Davies, 2001); eclogite also has a higher water content than the MORB source, which will influence the melting and rheological behaviour of the plume as well as the chemical signature of the resulting OIB (Bell and Rossman, 1992). The underlying assumption is that eclogitic melt does not react to a large extent with the peridotite matrix, but this is only justified if melt segregation is fast enough, *e.g.* due to transport in channels (see section II.4.2); otherwise, eclogite–peridotite reaction can refertilize the peridotite and lower its solidus (*e.g.* Yaxley, 2000). A limitation of this mechanism is the fact that eclogite has a greater density than peridotite in the upper mantle, so that a large eclogitic slab is expected to sink in spite of possible thermal expansion; however, if the crust slab is fragmented into pieces with a diameter of the size of its thickness, a sufficiently strong upwelling current is able to drag it upward, because the sum of diapiric upwelling and buoyant sinking velocities still points upward (Yasuda and Fujii, 1998); about 15–20 % of eclogitic material may be transported in a diapir this way.

I.2.2 Phase boundaries

A mineral is defined by its composition and its crystal structure, which is the energetically optimal configuration of atoms resp. ions at a given pressure and temperature; thus, under different p–T conditions another configuration may be more favourable, giving rise to a phase change of the material at a certain line in p–T space known as the Clausius–Clapeyron curve. There are several such solid–solid phase transitions in the mantle, some of which mark major structural boundaries; the most often mentioned are the olivine–modified spinel (α-ol $\rightarrow$ β-ol [wadsleyite]) and the spinel–perovskite (γ-ol [ringwoodite] $\rightarrow$ pv+mw) boundaries, which are related to the 410 km and 660 km seismic discontinuities, respectively. Apart from the density change due to the reconfiguration, latent heat changes in conjunction with the phase change cause shifts in the geotherm ("Verhoogen effect", Verhoogen, 1965) and thus have an important effect on the temperature and density of material trying to cross the phase boundary (Christensen and Yuen, 1985; Christensen, 1995): if the pressure of the flow on the rw–pv phase boundary is smaller than the excess pressure of the deformed boundary, an upwelling current will not be able to penetrate the upper mantle, so that the effect of the ringwoodite–perovskite transition will result in layered convection; this might be of particular importance for the dynamics of plumes of different sizes (Marquart *et al.*, 2000). If the phase boundary does separate convection in the upper and lower mantle, but is permeable to plumes, it offers an explanation for the geochemical differences in the MORB and plume reservoirs; numerical experiments support this case for $O(Ra) = 10^7 \ldots 10^8$, which are estimated for the mantle (*e.g.* White and McKenzie, 1995). If, however, a thermally anomalous flow crosses the phase boundary, it will induce a deflection of it: a hot upwelling plume causes a rise of a boundary with a negative Clapeyron slope and a depression in one with a positive. In the olivine-dominated mineralogy of the upper mantle, this results in a thinning of the transition zone, which is also detectable seismologically (see section IV.2.1 for an application to Iceland), whereas a subducted slab would have the opposite effect.

Other important, although not isochemical, phase transitions in the upper mantle are those involving the aluminous phase and the pyroxenes[5], *e.g.* the plagioclase–spinel[6] and the spinel–garnet transition at about 25...30 and 75...80 km, respectively. Opinions seem to differ whether the contributions from the pl–sp and the sp–gt transitions are important; Scott and Stevenson (1989) state that it does not seem that they have a substantial effect on the total buoyancy, while Niu and Batiza (1991) think that the effect on the density of melt-free solid mantle is relevant; models of this study tend to support the former conclusion

---

[5]Throughout this work, "Al-bearing" or "aluminous phase" will be used for one or all of the minerals plagioclase, spinel, and garnet, but not for minerals with Al-bearing endmembers like jadeite in clinopyroxene.

[6]Here it is indeed spinel, while in the olivine case, "spinel" rather refers to the crystal structure. For the rest of this work, the term "spinel" will be used for this aluminous phase, while for the high-p olivine polymorphs the names wadsleyite and ringwoodite, or β-olivine and γ-olivine, respectively, will be used.

(see sections V.1.2, V.1.6). They might be at least of some importance because the solidus of peridotite is distorted at its intersections with them and because of the effect of their phase change enthalpy on melting (see section II.3.1, II.3.2). – Apart from this, pyroxenes and garnet undergo a phase transition to majoritic garnet in a broad pressure range roughly coinciding with the stability field of wadsleyite, but with a negative Clapeyron slope (*e.g.* Yusa *et al.*, 1993; Weidner and Wang, 2000), which will partly compensate the effect of the olivine transitions; however, because of their irrelevance to melting processes in the shallow mantle, these transitions will not be considered in the convection models presented here, although this (common) practice is likely an oversimplification (Weidner and Wang, 2000).

For the purpose of this study it is assumed that phase changes occur linearly over the pressure range bounded by the stability fields of the pure low- or high-p phases, *i.e.* by the width of the binary loop in binary systems. In reality, though, the molar fraction of the phases changes non-linearly in the transitional regime, so that the phase transition is concentrated at one boundary of the phase loop, its effective width being much smaller than the width of the loop (Stixrude, 1997) and approaching common numerical grid resolutions of a few kilometers; further narrowing can be caused by non-transforming phases, such as pyroxenes and garnet in the case of the olivine–wadsleyite transformation. Stixrude (1997) also shows that this sharpening also increases the seismic observability *e.g.* of the 410 km transition.

The effect of phase changes on buoyancy can also be expressed by respective Rayleigh numbers, so that eq. I.8 can be extended as follows:

$$-\nabla \tilde{p} + \nabla \cdot \left\{ \eta [\nabla \vec{v} + (\nabla \vec{v})^{\mathrm{T}}] \right\} + \left(-Ra\,T + Rc_f\,f + Rc_\varphi \frac{\Delta \varrho_{\mathrm{f}}'}{\Delta \varrho_{\mathrm{f}0}'} \varphi + \sum_{\text{all phases}} Rc_{X_k}\,X_k \right) \vec{e}_z = 0,$$
(I.9)

where Rc_{X_k} is the phase Rayleigh number and X_k the fraction of phase k; the sum is taken over all relevant phases. The energy equation eq. I.3b also has to be generalized:

$$\frac{\partial T}{\partial t} = \nabla^2 T - \vec{v} \cdot \nabla T + Di \left\{ \underline{T} v_z + \frac{\eta}{2Ra} [\nabla \vec{v} + (\nabla \vec{v})^{\mathrm{T}}] \cdot [\nabla \vec{v} + (\nabla \vec{v})^{\mathrm{T}}] \right\}$$

$$+ \underline{T} \left[-\Delta S_{\mathrm{m}} \Gamma + \sum_{\text{all phases}} \Delta S_{X_k} \left(\frac{\partial}{\partial t} + \vec{v} \cdot \nabla \right) X_k \right]; \quad (I.10)$$

the T-dependence of the phase change contribution is taken into account by using the entropy change ΔS_{X_k} instead of a constant latent heat.

It is worth noting that for a convection current passing through the phase boundary, continuity conditions for the temperature apply and the appropriate solution of the conduction–advection equation for the temperature includes a precursor with the length scale κ/v on the upstream side of the boundary, *i.e.* before the phase transition itself takes place. However, detailed analysis of the temperature equation shows that this precursor is negligible for v higher than a critical value, which depends on the thermodynamic parameters of the phase change and is of the order of a few millimeters per year for the olivine phase changes (*e.g.* Schubert *et al.*, 2001).

I.2.3   Thermal expansivity

Although the thermal expansivity α is assumed constant in most numerical models of mantle convection, it is in fact notably pressure- and also temperature-dependent even in the upper mantle (see section V.1.3 and appendices D.1.3 and E). If a decrease of α with depth is included, the buoyancy of deep-mantle material is diminished; numerical models show that this effect, similar to a downward increase in viscosity, leads to the formation of a small number of large and stable hot plumes emerging from a thick basal boundary layer and of a multitude of small cold plumes descending from the cool top (Hansen *et al.*, 1993). If a compositional boundary is present, there is an episodic pattern of hot plume breakthrough in otherwise layered convection, which eventually leads to chemical heterogeneity in the upper mantle while leaving the lower mantle essentially uncontaminated (Hansen and Yuen, 1994). On the other hand, if temperature dependence and the effect of phase changes on α are also considered, the tendency of the mantle to develop layered convection is weakened (Ita and King, 1994).

In this study, in some instances another recent approach by Schmeling *et al.* (2003) will be followed in taking into account both p and T dependence of α (see sections III.4.4 and V.1.3). In this case, Ra and Di are not constant anymore, which affects both the Stokes equation I.9 and the temperature equation I.10. In particular, the adiabatic heating term in the latter becomes nonlinear, but it is still possible to handle it properly numerically; the dissipation term is not affected, because α cancels out. In the isoviscous, single-phase models of the upper mantle of Schmeling *et al.* (2003), where adiabatic heating was not included, the additional inclusion of the T dependence of α reinforced plume flow and retarded slab downwelling more strongly than with an only p-dependent α; in both cases, the buoyancy is less pronounced at the depth of the transition zone, but greater in the shallower mantle, than in constant-α models.

I.2.4 Viscosity and rheology law

Under high pressures and temperatures, the solid rock of the earth's mantle begins to creep in response to applied stress. The viscosity of the rock, which controls the extent of creep, depends in a complex way on several factors, such as temperature, pressure, stress and water content, and varies over several orders of magnitude throughout the mantle; for instance, a change in T of 100 K is estimated to result in a change of η by roughly a factor of 10 in the upper mantle. On the other hand it can be seen from eq. I.11 below that p and T have concurring effects on the viscosity, which *e.g.* probably leads to an approximately constant η for the depth range between 150 km and 400 km (Hirth and Kohlstedt, 1996). As the effect on convection flow patterns of the T dependence of η is strongest at lower Ra, it is probably less significant for the shape of large-scale convection in the mantle, though (White and McKenzie, 1995).

Creep occurs by various mechanisms, which become important under different conditions. Basically, the following creep regimes are distinguished (Karato and Wu, 1993; Paterson, 2001):

Diffusion creep. Deformation by diffusion creep occurs by diffusion of atoms along grain boundaries (Coble creep) or through the grains (Herring–Nabarro creep) and takes place under relatively low stress and/or in rock with small grain size and is probably characteristic for cold, shallow as well as deeper upper mantle regions. The strain rate is proportional to the stress, *i.e.* the rheology is Newtonian.

Dislocation creep. In dislocation creep, which dominates under high stresses and large grain sizes, the rock deforms by propagation of dislocations inside grains; it seems to prevail in hot, shallow parts of the upper mantle, *e.g.* under MOR or in plume heads. The stress–strain relation is nonlinear, but the rheology is independent of the grain size *a.* In contrast to diffusion creep, dislocation creep induces lattice-preferred orientation (LPO) in the crystal grains and thus causes anisotropy in several physical parameters such as seismic velocities.

Grain boundary sliding. This mechanism is just the relative motion of grains without significant shape changes in the grains themselves, although some accommodation by diffusion or dislocation creep or pressure melting and melt transfer along stress gradients can be involved locally to counteract bulk volume increase. It can be significant in the presence of melt in the form of a fairly contiguous melt network on grain interfaces, *i.e.* at moderate melt fractions, but below the threshold of rock dismembering (see section II.2.1).

The location of the transition between the diffusion and dislocation regimes in the mantle is not known very well, but examination of seismic anisotropy in the shallow upper mantle and of the presence of the Lehmann discontinuity at 220 km depth as well as postglacial rebound models with linear and nonlinear layers let Karato and Wu (1993) suggest that the rheology of the upper mantle is mostly linear, with at most a rather thin nonlinear layer between the thick lithosphere and $z = 200 \ldots 300$ km; thus, diffusion creep characterizes most of the upper mantle, and dislocation creep indeed is restricted to the shallow, hot regions beneath ridges or in plumes.

The properties of the different creep mechanisms can be described in a general form by an extended Dorn equation using an activation energy $\mathcal{E}$ and activation volume $\mathcal{V}$:

$$\dot{\varepsilon} = \mathcal{A} \left(\frac{\sigma}{\mu} \right)^n \left(\frac{u_\mathrm{B}}{a} \right)^q f_{\mathrm{O}_2}^m \, e^{-\frac{\mathcal{E}+p\mathcal{V}}{RT}}, \qquad \begin{cases} n = 1, q = 1 \ldots 3 & \text{for diffusional creep} \\ n = 3 \ldots 5, q = 0 & \text{for dislocational creep} \end{cases} \qquad (\mathrm{I}.11)$$

(*e.g.* Kohlstedt and Zimmerman, 1996; Karato and Wu, 1993), where ε is the strain, σ is the stress, a is the grain size, f_{O_2} is the oxygen fugacity, and $\mathcal{A}$ is a material parameter; under hydrous conditions, which have to be assumed at a concentration of OH^- ions of $C_{\mathrm{OH}} \gtrsim 50\,\mathrm{H}/10^6\,\mathrm{Si}$, the RHS has to be multiplied with C_{OH}, which for olivine and clinopyroxene is proportional to $f_{\mathrm{H_2O}}$, and the other parameters change (Kohlstedt *et al.*, 2000; Hirth and Kohlstedt, 1996; Kohlstedt *et al.*, 1996). The effect of melt is not included in this formulation and will be described in some more detail in section II.2.1. Kohlstedt *et al.* (2000) measured several parameters of eq. I.11 for an olivine–basalt aggregate for both diffusion and dislocation creep; they obtained $m = 1/7$ for diffusion creep and $n = 3.7$, $m = 0$ for dislocation creep, with an activation energy $\mathcal{E} = 530\,\mathrm{kJ}/\mathrm{mol}$ and suggest that in the high-stress regime a diffusion and a dislocation contribution to the strain rate be added, because both mechanisms are actually present. In the presence of water, the viscosity is reduced significantly in both regimes: for olivine aggregates, Hirth and Kohlstedt (1996) estimate that in the MORB source region the viscosity in the presence of water is 500 ± 300 times lower than in dry rock, and Mei and Kohlstedt (2000a,b) also report an increase of strain rates by a factor 5–6 in both deformation modes. For hydrous aggregates, the transition from diffusional to dislocational creep takes place at differential stresses exceeding 100 MPa. – The choice of $\mathcal{E}$ was shown to have substantial impact on the dynamical behaviour of plumes: in numerical experiments on whole-mantle convection (Thompson and Tackley, 1998), $\mathcal{E} = 500\,\mathrm{kJ}/\mathrm{mol}$ produced a huge, rapidly upwelling plume, while a value only half as large resulted in a conventionally shaped plume with head and tail.

Bystricki *et al.* (2000) remark that one of several imaginable mechanisms for focussing convective upwellings is due to the nonlinear character of the dislocation version of eq. I.11 associated with strain weakening, as suggested by laboratory experiments on olivine under high shear strain. It can also result in more complex behaviour such as the detachment of a plume head and formation of subsequent heads during ascent, as observed in numerical experiments (*e.g.* van Keken, 1997); this might be a possible mechanism for the apparent generation of plume "pulses". However, in numerical convection models, it is quite common use to include an exponential p–T dependence of the viscosity, but assume a linear stress–strain rate relation, because a nonlinear (non-Newtonian) rheology is less well tractable. Non-Newtonian models tend to develop structures on several length scales, and plumes in such models can reach very high velocities due to the positive feedback between velocity and stress; this will be further enhanced by viscous shear heating (*e.g.* Larsen *et al.*, 1999). In this study, all models are restricted to Newtonian rheologies.

It should be noted that many of the cited studies investigated the behaviour of essentially monomineralic systems, while peridotite consists of two to four major phases with possibly significantly different rheological characteristics; *e.g.* judging from experiments on forsterite–enstatite composites (Ji *et al.*, 2001), olivine has a higher flow strength and is dominated by other creep mechanisms than pyroxene. Ideally, it should be possible to calculate the viscosity of the rock from the proportions and viscosities of the constituents, as it seems that the deformation mechanism of a constituent does not depend on the presence of other minerals (Ji *et al.*, 2001); for application in a convection model, a closed-form mixing rule would be most convenient. While it is common use and probably acceptable to use an overall rheology in such models, *e.g.* equating the parameters of the mantle with those of olivine, it is a fact that the minerals undergo several internal structural changes and vary their relative proportions even in the p–T range of the upper mantle; changes like the depletion upon melting or the solid-state phase transitions might introduce a certain variability in the viscosity structure of the upper mantle. Ji *et al.* (2001) compared their results from experiments on fo–en composites with several binary mixing rules and concluded that from the three domains – weak-phase supported, transitional, and strong-phase supported – the first and the last can be described fairly well by (different) existing formulae, while the transitional ($X_{fo} = 0.4 \ldots 0.6$ in their case) is poorly tractable. This is unfortunate, if one considers *e.g.* the modal composition of pyrolite (table D.1), where olivine makes up about 55 % of the rock, so that the application of mixing rules will introduce some errors for fertile mantle. On the other hand, experimental findings are not yet unambiguous in that several authors presume that mixing has only a minor effect on rheology (*e.g.* Hirth and Kohlstedt, 1996). In conclusion, the rheology of the mantle, and in particular of several high-pressure phases, is not very well determined yet, so that large uncertainties are an inherent deficiency of probably every existing model.

A further aspect of the rheology of a heterogeneous mantle to be mentioned is the influence of a small component of eclogite (see section I.2.1) as in a "marble-cake" mantle. Allègre and Turcotte (1986) remark that old subducted oceanic lithosphere would be stretched out into thin layers, which would possibly be arranged preferredly in the horizontal direction and increase the vertical creep strength of the bulk mantle rock, *i.e.* result in anisotropic viscosity.

I.3   ASCENT AND NEAR-SURFACE EFFECTS OF A PLUME

I.3.1   *Ascent of a plume through the mantle*

Many plumes are assumed to originate from the CMB, which is the most important thermal (and also chemical) boundary layer in the earth's interior, and rise through *ca.* 2800 km of mantle rock of changing composition and physical properties, undergoing several phase changes which influence the plume temperature. Obviously, a plume which starts with an excess temperature of several hundred kelvins will lose a certain amount of its heat to its environment by several processes, *e.g.* by conductive cooling; important questions in this context are if an ascending plume head entrains normal mantle material, and if so, how much of it and how much this contributes to cooling, and if there are limitations with regard to size and volume flux which prevent plumes below a certain magnitude from reaching the lithosphere. Furthermore, if plumes encounter a steplike viscosity drop during ascent, *e.g.* at the boundary between the upper and the lower mantle, they accelerate and thin, which might result in pulsations due to stem disruption if there is an additional background shear flow and if the viscosity contrast is sufficiently large. However, cylindrical plumes were found to be rather robust and massive for viscosity contrasts of the magnitude expected for the upper and lower mantle, so that they do not develop pulsations (van Keken and Gable, 1995); it should be noted, though, that those models did not consider a temperature-dependent rheology, which is likely to lead to stronger thinning of plumes.

Albers and Christensen (1996) conducted numerical experiments on plumes with p-, T-, and σ-dependent viscosity starting with a given excess temperature, but varying volume flux and observe that plumes with small flux lose almost their entire excess temperature when arriving at the base of the lithosphere; this conclusion is valid mostly irrespective of the viscosity or other single material parameters and of the rheological law. At the base of the mantle, the low thermal expansivity and high thermal conductivity led to comparatively low buoyancy and result in a broad plume base to accommodate the flux while at higher levels the plume is thinner.

The importance of entrainment of normal mantle material into the plume head is still unresolved. The ascent of thermal plumes from a point source at the CMB was investigated analytically and in laboratory experiments by Griffiths and Campbell (1990) and by numerical modelling for both Newtonian and non-Newtonian rheologies and depth-dependent viscosity and thermal expansivity by Hauri *et al.* (1994b). They find significant degrees of entrainment in all of their models, most of it leading to the incorporation of lower mantle material into the plume resp. its thermal halo; this is due to the heating of ambient mantle by conduction, which increases its buoyancy and lowers its viscosity, thereby facilitating its inclusion into the upwelling plume. The importance of thermal interaction with the ambient causes plumes from the CMB with a small radius resp. a low buoyancy flux to entrain very large amounts of the surrounding mantle, so that the plume heads possibly double their diameters, whereas heads of plumes starting from the 660 km boundary will remain relatively small. According to Griffiths and Campbell (1990), entrainment takes place in the plume head in the first place and cools it considerably, whereas the trailing conduit contains mostly material from the source region. Laboratory experiments with chemically buoyant plumes having varying viscosity contrasts with the ambient fluid also showed some sort of entrainment, with the internal structure changing from a layered one at low η ratios to a more chaotic one at higher ratios, which evolves from an initial two-layer to a thoroughly stirred mode (Kumagai, 2002). Farnetani and Richards (1995) and Farnetani (1997) investigated the ascent of plumes from the CMB in numerical models with respect to different rheologies and the rw–pv phase boundary and did not find a significant degree of entrainment in any of their models, so that they conclude that the related cooling effect is not important. Although Farnetani and Richards (1995) conjecture that stronger $\eta(T)$ variations than theirs of at most 30 would not enhance entrainment significantly, it should be noted that their maximum corresponds to a temperature contrast of only about 150 K according to the rule of thumb on p. 10, which is much less than the initial temperature contrast prescribed in their model; apart from this limitation, the role of stress-dependent rheologies has not been considered. They explain the discrepancy with the analogue experiments, and actually also with the observations by Hauri *et al.* (1994b), with the difference in the source region – boundary layer vs. injection – and with the much shorter ascent path in their numerical experiments, which does not let the plume evolve enough to let entrainment happen; this argument, however, was countered by a numerical reproduction of laboratory experiments as well as models starting from a thermal boundary layer by van Keken (1997). The discrepancy with the results of Farnetani and Richards (1995) may thus be due to an underestimate of the temperature effect in their models.

I.3.2   Interaction of a plume with the lithosphere

In its latest stage of ascent, the plume impinges on the base of the cold, depleted, highly viscous lithosphere, which is an insuperable obstacle to further upwelling, and begins to spread out beneath the plate, losing heat to the overlying lithosphere by conduction in the process; however, in laboratory experiments (Griffiths and Campbell, 1990), the heads of small plumes already begin to spread early after starting from their source. The dynamic pressure of the rising material pushes upward the lithosphere directly above the plume conduit, causing a positive geoid and topography anomaly (*e.g.* Condie, 2001); additionally, the lateral spreading of plume material into a sublithospheric channel causes a large-scale uplift of the plate known as a hotspot swell, which is a very prominent feature in a number of hotspot regions. Much of the swell topography – except the surroundings of the conduit – is in isostatic equilibrium and thus gives an impression of the shape of the plume head. The shapes of the swell and the plume head result from the combined effect of plume head spreading and the motion of the overriding plate (Olson, 1990; Ribe and Christensen, 1994). Bercovici and Lin (1996) have investigated the spreading of plume heads with temperature-dependent buoyancy and viscosity and find that heat loss to the lithosphere causes them to slow down, deform and develop flat tops and steep lateral fronts, which might eventually evolve into swells. These plume heads grow both by thickening, especially of their outer parts, and by spreading, the former mechanism distinguishing them from isothermal, isoviscous plumes. Another remarkable feature in spreading plume heads with T-dependent viscosity is the possibility of oscillatory instabilities due to small, self-enhancing flow perturbations, which propagate outward as waves; the necessary critical flow velocity is likely to be exceeded in mantle plumes (Bercovici, 1992).

Several authors have already investigated the interaction of an upwelling plume with a ridge. Ribe *et al.* (1995) and Sleep (1996) have developed an analytic thin-layer flow ("lubrication") model and conducted numerical experiments on a plume–MOR setting with variable viscosity to deduce a relation for the width of a plume head, in particular for the flow along the ridge; they also applied their model to Iceland (see section IV.2.3). Feighner and Richards (1995) performed laboratory experiments with a chemical plume to examine the spreading and shape changes of a plume impinging on a fixed plate or on a simplified ridge analogue; these experiments were later complemented with numerical investigations of isoviscous chemical plume–ridge convection, which confirmed earlier findings, *e.g.* that the plume head already starts to spread before reaching the plate (Feighner *et al.*, 1995). The plume waist width y_{P0}, *i.e.* the along-strike width of the plume at the ridge, which is a steady-state feature, was found to be approximately proportional to the characteristic length scale $\sqrt{q/v_r}$ in all these studies. Both models of Ribe *et al.* (1995) do not indicate that y_{P0} is affected by upslope flow of plume material into the ridge; the analytical model, however, predicts a dependence of the waist width on the viscosity contrast. Recent numerical models by Albers and Christensen (2001), which could handle large viscosity contrasts, proved that plumes with viscosities several orders of magnitude below that of the adjacent lithosphere are strongly channelled into the ridge and propagate in it rapidly over large distances if the spreading velocity is low, thus confirming a prediction by Sleep (1996); for small viscosity contrasts or high spreading rates the plume head is much more radially symmetric instead, as seen in earlier studies, and axis-perpendicular drag by plate motion is more important in shaping the plume head. The aspect ratio of the plume head, y_{P0}/x_{P0}, was empirically found to be approximately proportional to $\sqrt{1/v_r}$ and independent of the volume flux. Sleep (1996) remarks that the length of the ridge interval affected by the plume depends on how much of the plume material is integrated into the newly formed lithosphere. Furthermore, the flux, an assumed T anomaly and the isostatically equilibrated excess topography above the ridge, Δt, were shown to be related by

$$q = \frac{y_{P0} v_r \Delta t (\varrho - \varrho_{H_2O})}{2 \varrho \alpha \Delta T} \qquad (I.12)$$

(Schilling, 1991; Sleep, 1990). – The relations between a plume and a ridge are even more variable if a plume and a ridge move relative to each other. In this case, the interaction is small and restricted to short ranges when the ridge moves towards the plume, because plate and ridge motion cooperate in dragging plume material away from the spreading center; it is quite large and far-reaching when the ridge moves away from the plume, as in this case the overall ridge motion and the spreading of the overriding plate act in opposite directions (Ribe and Delattre, 1998). Similar to on-ridge plumes, the material transferred from off-ridge plumes to the ridge also propagates along it and causes a chemical and a topographical anomaly (*e.g.* Schilling, 1991).

Phipps Morgan (1997) noticed that, with the exception of on-ridge hotspots, the melting rate of hotspots is independent from the age resp. thickness of the overriding oceanic lithosphere; this is in contradiction to the assumption that the base of the thermal lithosphere determines the upper boundary of the melting column, which should follow the well-known $\sqrt{age}$ law. Similar to other authors (*e.g.* White, 1993), he invokes the concept of a "compositional lithosphere" generated by depletion and dehydration due to melting beneath the mid-oceanic ridge and propagated sidewards by seafloor spreading; the base of this compositional lithosphere is assumed to lie at a more or less constant depth, at least beyond a certain distance from the spreading center.

Plumes have several direct or secondary expressions which are observable at or near the surface, *e.g.* flood basalts (*e.g.* Coffin and Eldholm, 1992; Olson, 1994), seismic dipping reflectors (Hinz, 1981), topographical, geoid, and gravity anomalies, and heat flow anomalies (*e.g.* Sleep, 1990), and a distinct geochemical signature. The topographic anomaly of a plume, which can be of the order of kilometers, is in principle well visible and separable from the reference topography in oceanic environments, in contrast to continents with their diverse topography, although some distortions are introduced through effects like flexure from volcanic loads; this effect is mostly reversible, though, so that the uplift will decay when the affected region has been moved away from the plume by plate drift, but the variations are likely to be preserved in the sedimentary record. Combination with gravity data from satellite altimetry provides a possibility to determine the depth of the anomaly and distinguish a plume from shallow, crustal sources (White and McKenzie, 1995).

CHAPTER II

PROPERTIES, GENERATION AND TRANSPORT OF MANTLE MELTS, AND THE FORMATION OF THE CRUST

II.1 SOME PHYSICAL PROPERTIES OF MELT

The physical properties of melt depend in a significant manner directly or indirectly on the p–T conditions of its current environment and its chemical composition and also have backlash on melt chemistry; some parameters, such as the compressibility and the density, are much more sensitive to p–T changes in melts than they are in the corresponding solid phases. Therefore, the dynamical characteristics of melt may change considerably over the depth interval of melting. In this section, some facts concerning the p–T dependence of important properties of melt are summarized with regard to their inclusion in numerical models in chapter V. The role of volatiles, which is not so well constrained anyway, will not be discussed in great detail, because much of this study essentially focusses on anhydrous melting; nonetheless, volatiles are certainly a significant component of melts, *e.g.* in the earliest stages of suboceanic melting.

II.1.1 Thermal expansivity and bulk modulus

The isobaric thermal expansivity,

$$\alpha = \frac{1}{V}\left(\frac{\partial V}{\partial T}\right)_p, \tag{II.1}$$

results from the combination of several effects on the molecular level and is a function of temperature and pressure. However, data on the isobaric temperature dependence of α is scarce, especially for silicate liquids, but as it is apparently small (Bottinga, 1985), α has traditionally been regarded as T-independent, although recent studies suggested that there is a significant T dependence *e.g.* in anorthite–diopside melts especially near the glass transition temperature (*e.g.* Toplis and Richet, 2000; Gottsmann and Dingwell, 2002). It also depends on composition: alkali- and Fe_3O_2-rich liquids have large α, which is especially important for basaltic lavas such as MORB (Lange and Carmichael, 1990). However, at the temperatures prevailing in the melting region and given the uncertainties *e.g.* in melt composition, it is sufficient to use a T-independent expansivity for the purpose of this study.

The isothermal compressibility of silicate melts,

$$\beta_T = \frac{1}{K_T} = -\frac{1}{V}\left(\frac{\partial V}{\partial p}\right)_T, \tag{II.2}$$

is significantly higher than that of solids, but also a function of p and T, and is also influenced by structural properties such as *e.g.* the coordination number of Si. It decreases with increasing pressure (Bottinga, 1985), but Lange and Carmichael (1990) assessed that for $p < 2\dots3\,\mathrm{GPa}$ it can be assumed constant. The p dependence varies greatly between different melts, and the p derivative, K_T', exceeds the range of $6\dots7$ suggested by Stolper *et al.* (1981) by a factor 4 for some liquids (see table D.10). Many experiments indicate that several compositional aspects, such as variations in silica, Al, and Fe content, have a strong, but complex influence on β (*e.g.* Walker *et al.*, 1988; Bottinga, 1985; Lange and Carmichael, 1990), which is difficult to quantify for use in melt-dynamical models.

II.1.2   Density

Many experiments suggest that melts are so compressible, that there is a certain pressure at which they become denser than the coexisting solid phases (Stolper *et al.*, 1981); Rigden *et al.* (1984) estimated a crossover pressure of 6…10 GPa from shock-wave experiments on a model basaltic ($an_{36}di_{64}$) liquid, which is shifted to 9–14 GPa by inclusion of a fayalite component ($di_{54.4}an_{30.6}fa_{15}$) (Chen *et al.*, 2002a). This possibility is not only of importance for the early history of the earth and the question if minerals, *e.g.* olivine, can float in the melt, but also for present-day segregation processes, because it would mean that there is a depth below which melt would not ascend due to buoyancy and finally erupt, but would stagnate or even segregate *downwards* and maybe form melt bodies at greater depths; it may also be responsible for the putative retention of small amounts of melt in the seismic low-velocity zone in the upper mantle under certain circumstances. In any case, the high compressibility reduces the density contrast between melt and matrix even in the normal melting regime, where a density inversion does not take place, so that melt ascent from greater depths could be slower than from shallower parts of the mantle; on the other hand, the increasing density contrast between melt and residue at shallow depths will enhance segregation in the higher parts of the melting column (Niu and Batiza, 1991)

Basically, a melt can be regarded as a mixture of several oxides, whose individual molar volumes combine according to their respective mole fraction to give the bulk molar volume of the liquid, whereby iron-bearing natural melts show anomalous behaviour (Lange and Carmichael, 1990, see app. D.2.1). It should also be noted that volatiles, chiefly water and CO_2, lower the density of a melt significantly, but not much is known about the p–T–X dependence of their molar volume in silicate melts. As the chemical variability of the melts is not included here, it should suffice to assume a constant composition and use laboratory measurements on a natural lherzolitic melt.

The pressure dependence of the volume or density is given by equations of state and parameterized by the bulk modulus at normal pressure, K_0, and its pressure derivative K_0', while the explicit temperature dependence is described independently by the thermal expansivity (see section II.1.1). The most widely used equation of state is the third-order Birch–Murnaghan equation

$$p(\varrho) = \frac{3K_0}{2}\left[\left(\frac{\varrho}{\varrho_0}\right)^{\frac{7}{3}} - \left(\frac{\varrho}{\varrho_0}\right)^{\frac{5}{3}}\right]\left\{1 + \frac{3}{4}(K_0' - 4)\left[\left(\frac{\varrho}{\varrho_0}\right)^{\frac{2}{3}} - 1\right]\right\} \qquad \text{(II.3a)}$$

(*e.g.* Stolper *et al.*, 1981). However, this equation of state is known to be increasingly inexact for very large strains. Therefore a number of alternative equations of state have been proposed, *e.g.* the third-order logarithmic equation of state of Poirier and Tarantola (1998),

$$p(\varrho) = K_0\frac{\varrho}{\varrho_0}\left[\ln\frac{\varrho}{\varrho_0} + \frac{K_0' - 2}{2}\left(\ln\frac{\varrho}{\varrho_0}\right)^2\right], \qquad \text{(II.3b)}$$

which is derived from total finite strain instead of Eulerian strain. The one by Vinet *et al.* (1987),

$$p(\varrho) = 3K_0\left[1 - \left(\frac{\varrho_0}{\varrho}\right)^{\frac{1}{3}}\right]\left(\frac{\varrho}{\varrho_0}\right)^{\frac{2}{3}}\exp\left\{\frac{3}{2}(K_0' - 1)\left[1 - \left(\frac{\varrho_0}{\varrho}\right)^{\frac{1}{3}}\right]\right\}, \qquad \text{(II.3c)}$$

is based on binding energy considerations and was regarded as the most accurate in a comparative study by Cohen *et al.* (2000). Although eqs. II.3a and II.3b are in quite good agreement for small strains, they start to differ notably for $\varrho/\varrho_0 > 1.5$, especially if $K_0' > 4$ (Poirier and Tarantola, 1998). On the other hand, the Vinet equation II.3c has also been criticized for leading to distortions in the pressure derivatives of K by Stacey (2000), who proposed another equation of state (in the form $\varrho(p)$); from a practical point of view, the problem with his equation though is, that it needs information on either K_∞' or K_0'', which is hardly available in literature.

All three eqs. II.3 have the disadvantage to express p as a function of ϱ instead of the other way round, which would be more useful for an application in a convection–segregation algorithm; unfortunately, the inverse function cannot be obtained analytically, so that an iterative numerical root-finding routine would be necessary to get $\varrho(p)$. If this value has to be obtained hundreds or thousands of times for each time step, it is too time-consuming. As a workaround, one can derive a simple equation of state from the definition of the bulk modulus based on the assumption of a linear K_T–p relationship:

$$\frac{1}{K_T} = -\frac{1}{V}\left(\frac{\partial V}{\partial p}\right)_T = \frac{1}{\varrho}\left(\frac{\partial \varrho}{\partial p}\right)_T; \qquad K_T(p) = K_0 + K_0'p$$

$$\Rightarrow \int_0^p \frac{\mathrm{d}\hat{p}}{K_0 + K_0'\hat{p}} = \int_{\varrho_0}^{\varrho} \frac{\mathrm{d}\hat{\varrho}}{\hat{\varrho}} \;\Leftrightarrow\; \varrho(p) = \varrho_0\left(\frac{K_0 + K_0'p}{K_0}\right)^{\frac{1}{K_0'}}. \tag{II.4}$$

Strains of a magnitude at which significant deviations between the different equations of state appear are not likely to occur in solid phases at upper or intermediate mantle depths due to their low compressibility (see table D.5), but might be more important for melts (see table D.10). This is shown in figure II.1, where four different isothermal equations of state, eqs. II.3 and eq. II.4, are plotted for solid forsterite and diopside melt, which can be regarded as rough approximations for upper mantle material and basaltic melt, respectively. Obviously, under upper-mantle conditions, the choice of the equations of state does not matter for solids, but for melts, there are significant differences at pressures larger than 4–5 GPa, which might be important for very deep melts as those expected in very hot plumes or in the hydrous melting region. Therefore, it must be emphasized that eq. II.4 is only valid for small compressions of at most 1.1 . . . 1.2, which are to be expected for $p \lesssim 5\,\mathrm{GPa}$ for common mantle melts; at greater depths, it yields too low densities. It should also be noted that the above equations of state are possibly inappropriate for extrapolation in cases like alumosilicate liquids, where K_0' is not constant, but varying continually.

The actual density does of course not only depend on the compressibility, but also on temperature: the effect of thermal expansion counteracts the pressure effect and shifts the critical depth downwards. Taking the definitions of α and K_T together, the general p–T dependence of ϱ can be written as

$$\varrho(p,T) = \varrho_0 \exp\left(\int_0^p \frac{\mathrm{d}\hat{p}}{K_T} - \int_{T_0}^T \alpha\,\mathrm{d}\tilde{T}\right) \tag{II.5}$$

$$\approx \varrho_0(1 - \alpha\Delta T)\left(\frac{K_0(T) + K_0'(T)p}{K_0(T)}\right)^{\frac{1}{K_0'}}.$$

The density difference between melt and (depleted) matrix at a pressure p and a temperature

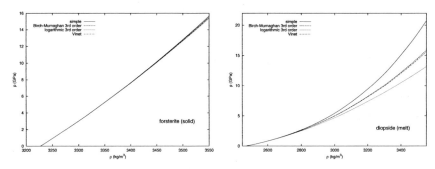

Figure II.1: Isothermal equations of state (eqs. II.3 and II.4) for solid forsterite (left; $\varrho_0 = 3227$ kg/m³, $K_0 = 128$ GPa, $K_0' = 5$) and diopside melt (right; $\varrho_0 = 2471$ kg/m³, $K_0 = 13.8$ GPa, $K_0' = 6.6$). Note the different axis scales.

$T = T_0 + \Delta T$ which causes buoyancy-driven segregation is hence

$$\Delta\varrho_{\mathrm{f}}(p, T_{\mathrm{f}}, T_{\mathrm{dp}}) = \varrho_{\mathrm{f}}(p, T_{\mathrm{f}}) - \varrho_{\mathrm{dp}}(p, T_{\mathrm{dp}}) = \varrho_{\mathrm{f}}(p, T_0)(1 - \alpha_{\mathrm{f}}\Delta T_{\mathrm{f}}) - \varrho_{\mathrm{dp}}(p, T_0)(1 - \alpha\Delta T_{\mathrm{dp}}); \quad (\mathrm{II.6})$$

usually the solid is assumed incompressible $(\varrho_{\mathrm{dp}}(p, T_0) = const.)$, and for the pressure dependence of ϱ_{f} eq. II.4 or any other equation of state can be used. The distinction between the temperatures T_{f} and T_{dp} is made, because the melt is not necessarily in thermal equilibrium with the solid, *i.e.* it can have a different temperature. In this study, the T difference is though neglected in this context.

Assuming a melt whose compressibility is quantified according to the previous paragraphs and an incompressible solid, it is possible to include the spatial variability of melt buoyancy throughout the melting region in numerical models. Melt compressibility is included in most models of this study and compared to the ascent of incompressible melt in section V.1.8.

II.1.3   Heat capacity and melting enthalpy

Richet and Bottinga (1986) stated that c_p does not vary strongly with T at temperatures above the Debye temperature, although variations are not negligible in precise thermodynamic calculations, but for the aluminous liquids considered here, the increases are not larger than 0.01 %/K. Similarly, $(\partial c_p/\partial p)_T$ is expected to be relatively small, so that it can be neglected (Bottinga, 1985). The c_p of multicomponent liquids can be approximated as the sum of the molar heat capacities of the oxide components (Navrotsky, 1995).

To determine the melting enthalpy L_{m} of peridotite, experimental values such as those of table D.11 must be corrected for the p–T conditions of melting and the effects of solid solution and mixing, and have to be weighted with the proportions of the melting phases. Hess (1992) estimates the effect of temperature on L_{m} to be ≈ 0.42 J/g K and the effect of pressure ≈ 29.4 J/g GPa under the conditions of suboceanic mantle. Comparison of experimental results from multicomponent systems with the weighted sums of the respective endmember enthalpies indicates that summation and cancellation of the enthalpy of mixing are justified, and that the minor components contribute only 5 . . . 6 % to the total latent heat change (Kojitani and Akaogi, 1997); experiments also confirm the significance of p and T effects on L_{m} (Kojitani and Akaogi, 1995). As the temperature-corrected L_{m} for the main components of peridotite are quite similar at upper mantle temperatures, Hess (1992) concluded that $\Delta L_{\mathrm{m}}(T)$ does not change significantly for varying degrees of depletion. As a working estimate for L_m along a 1430 °C adiabat in the pressure range of 1...3 GPa, he calculated 712...795.5 J/g. This, however, is a rather high value compared with a variety of experimental results for both endmember minerals and more complex systems as summarized in table D.11, and other authors argue for entropy values between 0.3 J/gK (Hirschmann *et al.*, 1999) and 0.4 J/gK (McKenzie and O'Nions, 1991; Kojitani and Akaogi, 1997) for the melting entropy, which corresponds to an enthalpy of 510 and 680 J/g at the same temperature, respectively.

The large uncertainty in melting enthalpy is unfortunate, because varying this value has considerable impact on melt production in the mantle and hence on the amount of material available for crust production. Therefore, the effect of latent heat is highlighted in some numerical models in sections V.1.4 and V.1.5.

II.1.4 Viscosity

Several experiments have established that the viscosity η_f of most silicate melts, in particular of those which are derived from partial melting of upper mantle rocks, decreases with increasing pressure, as well as with increasing temperature and water content; the reason for this behaviour probably is depolymerization and a change of the coordination number for certain atoms in the melt. Different attempts have been made to quantify the temperature dependence of η_f, *e.g.* by Arrhenius-type laws, but simple relations frequently fail to describe the physical behaviour for a large range of temperatures resp. viscosities (*e.g.* Richet, 1984). More sophisticated approaches try to capture the temperature dependence by using a configurational entropy related to the number of system configurations accessible in structural rearrangements during flow (Adam and Gibbs, 1965; Richet, 1984; Bottinga and Richet, 1995), and attempts have also been made to introduce pressure dependence into this or other theoretical models (*e.g.* Bottinga and Richet, 1995; Taniguchi, 1995).

Kushiro (1986) has measured the viscosities of different melts in the pressure interval from 0.7 to 1.5 GPa and stated that the decrease is of about a factor 2 for alkali-rich melts, but less in olivine tholeiites; at lower pressures, $\partial\eta/\partial p$ is even smaller. He also estimated the viscosity for partial melts along the anhydrous solidus of peridotite and concluded that it decreases from 2.5 Pa s at (0.8 GPa, 1220 °C) to 0.2 Pa s at (3.5 GPa, 1580 °C), which of course involves compositional changes; the results of Kushiro *et al.* (1976) for olivine tholeiite melt are roughly of the same order of magnitude. It then depends on $\varrho(p)$ whether the viscosity drop results in an enhancement of melt mobility or if low viscosity/high pressure melts remain at great depths and influence the rheology of the rock at the base of the melting zone. In fact, Bagdassarov (1988) deduced from a simple buoyancy-driven, one-dimensional model of melt segregation for melts from spinel lherzolite or pyrolite that the viscosity exerts the principal control on segregation rate due to its large changes over the depth range of melting, so that deep melts might actually ascend faster if appropriate driving forces act. Taking into consideration changes in chemical composition, in particular the amount of network-forming components, he derived the empirical relation[1]

$$\ln \eta_f = \frac{A_1 \exp(-A_2 \mathcal{P}_{SC}) - A_3 \mathcal{P}_{SC} + \mathcal{E}_0}{RT} - A_4 p - A_0, \qquad (II.7)$$

with η_f in Pa s, p in GPa, T in °C, $A_1 = 4.9056 \cdot 10^4$ J/mol, $A_2 = 5.8 \cdot 10^{-3}$, $A_3 = 85.393$ J/mol, $A_4 = 0.502$ 1/GPa, $A_0 = 5.803$ and $\mathcal{E}_0 = 7.6672 \cdot 10^4$ J/mol. The Saucier–Carron parameter $\mathcal{P}_{SC}$ is a measure of the ratio of oxygen to network-forming components (*e.g.* Si^{4+}, Al^{3+}, Fe^{3+});

---

[1]Here and in several of the following formulae quoted from other studies, the units were transformed to SI if necessary.

Figure II.2: Viscosity η_f of mantle melts (in Pas) for $\mathcal{P}_{SC} = 150$
(solid lines) and $\mathcal{P}_{SC} = 100$ (dashed
lines) according to eq. II.7. Note
that the assumption of a constant
$\mathcal{P}_{SC}$ is a simplification, because melt
compositions change with f, espe-
cially at incipient melting. In reality, of course, melts are only present
in the region above the solidus
(Hirschmann, 2000, see eq. II.25c).

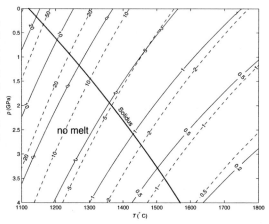

it was found to range between *ca.* 50 at low p and f and 150 at high p and f in experimental
batch melts (Bagdassarov, 1988), but can probably be assumed to be > 100 for all but the
initial melts in the case of fractional melts. As figure II.2 shows, eq. II.7 agrees roughly
with the estimates by Kushiro (1986), although the original values were derived for spinel
lherzolite up to only 2 GPa.

As already mentioned, the composition also affects the viscosity. For example, for hydrous
melts, miscibility of water and silicate melts may be an issue: Bureau and Keppler (1999)
observe complete miscibility for $p \gtrsim 2$ GPa and expect unmixing to occur between 30 and
60 km depth; it is reasonable to assume a change in melt viscosity due to this mechanism if
the melt contains a significant amount of water.

Because of its potential interest for the segregation of deep melts, the effect of variable
viscosity will be given consideration in section V.1.8, but the effect of volatiles will not be
included here.

II.1.5   Melt geometry and the permeability–porosity relation

As will be detailed further in section II.2, melt has a substantial effect on several physical
properties of mantle rock such as density and viscosity, on its behaviour in convection, its
chemical signature, and its relation to MORB. This effect does not only depend on the mere
presence of melt, but also on its geometry, which also is essential for its mobility in the
mantle. Field studies suggest that melt is distributed in vein-like structures of a certain
orientation (Boudier and Nicolas, 1972; Menzies, 1973) allowing for efficient segregation, but
that a certain amount of melt is retained in parts of the mantle rock, with melt fraction
estimates ranging from about 2 % (Seyler *et al.*, 2001) to 6–7 % (Dick, 1977). Only the lower
end of this range seems to be in accordance with results of several laboratory experiments
as well as geochemical investigations and estimates from seismic and electromagnetic data
acquired at MOR, which are discussed in the following and indicate melt contents between
0.1 % for low-degree volatile-rich melts and 1–3 % (*e.g.* McKenzie, 1985a,b, 1987; Johnson
et al., 1990; Faul *et al.*, 1994; Hellebrand *et al.*, 2002). Furthermore, the amount of retained
melt, resp. the fraction of melt which leaves the melting region at some point and erupts or
intrudes the overlying thermal boundary layer, is also crucial for the thickness of the crust.

Two idealized models are usually considered when discussing the process of melt generation and migration and its chemical and petrological consequences (*e.g.* Wilson, 1989):

Equilibrium or batch melting: All of the generated melt remains in its source region until it possibly segregates as a whole. In this melting mode, thermal and chemical equilibrium exists during the whole time interval of melt production, and the bulk composition of the melt–matrix system does not change.

Fractional or Rayleigh melting: The melt is continuously and instantaneously removed from its source region, so that it cannot equilibrate with the matrix. The bulk composition of the system changes throughout the melting process.

Clarification of the question which of these models is closer to reality for mantle melting is crucial for assessing the mobility of the melt at a given porosity and explaining measured physical effects of melt and the development of the observed chemical signatures.

It is generally assumed that the amount of melt present in the rock, *i.e.* the porosity, and the permeability are related by an equation of the form

$$k_\varphi = \frac{a^2}{b} \varphi^n, \tag{II.8}$$

whereby a is of the order of millimetres for mantle peridotite; Toramaru and Fujii (1986) give sizes of 2–3 mm for olivine and orthopyroxene and of 0.5–1 mm for clinopyroxene in the spinel lherzolite KLB-1 (see appendix D.1.1), which would give a weighted average grain size of about 2.2 mm, with the small fraction of spinel being neglected. However, experiments have shown that porosity and permeability cannot simply be linked by an equation like eq. II.8, but that additional aspects of melt geometry, crystal structure and modal composition have to be taken into account in order to determine under which conditions porous flow is possible at all.

At the very first stages of melting, the melt occurs as isolated droplets at the grain corners of different minerals. The shape of these droplets strives to minimize energy and depends on the ratio of grain–grain and grain–melt surface tension, γ_{ss}/γ_{sf}, which is of the order of $0.3 \ldots 0.9 \, \text{N/m}$ for partially molten upper mantle materials (Laporte and Provost, 2000) and related to the dihedral angle Θ by

$$\cos \Theta_{s_1 s_2 f} = \frac{\gamma_{s_1 s_2}^2 - \gamma_{s_1 f}^2 - \gamma_{s_2 f}^2}{2 \gamma_{s_1 f} \gamma_{s_2 f}} \tag{II.9}$$

(*e.g.* Toramaru and Fujii, 1986) in the case of two different isotropic solids. The value of Θ is crucial for the formation of an interconnected melt network. For one solid phase, the connectivity criteria are $\Theta_{ssf} < 60°$ for edges resp. $\Theta_{ssf} < 70.5°$ for corners, whereas for larger dihedral angles, a critical melt fraction φ_c must be exceeded for interconnection to take place.

The dihedral angles in mantle rock depend considerably on petrological composition and water content. Experimental results with olivine–basalt mixtures (*e.g.* Waff and Bulau, 1979) and numerical investigations (von Bargen and Waff, 1986) indicated that the connectivity criteria for eq. II.9 are always fulfilled and that the melt is interconnected if $\varphi > 0.01 \ldots 0.02$, so that it can be extracted; these results have had a great influence on many later studies, among them those of McKenzie (1984) and Spiegelman and McKenzie (1987). However, calculations and experiments with more complex systems (Toramaru and Fujii, 1986; Bussod and Christie, 1991; Kohlstedt and Zimmerman, 1996), which are more appropriate for the deeper parts of the melting domain, showed that the presence of pyroxenes hinders the formation of a melt network. In such assemblages, edges with different solids and melt also have to be taken into account; in an ol–opx–cpx system 25 different types of edges or corners are possible, of which only the ol–ol–melt edges and the ol-only and ol–ol–ol–opx corners seem to establish interconnection. Toramaru and Fujii (1986) deduced from a statistical connectivity model that the rock will become permeable when a critical value is exceeded; from this follows that connectivity controls the type of melting (Maaløe, 1982) and is related to a certain critical solid modal composition, which depends on the state of melting and could be determined on the basis of melting proportions (*e.g.* section II.3.3). Although their unrealistically high threshold for full interconnection of 29 % is considerably relaxed in a later model (Nakano and Fujii, 1989), this latter study confirmed the existence of a modally controlled threshold and predicted sudden establishment of full melt connectivity in a natural three-phase peridotite for olivine fractions above 63 vol. % at $\varphi \geq 0.008$.

Water is known to lower the dihedral angles in partially molten olivine–pyroxene systems, thereby supporting interconnection of melt (Fujii *et al.*, 1986; von Bargen and Waff, 1988; Daines and Kohlstedt, 1993). These experiments indicate that the presence of orthopyroxene is of only minor importance *in wet peridotite*. The estimates of Nakano and Fujii (1989) and the modal compositions listed in table D.1 thus suggest that the low-connectivity interval ranges at most from the end of the hydrous melting stage to the melting of the first few percent of pyroxenes.

A shortcoming of the above considerations is that crystalline anisotropy, which results in a direction dependence of the surface tensions, is not accounted for. In the realistic case of anisotropic crystals, eq. II.9 is not valid anymore, but nonetheless, the average equilibrium dihedral angle in monomineralic systems is only slightly different from the isotropic reference case (Laporte and Provost, 2000). More recent experiments on olivine–basalt mixtures (Waff and Faul, 1992; Cmíral *et al.*, 1998; Jung and Waff, 1998), however, indicated that already at $\varphi < 5\%$ many flat grain boundaries (F-faces) of olivine are also wetted by melt and that its topology is quite variable and cannot be described by a single dihedral angle; on the other hand, F-faces might form more easily in the small grains used in experiments than in the mantle (Cmíral *et al.*, 1998). If they exist in nature, the presence of melt on them at melt fractions of a few percent will have a strong impact on the large-scale permeability of the melt network at increasing melt fractions (Laporte and Watson, 1995; Schäfer and Foley, 2002), especially if the grains are aligned, *e.g.* by lattice preferred orientation (LPO). In fact, petrological and seismological observations indicate that the olivine crystals are oriented such that the horizontal permeability would be enhanced and the vertical reduced, thus blocking the buoyant ascent of melt in mantle regions far from the spreading center (Waff and Faul, 1992). In contrast, deformation-induced LPO near the ridge can increase k_φ in direction to the spreading center (Ribe, 1989). On the other hand, Wark *et al.* (2003) did not find substantial amounts of melt on F-faces, but rather confirmed the tubule network model.

In all these experiments only the behaviour of melt–rock systems under hydrostatic pressure has been investigated; in contrast, partially molten rock experiences considerable shear deformation in the mantle beneath a spreading MOR. The distribution of melt changes drastically if the rock is deformed at stresses of a certain magnitude, although it is still now very clear at which stresses the transition takes place (Bussod and Christie, 1991; Jin *et al.*, 1994; Kohlstedt and Zimmerman, 1996; Daines and Kohlstedt, 1997; Karato *et al.*, 1998; Zimmerman *et al.*, 1999). Several of these experiments showed that under deformation, the melt concentrates in planar pockets with a length of several grain diameters which are oriented at about 20–30° from the maximum principal stress σ_1 under mantle-like conditions; as the degree of orientation depends, among other factors, on the applied differential stress and on a, Daines and Kohlstedt (1997) estimated that in coarse-grained real mantle rock stresses of a few kilopascals, which are easily reached in the upper mantle below spreading ridges, are sufficient to induce a melt-preferred orientation (MPO) similar to that observed in the experiments. The differential stress dominates melt distribution, and the strong preference of a certain orientation arises because the tips of favourably oriented melt pockets are sites of high tensile stress and low mean pressure, so that melt from pockets of non-optimal orientation will follow the pressure gradient and be drained into the optimum pockets (Daines and Kohlstedt, 1997; Zimmerman *et al.*, 1999). The formation of such a pronounced MPO is thought to have a measurable effect on seismic anisotropy and also leads to a strong anisotropic permeability, thereby greatly enhancing directional melt flow and possibly supporting the concentration of melt in veins as suggested by Stevenson (1989) (Zimmerman *et al.*, 1999, see sect. II.4.2); Daines and Kohlstedt (1997) calculated $k_\parallel / k_\perp$ ratios between 3.2 and 15.1 for their experiments.

Although deformation experiments are closer to the conditions in the mantle, the pyroxene problem was neither resolved by the experiments on olivine–basalt aggregates, where pyroxene was not present as a significant solid phase at all, nor by the experiments on lherzolite of Bussod and Christie (1991), who had added 1.6 % water to their samples for technical reasons, which may have masked the pyroxene effect. However, as the experiments of Jin *et al.* (1994) have been performed on dry samples and also resulted in coverage of olivine and pyroxene grains with melt, it may be concluded that their observation is applicable to the mantle beneath ridges and that indeed connectivity is already established at rather low melt fractions of 4–7 % as achieved in these experiments, especially if one assumes initial melting to occur in the presence of volatiles.

It should though be emphasized that mere connectivity is not a sufficient condition for efficient melt drainage. As can be seen from eq. II.8, the exponent n and the geometry factor b also have to be taken into account and are likely to change significantly over the first few percent of melt generation in the course of the geometrical reconfiguration of the melt. Most common choices for n are between 2 and 3 (*e.g.* Schmeling, 2000; Tanimoto and Stevenson, 1994), although Faul (2000) stated that n can be greater than 2 or even 3 for very small φ, so that k_φ will be extremely low, and no relevant segregation will take place in spite of interconnectivity; in addition, Faul (2001) estimated on the basis of experiments on olivine–basalt systems that $\varphi_c = 0.02 \ldots 0.03$ and that k_φ will then increase by at least four orders of magnitude. In contrast, Wark *et al.* (2003) claimed that the tubule network is adequate also at high φ (with $b = 270$ in eq. II.8) and that a threshold permeability separating a low-k_φ and a high-k_φ domain as proposed by Faul (1997) does not exist. Their tubule network model would have a higher permeability at very low porosities.

So far, a uniform porosity structure has been assumed for the mantle; however, the larger-scale structural features discussed in more detail in section II.4.2 and the chemical characteristics outlined in section II.2.5 suggest that there is more than one porosity scale in the mantle. As an example, Lundstrom *et al.* (2000) proposed the existence of two such scales: one provided by highly permeable dunite channels for fast non-equilibrium flow (see section II.4.2), which leads to the fractional-melting geochemical signature, and another one with low-porosity ($\varphi \leq 0.001$) lherzolite or harzburgite, where equilibrium porous flow takes place.

II.2 THE INFLUENCE OF MELT ON MATERIAL PARAMETERS AND CONVECTION

Melting processes, the presence of melt (as well as that of another fluid phase) and its geometry and degree of connectivity have a considerable and complex impact on many physical properties of the mantle, *e.g.* on the viscosity η, the seismic velocities $v_{P,S}$ and the attenuation $Q_{P,S}^{-1}$, the electrical conductivity σ, the density ϱ and also on the volatile content and chemical composition. Therefore, a brief overview of its influence on these parameters is given in this section.

II.2.1   Viscosity and creep mechanisms

The presence of melt generally leads to a decrease of the overall shear viscosity of the mantle. To describe its effect on creep strengths, eq. I.11 has to be amended by a term describing the dependence of the strain rate on the melt fraction φ for the different creep regimes. As a unified description for both diffusion and dislocation creep, Mei *et al.* (2002) suggest

$$\dot{\varepsilon}(\varphi) = \dot{\varepsilon}(0)\, e^{s\varphi} = \mathcal{A}' \, \frac{\sigma^n}{a^q} f_{O_2}^m \, e^{-\frac{\mathcal{E}+pV}{RT}+s\varphi} \tag{II.10}$$

(cf. eq. I.11); for the dislocation regime, $s = 31$, and for the diffusion regime, $s = 26 \ldots 28$ (Kohlstedt *et al.*, 2000). The distribution of melt in pockets resp. on grain surfaces (see section II.1.5) under shear results in viscosities considerably below that of olivine (Bussod and Christie, 1991; Jin *et al.*, 1994) and, taking into account the melt-preferred orientation, in anisotropy, if a certain fraction is retained. As discussed in more detail in section II.1.5, melt seems to wet the grain boundaries to a considerable extent; this is also supported by the strong increase of strain rates with melt fraction in diffusion creep observed by Kohlstedt and Zimmerman (1996), which cannot be explained with the tubule model. However, a melt fraction of at least $7\,\%$ is necessary for a viscosity reduction of one order of magnitude (Kohlstedt *et al.*, 2000; Gribb and Cooper, 2000).

Kohlstedt and Zimmerman (1996) compared $\dot{\varepsilon}$–σ relations for olivine–basalt aggregates and spinel lherzolites at different melt fractions and observed a similar magnitude of the effective viscosities in both media; nevertheless, the strain rate seems to increase rather smoothly with φ in lherzolite, while there is an pronounced change in the behaviour of the olivine–MORB samples at $\varphi \gtrsim 4.5\,\%$, which they ascribe to different wetting characteristics of the mineralogies. In the lherzolite experiments, they also observed a change in the activation energy $\mathcal{E}$ in diffusion creep.

It is obvious and was confirmed by numerical experiments by Buck and Su (1989) and Su and Buck (1993) that the effect of melt should enhance mantle flow in supersolidus regions and can lead to a concentration of the upwelling near the spreading center and to small-scale convection, although it must be noted that melt fractions $> 10\,\%$ as assumed by Buck and Su (1989) are not realistic (see section II.1.5). Furthermore, it has been claimed that the concentration of melt in pockets observed in differential-stress melting experiments (see sect. II.1.5) would ensure low viscosities in the melting zone even at low porosities (Rabinowicz and Briais, 2002); it is though questionable whether the formation of this structure simply results in a decrease of the viscosity or whether it rather makes it anisotropic, so that the material is weaker in certain directions, which are though not necessarily favourable for mantle flow.

However, it is quite possible that the abovementioned effect of melt is counteracted by the influence of water and its extraction, at least at low melt fractions. In the deeper, subsolidus parts of the mantle a small amount of water is present (*e.g.* Bell and Rossman, 1992) which decreases the viscosity of rocks by at least two orders of magnitude (*e.g.* Hirth and Kohlstedt, 1996; Mei and Kohlstedt, 2000a,b, see section I.2.4). On the other hand, the strong incompatibility of water led Karato (1986) to the estimate that about 0.1 wt% of melt will mostly dehydrate the olivine and thus cause a substantial increase of viscosity in the supersolidus domain; Bell and Rossman (1992) and Hirth and Kohlstedt (1996) estimated a partition coefficient of 0.01 for water in peridotite, with which the water content in the partially molten rock can be calculated using the formulae of Shaw (1970) in eqs. II.24, resulting in a more moderate, but still large viscosity change. It is consistent with this notion that creep experiments on hydrous partially molten olivine–basalt aggregates ($\varphi =$ 0.02...0.12) lead to similar results as experiments on anhydrous ones (Hirth and Kohlstedt, 1995; Mei *et al.*, 2002). Phipps Morgan (1997) presented numerical models in which a layer of melting residue with a viscosity higher by a factor 50–500 and a reduced density is formed beneath a spreading center and is transported with the drifting oceanic plates, thereby forming a compositional lithosphere with a sharp viscosity jump at its lower boundary (see also Hirth and Kohlstedt, 1995, 1996). Braun *et al.* (2000) investigated the effect of incipient low-productivity "damp melting" under hydrous conditions in the garnet stability field as a prelude to the common anhydrous melting and the effect of dehydration, melt retention and rheological transitions. They and Ito *et al.* (1999) observed that water extraction indeed leads to high viscosities and inhibited buoyant upwelling in the dry melting regime, but hydrous melting enables grain boundary sliding in the deep melting interval, which results in low enough viscosities $\sim 10^{18}$ Pa s for buoyant upwelling. According to Braun *et al.* (2000), melt retention seems to be of minor importance unless the threshold is unrealistically high. – In this study, the effects of retained melt and dehydration are included in several models following the suggestions from Mei *et al.* (2002) and Hirth and Kohlstedt (1996), respectively, and confirm the general picture from the earlier investigations; a detailed discussion and results are given in sections V.1.4 and V.2.3.

The increasing depletion in pyroxenes during the progress of melting also shifts the relative proportions of mineral phases with different flow strengths towards the stiffer olivine rheology. Based on the experiments by Ji *et al.* (2001) and on the theoretical model of Wheeler (1992), it can be speculated that this will further increase the viscosity in the melting domain, although the extrapolation from the experiments, which were performed on a binary, iron-free system at normal pressure, to a four-phase mantle rock under high-p conditions can only be a qualitative assessment at best. As far as the application of viscosity mixing rules is concerned, however, the shift to higher olivine contents favours the use of strong-phase supported mixing models.

II.2.2 Seismic wave propagation

The influence of melt on seismic observables, *e.g.* on the seismic velocities $v_{\mathrm{P,S}}$ and on the attenuation $Q_{\mathrm{P,S}}^{-1}$, is quite complicated, among other reasons due to its dependence on the frequency of the seismic waves and the effect of variable melt pocket geometries and degrees of interconnection, which are partially controlled by the anisotropy of the crystals (see section II.1.5). In particular, wave propagation under mid-oceanic ridges or through plumes is notably influenced by the presence of melt and by the flow-induced orientation of olivine crystals, which results in seismic anisotropy.

Elastic moduli and seismic velocities — The elastic moduli are complex, frequency-dependent quantities, and their zero-frequency and infinite-frequency limits are defined as the relaxed and the unrelaxed elastic moduli, corresponding to isolated and connected, pressure-equalized inclusions, respectively. In the calculation of the effective elastic moduli of a medium with both connected and isolated, densely distributed melt inclusions, one has to take into account their geometry and the interactions between them; the unrelaxed moduli are independent of the degree of interconnection, while the relaxed moduli strongly depend on the connectivity. Gassmann (1951) described the long-wavelength limit of the elastic moduli of a saturated porous medium as the sum of the moduli of the frame and the fluid-filled pores.

The actual moduli are usually bracketed between the upper and lower Hashin–Shtrikman bounds

$$K_{\mathrm{HS+}} = (1 - c_K)K_{\mathrm{s}} + \frac{c_K^2}{\frac{\varphi}{K_{\mathrm{f}}} + \frac{c_K - \varphi}{K_{\mathrm{s}}}} \qquad\qquad K_{\mathrm{HS-}} = \frac{1}{\frac{\varphi}{K_{\mathrm{f}}} + \frac{1-\varphi}{K_{\mathrm{s}}}} \qquad (\mathrm{II.11a})$$

$$\mu_{\mathrm{HS+}} = \left(1 - \frac{5(4\mu_{\mathrm{s}} + 3K_{\mathrm{s}})\varphi}{(8\mu_{\mathrm{s}} + 9K_{\mathrm{s}}) + (12\mu_{\mathrm{s}} + 6K_{\mathrm{s}})\varphi}\right)\mu_{\mathrm{s}} \qquad \mu_{\mathrm{HS-}} = 0 \qquad (\mathrm{II.11b})$$

with $c_K(\varphi) = [(4\mu_{\mathrm{s}} + 3K_{\mathrm{s}})\varphi]/(4\mu_{\mathrm{s}} + 3K_{\mathrm{s}}\varphi)$ (Nolen-Hoeksema, 2000; Hashin and Shtrikman, 1963); the lower Hashin–Shtrikman bounds are equal to the Reuss bounds, but the upper Hashin–Shtrikman bounds are lower than the Voigt bounds, which are the porosity-weighted average of the frame and pore moduli. Arguing that the matrix moduli drop to zero above a critical disintegration porosity φ_{c}, these bounds have later been modified by replacing φ by $\varphi/\varphi_{\mathrm{c}}$ in eqs. II.11. Extending the general model, Schmeling (1985a) and other workers have derived a set of equations for the relaxed and the unrelaxed bulk and shear moduli and an ensemble of different melt inclusion geometries, in particular for film-shaped, tubular, and spheroidal inclusions. Numerical tests by Schmeling (1985a) have shown that for aspect ratios > 0.2 the effective moduli do not vary much as a function of the aspect ratio.

From the elastic moduli follow the seismic velocities, thus it is obvious from the above that the change from a purely solid body to a two-phase system consisting of a solid and a liquid phase will result in a change in seismic velocities, which makes the presence and content of melt in the earth observable. In particular, v_{P} and v_{S} are affected in different ways, because for v_{P}, the bulk modulus of the liquid phase is important. Therefore, $v_{\mathrm{P}}/v_{\mathrm{S}}$ is a quantity of special interest for the seismological detection of melt, but $\mathrm{d}\ln v_{\mathrm{P}}/\mathrm{d}\ln v_{\mathrm{S}}$ also contains information on pore geometry and is higher for disequilibrium structures (Takei, 2002); hence, it should be possible to derive constraints on melt transport mechanisms from seismological measurements. – Tanimoto and Stevenson (1994) used S-wave velocities from ridge areas and simple plate cooling and melt percolation models with a power-law k_{φ}–φ relation similar to eq. II.8 to show that the S-wave anomalies in the upper few dozen kilometers beneath a spreading center cannot be explained by the thermal effect alone, but must be caused by a few percent melt. P-wave travel times under MOR are not very sensitive to the microstructure of the melt and are mainly controlled by the texture effect, whereas S-waves are more noticeably influenced by the melt and its geometry, especially the vertically travelling ones (Blackman and Kendall, 1997; Karato *et al.*, 1998). It seems that in the case of buoyant upwelling, which is also of particular interest for plumes, large effects will occur in a narrow region around the center of the upwelling, whereas passive upwelling is characterized by a lower-amplitude anomaly in a broader domain (Blackman and Kendall, 1997). Apart from this direct effect, Braun *et al.* (2000) proposed that the reinforcement of finite strain due to elevated viscosities in mantle dehydrated by melting will enhance LPO of mantle minerals and thus seismic anisotropy; the seismically fast $\mathfrak{a}$ axis of olivine aligns in the flow direction.

Attenuation — The attenuation of seismic waves, which is measured in terms of the quality factor Q (or its inverse Q^{-1}), varies significantly throughout the mantle and is mostly related to the response to shear; it is strongly correlated with v_{S}, low-v_{S} regions corresponding to high-Q_{S}^{-1} ones. Strong attenuation coincides at least partially with the low-velocity zone, with $Q_{\mathrm{S}}^{-1} > 0.01$ between 80 and 220 km depth, whereas above and below it drops to < 0.003 (*e.g.* Tan *et al.*, 2001).

Q is composed of an anharmonic and an anelastic contribution (see appendix D.1.8); the latter is a function of frequency and possibly important at seismic and sub-seismic frequencies (Karato, 1993). There are different physical mechanisms which possibly can cause attenuation. Viscous shear is one such effect, but it was shown to be of importance for seismic waves only at viscosities which are much higher than those expected for melts in the asthenosphere. Melt squirt, *i.e.* pressure-driven fluid flow in or between pores deformed during the transition of the seismic wave could be important at seismic frequencies for small aspect ratios of the melt inclusions. Reduced interconnection leads to a reduced absorption at melt fractions which cause a moderate modulus decrease, whereas superposition of several melt geometries lets the bulk relaxation strength and the corresponding absorption increase significantly (Schmeling, 1985a).

Experimental data for seismic and sub-seismic frequencies (Gribb and Cooper, 2000; Bagdassarov, 2000) show a relatively weak power-law relation between Q_S^{-1} and frequency and an Arrhenian dependence on T and suggest that a melt phase equilibrated with the matrix on the grain scale does not enhance attenuation significantly and has the same effect as a temperature increase of about 50 K; this is basically in agreement with numerical models of Hammond and Humphreys (2000). For Q_P, Sato *et al.* (1989) found an even weaker frequency dependence under p–T conditions below and above the solidus and stated that it depends rather on the homologous temperature than on φ; it must though be noted that their experiments were performed at ultrasonic frequencies and that they possibly did not measure Q_P under relaxed conditions.

Although it is a fairly wide-spread opinion that the low v/high Q^{-1} depth interval corresponding to the asthenosphere is characterized by the presence of a small amount of melt or at least by temperatures near the peridotite solidus, it is worth mentioning that the presence of water in important minerals such as olivine could produce this effect as well; in this case, the asthenospheric features would rather be due to the effect of water and the *absence* of melting, while onset of melting at shallower levels would dry out the rock quite efficiently and lead to a pronounced increase of velocities, under the assumption of fractional melting (Karato and Jung, 1998).

Application to results of dynamical models — In sections V.2.3 and V.2.4, the results of mantle and melt dynamical models – the temperature, porosity and depletion fields, and possibly the spatial distribution of water content – are taken as a starting point for the computation of seismological observables. The basis for these calculations are partly the models and mechanisms outlined above; Kreutzmann *et al.* (2004) have extended the method and updated it with newer experimental results, and their procedure is followed in this study with small modifications.

The anelastic contribution to the T effect on seismic velocities includes the frequency-dependent attenuation of the form

$$Q^{-1} = \frac{B_Q}{a^m}\left(\frac{\mathfrak{c}}{\omega}\right)^{\mathcal{Q}} e^{-\frac{\mathcal{Q}\mathcal{E}}{RT}}, \qquad \mathfrak{c} = \begin{cases} 0.56 C_{\mathrm{OH}} & C_{\mathrm{OH}} \geq 1.7857\,\mathrm{ppm\,^H/Si} \\ 1 & \text{else,} \end{cases} \tag{II.12a}$$

where $\mathfrak{c}$ is a dislocation mobility increase factor; in the following, $B_Q = 12.6\,\mathrm{^{m^m}/s^Q}$, $m = 0.33$, $\mathcal{Q} = 0.25$, $\mathcal{E} = 430\,\mathrm{kJ/mol}$ (after Jackson *et al.*, 2000; Karato and Jung, 1998), $\left(\partial \ln v_P/\partial \ln \varrho\right)_p = 2.1$, $\left(\partial \ln v_S/\partial \ln \varrho\right)_p = 2.5$ (Christensen, 1989) and $a = 1\,\mathrm{mm}$ are used, and a typical seismic frequency of 2 Hz is assumed throughout. The pressure dependence is calculated with the empirical law

$$Q(p,T) = Q(0,\hat{T})\frac{Q(0,\hat{T}(0) = 1) + \frac{p}{p_0}}{Q(0,\hat{T}(0) = 1)}, \qquad \hat{T}(p) := \frac{T}{T_s(p)}, \tag{II.12b}$$

with $p_0 = 1\,\mathrm{GPa}$ (Kreutzmann *et al.*, 2004; Sato *et al.*, 1989); T_s in this case is the solidus of olivine fo$_{90}$, which is estimated with the method detailed in section II.3.3 from the forsterite solidus eq. II.27 using eq. II.30. Following these authors, the dependence of Mg# on f is taken as $\partial \mathrm{Mg\#}/\partial f = 0.12$ (after Jordan, 1979), which can be used to include the upward shift of T_s with increasing depletion; it is implicitly assumed here that this is the only way in which depletion affects Q over the Mg# range of interest (about 0.89–0.925), but the possible error thus introduced is likely to be minor compared to the much greater simplification of equating the seismic properties of peridotite with those of olivine. Furthermore, the relation $Q_P^{-1} = 0.4 Q_S^{-1}$ is applied. Instead of determining the change of $v_{P,S}$ with melt content with the formalism mentioned above, the experiment-based values

$$\left(\frac{\partial \ln v_P}{\partial \varphi}\right)_T = -1.23, \qquad \left(\frac{\partial \ln v_S}{\partial \varphi}\right)_T = -2.04 \qquad \text{(II.12c)}$$

are applied, which give results similar to those for ellipsoidal inclusions of about the aspect ratio expected here for $\varphi_{\max} = 0.01$ (Kreutzmann *et al.*, 2004). The effect of petrological changes of the rock on seismic velocities had been found to be smaller than $1\,\%$ even for substantial melting (Jordan, 1979) and is therefore neglected.

II.2.3   Electrical conductivity

Similar to the influence on seismic properties, melt also affects the electrical properties, *i.e.* the conductivity σ of a rock. The conductivity of melt exceeds that of the solid phase by some orders of magnitude; thus, it is a common approach to model a partially molten rock as a network of possibly different resistors representing melt bodies with a varying degree of interconnection, while the solid is assumed to be an insulator. The degree of interconnection can be estimated statistically from the melt fraction and the aspect ratio of the inclusions (Schmeling, 1986).

For isotropic melt distributions, the Hashin–Shtrikman bounds (Hashin and Shtrikman, 1963)

$$\sigma_{HS-} = \sigma_s + \frac{\varphi}{\frac{1}{\sigma_f - \sigma_s} + \frac{1-\varphi}{3\sigma_s}} < \sigma < \sigma_f + \frac{1-\varphi}{\frac{1}{\sigma_s - \sigma_f} + \frac{\varphi}{3\sigma_f}} = \sigma_{HS+} \qquad \text{(II.13)}$$

are the best possible bounds for the effective conductivity. On the other hand, empirical data indicate a relation of the type $\sigma \sim \varphi^2$ known as Archie's law between the conductivity and the porosity. These relations, however, do not account for the geometry and the degree of connection of the melt. To account for the fact that the matrix in partially molten rock has a finite conductivity and including the connectivity and shape of the melt bodies, Schmeling (1986) derived a model for the effective conductivity of randomly oriented melt ellipsoids as a function of porosity and interconnectivity. σ_{eff} can then be approximated by the geometric mean of the conductivity of isolated inclusions and of the upper Hashin–Shtrikman bound,

$$\sigma_{\mathrm{eff}} = \sigma_{HS+}^{P}\sigma_{\mathrm{iso}}^{1-P}, \qquad \text{(II.14)}$$

where the connectivity P depends on the aspect ratio of the inclusions and the porosity, but is 1 when full connectivity is achieved (see Schmeling, 1986; Kreutzmann *et al.*, 2004, for details). As is obvious from section II.1.5, the geometry resp. connectivity parameters are functions of φ and can vary between different domains of melt generation and transport. Numerical modelling for different melt fractions, geometries and degrees of interconnection show that a reduced connectivity can cause significant deviations from values suggested by the upper Hashin–Shtrikman bound at porosities which are also related to moderate negative seismic anomalies. It was also shown that for $\varphi > 5\%$ Archie's law is an acceptable rule of thumb (see Schmeling, 1986), but it seems unlikely that such high melt fractions are reached in large volumes of the mantle (see sections II.1.5 and II.4.2).

σ_s is also variable, being a function of T, p and composition, and is composed of several different mechanisms; in the earth, grain interior conductivity seems to be the dominating term, while grain boundary processes are negligible as long as no melt is present (Huebner and Dillenburg, 1995). For a given mineral, $\sigma_s(p, T)$ can be expressed by an Arrhenius-type law (see eq. D.20). For a mixture of two mineral phases, the effective conductivity is bracketed by the Hashin–Shtrikman bounds and can be approximated by mixing rules (Xu et al., 2000; Berryman, 1995).

It should be mentioned that in a similar way that water dissolved in olivine mimics the effect of melt on the propagation of seismic waves, it can also serve as an alternative explanation of the high conductivity in parts of the upper mantle (Karato, 1990), even at a water content much lower than that necessary to produce a significant downward shift of the solidus, so that it would be compatible with a melt-free, well-conducting mantle.

Application to results of dynamical models — In analogy to the prediction of seismological observables, a conductivity model can be derived from the geodynamical model using the above models. Again, the respective method of Kreutzmann et al. (2004) will be applied to some models of sections V.2.3 and V.2.4 in a modified and extended form, which is shortly described here.

For calculating the electrical conductivity, the effects of p, T, composition, and melt content have to be taken into account. The melt content is treated after the theoretical model by Schmeling (1986). For the composition of the solid matrix, the p–T data from Xu et al. (2000) for olivine, orthopyroxene, and clinopyroxene (see appendix D.1.9), recalculated for Mg# = 0.88 with $\partial \log \sigma_s / \partial \mathrm{Mg\#} = -12$ (Kreutzmann et al., 2004; Tyburczy and Fisler, 1995) for all three minerals, is used; garnet resp. spinel are considered to be insignificant, so that mineral fractions for the standard modal mineralogy have to be renormalized for this calculation. The effective conductivity of the solid is then determined as the geometric mean of the three constituent conductivities, $\sigma_s = \Pi_{i=1}^3 \sigma_i^{X_i}$, (Shankland and Duba, 1990). As the mineralogy changes due to melting, σ_s will also change as a function of f; this is approximately taken into account by using a simplified parameterization of modal composition as a function of f (cf. section II.3.3). The effect of iron loss by depletion can be expressed as

$$\frac{\partial \log \sigma_s}{\partial f} = \frac{\partial \log \sigma_s}{\partial \mathrm{Mg\#}} \frac{\partial \mathrm{Mg\#}}{\partial f} = -1.44 \Rightarrow \sigma_s(p, T, f) = \sigma_s(p, T, 0) \cdot 10^{-1.44f}, \qquad (\text{II.15})$$

where $\sigma_s(p, T, 0)$ is given by the Arrhenian-type relation eq. D.20.

As mentioned, the presence of hydrogen ions is expected to cause a strong increase in σ_s in hydrous models. An attempt can be made to estimate the corresponding effect by adding the conductivity for the anhydrous material and the conductivity caused by H^+ using the Nernst–Einstein equation for monovalent ions,

$$\sigma_{H^+} = \frac{\kappa_H e_0^2 C_H}{k_B T}, \qquad (\text{II.16})$$

(after Tyburczy and Fisler, 1995); the diffusion coefficient κ_H obeys an Arrhenius equation (Brady, 1995). As only diffusion data for olivine have been available, it is assumed that the other minerals behave like olivine for this matter, and the parameters from Mackwell and Kohlstedt (1990), $\kappa_{H,0} = 5.998 \cdot 10^{-5}\,\mathrm{m^2/s}$ and an activation energy of $130\,\mathrm{kJ/mol}$, are used. For the calculation of the charge carrier concentration, C_H, the average density of the hot model

Figure II.3: Electrical conductivities resp. resistivities for some mineral components and assemblages of dry and hydrous peridotite at different pressures after eqs. II.15 and II.16, using the data of Xu *et al.* (2000). The curve used by Kreutzmann *et al.* (2004) is also shown for comparison.

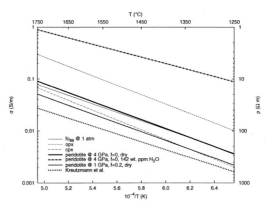

parts of *ca.* $3500 \, \mathrm{kg/m^3}$ is taken. The decrease of C_H in the solid with increasing depletion follows from the partitioning of water between melt and matrix (see section II.2.5) and is determined with equation II.24d (Shaw, 1970). As no data have been available for diffusion in melt, this contribution has to be neglected here, although it may well be significant at least at incipient melting. – Figure II.3 shows some example curves for solid conductivities in the temperature range relevant to the mantle in these models. The strong increase of σ_s when water is added corresponds approximately to the prediction of Karato (1990) for a hydrogen content of $1000 \, \mathrm{ppm} \, {}^H/_{Si}$. The curve for dry, depleted peridotite agrees quite well with the one by Kreutzmann *et al.* (2004).

II.2.4   Density and gravity anomaly

Melting changes the ratio of mineral components in the rock (see section II.3.3) and thus leads to a change in the density of the residue; besides, the melt itself has a density which is different from that of the parental material. The preferred loss of Fe from the residue (see section II.2.5) has long been assumed to enhance buoyant upwelling (Oxburgh and Parmentier, 1977) and possibly raise the melting rate by increasing material flux through the melting zone. However, this is not necessarily the case: if both thermal and chemical buoyancy are taken into account, thinning of the lithosphere can lead to trapping of the depleted material and counteract further upwelling and melting by formation of a depleted root, so that an initial peak in melt production is followed by a long period of unsteady decrease (Manglik and Christensen, 1997); this effect is particularly likely for small spreading velocities, while at larger v_r the depleted batch is carried away more easily.

As the depleted mantle has a lesser density than the unmolten (see section D.1.2), a mass deficit is to be expected in the mantle especially under large hot thermal anomalies and might have considerable influence on the regional gravity, geoid, and topography signal (Jordan, 1979). The most prominent topography signal at normal ridges, however, is well approximated up to an age of 80 Ma by a $\sqrt{\text{age}}$ law, assuming local isostasy and a cooling half-space, whose temperature distribution is also given by such a law. The geoid of a normal MOR follows from both topography and mantle T structure and hence is a linear function of crust age (*e.g.* Forsyth, 1992).

II.2.5   Chemical composition

The generation of melt has a complex influence on both the major and trace element characteristics of mantle rock. In section I.2.1 it was already mentioned that Fe enters the melt to a larger extent than Mg, which leads to an increase of Mg# in olivine and pyroxenes (*e.g.* Pickering-Witter and Johnston, 2000), causing *e.g.* a reduction of bulk density (Oxburgh and Parmentier, 1977) and possibly also of matrix electrical conductivity; similarly, Al is removed by the melt, leading to an increase of Cr# in spinel[2]. Removal of Na is known to take place in the early stage of melting and to have a significant effect on the position of the solidus (see sections II.3.1–II.3.3); most of the Na in spinel and garnet lherzolite is contained in the jadeite component of clinopyroxene, so that the evolution of Na in melting peridotite is essentially characterized by Na partitioning between cpx and melt (Blundy *et al.*, 1995). The ratios of radiogenic to stable isotopes of some elements are of particular interest, because they are conserved during the melting process and thus provide information about the chemical composition of the source rock.

Element fractionation and partition coefficients — When rock melts, some elements will be transferred to the melt phase more readily than others, because their ions do not fit so well into their particular site in the crystal lattice, *i.e.* they are incompatible with respect to the mineral; among them are the light rare-earth elements (REE). For a multiphase system such as mantle peridotite, the compatibility is quantified by the Nernst bulk partition coefficient and the exchange coefficient

$$D = \sum_x X_x D_x^a, \quad D_x^a = \frac{C_x^a}{C_f^a}; \qquad \mathcal{K}_{a-b}^{x-liq} := \frac{C_x^a}{C_f^a} \cdot \frac{C_f^b}{C_x^b} = \frac{D_x^a}{D_x^b} \qquad (II.17)$$

of the element, where D_x^a denotes the partition coefficient of component a with respect to phase x and C_{xf}^a the concentration of a in x and in the liquid, respectively (Wilson, 1989). The behaviour of incompatible elements during melting is only sensitive to the melting degree f as long as $f < D$ (McKenzie, 1985a). D_i varies for different minerals, of course, and also depends on the p–T conditions, on the major element composition and the water content of the mineral, and in the case of the bulk coefficient on the corresponding modal composition, although these dependencies are often neglected; recently, detailed lattice strain models of D_{cpx} and D_{gt} have been developed (*e.g.* Blundy and Wood, 1994; Wood and Blundy, 1997; van Westrenen *et al.*, 2001). Furthermore, recent investigations suggest that variations in melt structure also affect D and $\mathcal{K}$ (Kushiro and Mysen, 2002). The variability of D_{cpx} and D_{gt} plays a crucial role in the ongoing debate whether MORB melting begins in the garnet stability field (*e.g.* Johnson *et al.*, 1990; Shen and Forsyth, 1995; Ryabchikov, 1998; Ozawa, 2001; Salters *et al.*, 2002; Hellebrand *et al.*, 2002) or is restricted to the spinel stability field (*e.g.* Blundy *et al.*, 1998; Johnson, 1998; Chauvel and Blichert-Toft, 2001). – Following the logic that incompatible elements enter the melt easily and during the first steps of melting, enrichment of a volcanic rock can indicate low-degree melts; in the case of hotspots, where enriched lavas are common, this is obviously not reasonable, so that the contribution of an undepleted source, possibly from the lower mantle, which introduces volatiles and incompatible elements, is the preferred explanation here (*e.g.* Dick *et al.*, 1984).

When a rock melts and the melt is not removed immediately, it will react with the solid phases until a local chemical equilibrium is reached, which is particular for the composition of the host rock; for this reason, melt which migrates rapidly through a porous matrix is usually not in equilibrium with the solid, as observed *e.g.* in dunite channels (see section II.4.2),

---

[2]Mg# := $\frac{Mg}{Mg+Fe}$; Cr# := $\frac{Cr}{Cr+Al}$. These molar ratios can serve as measures for f.

and absence of equilibrium triggers melt–solid reactions such as dissolution or precipitation of phases.

Radionuclide disequilibria and incipient melting — Although chemical investigations cannot directly determine melt topology, they show if melt and matrix were isolated and thus can be expected to provide some hints for one or another melting type and segregation mechanism (Kelemen *et al.*, 1997). One feature frequently discussed in connection with melting rates and mechanisms is the relative abundance of certain radioactive members of the ^{238}U decay chain, in particular the excess of ^{230}Th relative to ^{238}U in the melt. ^{230}Th is an instable, but relatively long-lived ($t_{1/2} \approx 75.38$ ka) member of the ^{238}U decay chain and has the same activity as its parent in undisturbed mantle, *i.e.* $c_s^U/c_s^{Th} = t_{1/2}^U/t_{1/2}^{Th}$; this equilibrium can be disturbed by a melting event, in which Th will partition into the melt to a larger extent than U if $\mathcal{K}_{U-Th}^{x-liq} > 1$, as commonly assumed for spinel lherzolite, and will only gradually be restored some half-lives after its cessation (*e.g.* Elliott, 1997). The ratio of activities is thus an indicator of the degree of disequilibrium. A complication in the formation of radionuclide excesses is introduced by "ingrowth" during melting, by which the less incompatible parent in the residue, which has a relatively larger activity after a melting increment, produces additional amounts of the more incompatible daughter and thereby reinforces the excess of the latter in the melt if the melting process itself is slow enough; this effect is independent of f and thus acts also after initial melting, but becomes less important, because the concentration of the parent approaches zero fast as well (*e.g.* McKenzie, 1985b; Elliott, 1997).

The observed excesses of ^{230}Th and ^{226}Ra, which range from a few to some tens of percents, require $\mathcal{K}_{U-Th}^{x-liq} > 1$ and $O(D_{bulk}^U) \approx \varphi_c$ (*e.g.* Wood *et al.*, 1999; Landwehr *et al.*, 2001) and suggest that melt is erupted within less than 20 a after being isolated from the source; they have therefore been interpreted as indicators for threshold porosities less than 0.1 % under the assumption of dynamic melting, where melt is isolated and segregates very fast after a certain threshold is exceeded (*e.g.* McKenzie, 1985b, 2000; Hauri *et al.*, 1994a), but they can also be explained by reactive porous flow along a certain distance in a low-φ part of the melting region which contains some amount of a mineral phase which preferredly retains U and will then lead to even larger ^{230}Th excesses due to ingrowth (Elliott, 1997; Sims *et al.*, 1999). Knowledge of $D_{cpx,gt}^U$ and $D_{cpx,gt}^{Th}$ is therefore of great interest to put possible constraints on the depth of incipient melting. In fact, several workers have postulated that MORB melting begins in the garnet stability field, because garnet has been known to have $\mathcal{K}_{U-Th}^{x-liq} > 1$, in contrast to most other major mantle phases, so that Beattie (1993a,b) concluded that slow initial melting must take place in the garnet stability field at $\varphi \ll$ 1 %. Although more recently Wood *et al.* (1999) and Landwehr *et al.* (2001) showed that aluminous, Ca-poor cpx, which is formed at increasing pressures already in the spinel stability field, can also produce ^{230}Th excess, this does not discard the possibility that melting indeed begins in the garnet field, because the likely presence of small amounts of water in fertile mantle leads to early hydrous melting (*e.g.* Bell and Rossman, 1992), or that a small garnet-bearing component such as eclogite is present in the spinel stability field (*e.g.* Hirschmann and Stolper, 1996).

Based on the assumption of a fractal drainage system, Hart (1993) and Lundstrom *et al.* (2000) even argue that melt *is* in equilibrium at least on the smallest, near-source scales of the percolative network, but becomes more and more isolated at a certain distance due to increasing flow velocities. This is in agreement with the result of Faul (2001) that most of the observed U-series disequilibria can only be preserved if the melt is volatile-rich at low porosities and the permeability (and segregation velocity) increases rapidly beyond a certain threshold. Avoiding unrealistically high mantle temperatures ($\mathcal{T} > 1450\,°C$) also requires initial melting to take place in the presence of volatiles in the garnet stability field, as one possible alternative, because anhydrous melting would result in either too low melting degrees to account for observed chemical signatures or too large crustal thicknesses (Robinson and Wood, 1998).

The consequences of source heterogeneity, which in principle have already been considered for a long time (*e.g.* Hofmann and White, 1982; Allègre and Turcotte, 1986), were recently discussed from the U–Th point of view by Lundstrom *et al.* (2000), who confirmed that variations as those observed in near-ridge off-axis seamounts are due to source heterogeneity, and the ^{230}Th excesses observed in ridge lavas are best explained by mixing of melts from different sources, with a $\sim$5 % contribution of an enriched mafic source like pyroxenite. To reconcile the fractional-melting signature *e.g.* from REE abundances and the equilibrium signature of the U series, Lundstrom (2000) and Lundstrom *et al.* (2000) proposed that melt percolates in equilibrium with the matrix in the relatively homogeneous, low-φ, peridotitic regions of its generation and attains the disequilibrium only at the last stages of melting near and within a dunite channel; they note that only a few percent of near-fractional melting at this stage could leave the observed REE disequilibrium footprint on melt which had formerly been in equilibrium with the rock. For the porosity, they determine φ values of 0.1...0.5 % over travel distances of up to tens of kilometers for both the enriched and the depleted source endmember.

II.3   THE GENERATION OF MELT

II.3.1   Physical melting mechanism

Melting of a given material in the mantle mainly depends on two physical quantities: the pressure p and the temperature T. As has already been mentioned in section I.1.3, $Pe \gg 1$ in the mantle, so that a mass parcel in upwelling mantle can be considered isolated from its surroundings and will not exchange heat with them. The upwelling transports it to shallower parts of the earth, so that is decompressed; thus, it experiences a change of state which is isentropic (adiabatic and reversible) to a very good approximation (*e.g.* Turcotte and Schubert, 1982; McKenzie and Bickle, 1988; Langmuir *et al.*, 1992) and therefore can be expressed as

$$dS = 0 = \left(\frac{\partial S}{\partial T}\right)_p dT + \left(\frac{\partial S}{\partial p}\right)_T dp = \frac{c_p}{T}dT - \frac{\alpha}{\varrho}dp \qquad (II.18)$$

(*e.g.* Phipps Morgan, 2001), where phase transitions are not included. The change in temperature is then given by

$$\left(\frac{\partial T}{\partial p}\right)_S = \frac{\alpha T}{\varrho c_p}; \qquad (II.18a)$$

$$\left(\frac{\partial T}{\partial z}\right)_S = \frac{\alpha g T}{c_p}. \qquad (II.18b)$$

It should be noted that α, ϱ and c_p are functions of p, T and the composition and that g also varies in the earth, mostly with depth; the frequently made assumption of a linear adiabat is a simplification which may hold within a certain depth range and might be justified by the limited knowledge of the form of the material parameters. The temperature decrease caused by adiabatic decompression during ascent of the rock makes it difficult to compare the heat content of material at different depths. Therefore, McKenzie and Bickle (1988) introduced

$\mathcal{T}$ (°C)	reference
1260	Presnall *et al.* (2002)
1280	McKenzie and Bickle (1988)
1320	White (1997)
≈ 1330	Nicholson and Latin (1992)
1330	White *et al.* (1992)
1350	White and McKenzie (1995)
≈ 1400	Langmuir *et al.* (1992)
1400	Ryabchikov (1998)
1410	Asimow *et al.* (2001)
1450	Green and Falloon (1998); Green *et al.* (2001)

Table II.1: Estimates of the potential temperature of normal suboceanic mantle.

the potential temperature

$$\mathcal{T} := T(z)\, e^{-\frac{g\alpha z}{c_p}} \tag{II.19}$$

by integration of eq. II.18b; this is the temperature the rock would have if it were transported adiabatically to the surface. They estimated a potential temperature of *ca.* 1280 °C for normal suboceanic mantle, whereas some newer investigations suggest considerably higher values such as 1350 °C (White and McKenzie, 1995) or even 1430 °C (*e.g.* Green *et al.*, 2001) (see table II.1). These estimates are mostly based on geochemical and petrological arguments and therefore subject to certain restrictions imposed by underlying assumptions or experimental limitations; for instance, the estimate of McKenzie and Bickle (1988) stems from batch melting data, which is not an exact representation of mantle melting. In fractional melting, the productivity is lower than in batch melting (*e.g.* Langmuir *et al.*, 1992), thereby requiring potential temperatures higher by perhaps 50 K to fit relatively hard constraints such as the crustal thickness (Nicholson and Latin, 1992). Furthermore, there is also disagreement about the along-ridge variability of $\mathcal{T}$: estimates range from as much as 250 K (Langmuir *et al.*, 1992) down to less than 60 K (Shen and Forsyth, 1995), which also has implications for the fraction of melting taking place in the garnet stability field.

Multiphase systems as mantle rocks do not have a unique temperature at which they melt as a whole, but melt over a certain p–T range given by the solidus and the liquidus curve $T_s(p)$ and $T_l(p)$, respectively, which are usually determined by experiments (see appendix C); the chemical-petrological composition, especially the content of CaO and Al_2O_3, which is closely linked to the abundance of clinopyroxene and garnet, has a large impact on the melting history of the rock (also see sections II.3.2 and II.3.3). Besides, the presence of volatiles, in particular of water, and of alkalis lowers and deforms the solidus of mantle rock to a large extent (*e.g.* Thompson, 1992, see sect. II.3.2, app. C). In particular has it long been known (*e.g.* Green, 1973) that the water-saturated ("wet") solidus is some 500–800 K lower than the dry solidus for a large part of the upper mantle, reaching a minimum of about 900 °C around 2–3 GPa. The depression of the solidus is proportional to the mole fraction of water in the melt (Thompson, 1992; Davies and Bickle, 1991),

$$\Delta T_{s,dw} = b_{T_s} C_{H_2O}, \tag{II.20a}$$

the factor b_{T_s} being essentially independent of pressure at least between 2 and 3 GPa (Davies and Bickle, 1991); more recently, Katz *et al.* (2003) used a slightly more general equation,

$$\Delta T_{s,dw} = \tilde{b}_{T_s} X_{H_2O}^{\nu_s}, \tag{II.20b}$$

for fitting experimental data, with X_{H_2O} in wt %. The fraction of water in mantle peridotite is of the order of 0.1 % or less under MOR (*e.g.* Bell and Rossman, 1992), and water dissolves readily in melt or mixes with it, so that most of it will be extracted in the first stages of melting and generates only minor amounts of melt; the effect on major element chemistry is not expected to be substantial either at water contents of less than 1 wt % (Kushiro, 2001). Thus, for upwelling melting mantle increasingly dry solidi will hold, and the melting process under a spreading ridge as a whole can be considered as dry (Nicolas, 1990; Kostopoulos, 1991; Karato and Wu, 1993; Kushiro, 2001). Nonetheless, there might exist a "melting gap" between the early melting event and the main melting regime; its existence and dimensions are controlled by volatile and alkali content, but also by the (probably high) connectivity and mobility of the hydrous melt.

The mantle solidus has a steeper gradient $\partial T_s/\partial p$ than the mantle adiabat, so that upwelling mantle material will cross it at some depth which depends on its temperature. At this point, the rock begins to melt, thereby lowering its temperature by

$$\Delta T_m = \frac{\tilde{\varphi} L_m}{c_p} \tag{II.21}$$

(*e.g.* Hess, 1992) by consuming a part of its internal latent heat in breaking up the chemical bonds of the crystal lattice[3]. The cooling effect causes the melting path of the mantle parcel to have a steeper p–T slope $\tilde{\varphi} \, \partial T/\partial \tilde{\varphi}$ than the melt-free adiabat and reduces the temperature contrast between the melting region *e.g.* of a plume head and the surrounding solid mantle; besides, similar to solid–solid phase transitions (see section I.2.2), an upstream temperature precursor appears on the unmolten side and induces some cooling in an upwelling even before the rock has begun to melt (Turcotte and Phipps Morgan, 1992). As Johnson *et al.* (1990) remark, L_m will shift to higher values with increasing melting degree, because the melting enthalpy of forsterite is higher than that of pyroxenes (also see table D.11). If the assumption of isentropic ascent is maintained, an additional term for entropy change due to depletion has to be introduced into eq. II.18. This term describes the melting entropy ΔS_m, which is related to the melting enthalpy L_m, and in principle must also account for the compositional entropy changes in solid solutions, although these are minor effects (Phipps Morgan, 2001; Asimow, 2002); with regard to the considerable range of proposed values for ΔS_m resp. L_m (see section D.2.4) it does not seem reasonable to put much emphasis on this factor at the present stage. An important constraint related to the temperature drop in eq. II.21 is that it equals the change in T_s due to decompression and depletion (Scott, 1992; Phipps Morgan, 2001). It should be noted that fractional melting is neither isentropic nor reversible, because a part of the material is removed from the system and transports entropy with it; it can though be approximated by a succession of infinitesimal batch melting and melt removal steps in a continuously changing system (Asimow *et al.*, 1997), which is in principle the approach taken in the melt-dynamical models of this study.

The pressure-dependent change in melting degree depends on several material parameters and variables (Iwamori *et al.*, 1995; Langmuir *et al.*, 1992; McKenzie, 1984) and is an

---

[3]In thermodynamical relations, the relevant quantity is the amount of melt by weight, whereas for several observables, the melt fraction by volume has to be used (see sections II.2.2, II.2.3). For instance, in batch melting (which was implied in eq. II.21), the melt fraction by weight, $\tilde{\varphi}$, can be related to the melt fraction by volume, φ, through

$$\tilde{\varphi} = \frac{\varphi \varrho_f}{\varrho_s + \varphi(\varrho_f - \varrho_s)} \Leftrightarrow \varphi = \frac{\tilde{\varphi} \varrho_s}{\varrho_f + \tilde{\varphi}(\varrho_s - \varrho_f)}$$

(McKenzie, 1984), but it should be kept in mind here that in (near-)fractional melting, the total amount of melt generated in the melting process is not equal to the amount which is retained in the mantle and influences observations; note also that ϱ_s and ϱ_f are not constant at different melting degrees f.

essential factor concerning the influence of melt on convection or the generation of crust. Some of its constituents, *e.g.* $\partial T/\partial f$, are not well known; moreover, phenomena such as the exhaustion of certain phases due to progressive melting during ascent resp. with increasing f are responsible for this term not being constant. The most important factors in $(\partial f/\partial p)_S$ are the isobaric productivity $(\partial f/\partial T)_p$, which follows from mass balance,

$$X_0 = f X_f + (1 - f) X_s,\tag{II.22}$$

and the slopes of the melting degree isolines, $(\partial T/\partial p)_f$ (Asimow *et al.*, 1997). Predictions of its shape resp. the form of $f(T)$ are contradictory: while theoretical investigations (Asimow *et al.*, 1997; Hirschmann *et al.*, 1999) and some experiments (*e.g.* Walter and Presnall, 1994; Schwab and Johnston, 2001) find f to be a concave-up function of supersolidus temperature at constant pressure, others indicate a linear or concave-down shape (*e.g.* McKenzie and Bickle, 1988; Langmuir *et al.*, 1992; Iwamori *et al.*, 1995), *i.e.* they predict a decrease in melt productivity with progressing melting. In the concave-up f models, the productivity will be low near the solidus and increase with f until a certain phase is exhausted for most cases especially in batch ($f = \tilde{\varphi}$) melting, but probably also in fractional ($\tilde{\varphi} = 0$) melting. – As f is constrained by $T_l(p) - T_s(p)$, its gradient in p–T space steepens with increasing pressure, because the melting interval narrows; often, f is expressed as a function of the dimensionless homologous temperature

$$\tilde{T} := \frac{T - \frac{T_s + T_l}{2}}{T_l - T_s}\tag{II.23}$$

(McKenzie and Bickle, 1988), whose purpose is to avoid the difficulty of expressing it as an explicit function of both p and T.

II.3.2   Petrological and chemical characterization of melting

As already pointed out in section II.3.1, the modal and chemical composition of the rock is of potentially substantial influence, in particular the transitions of the Al-bearing phases plagioclase, spinel, and garnet (section D.1.6), which happen in or near the usual pressure interval of decompression melting, and the general modal evolution of the rock due to depletion upon melting.

Mantle and melt composition — As the transitions from plagioclase to spinel and from spinel to garnet both have positive Clapeyron slopes, *i.e.* the lower-pressure phase has a higher entropy, melting must cease or even turn into freezing to ensure that the whole process remains adiabatic, because the liquid phase also has a higher entropy than the solid. Therefore, Asimow *et al.* (1995) argued that the transition zones, which would coincide with cusps in the solidus curve in a pure endmember system like CMAS, will be regions of reduced melting or even re-solidification and may be correlated with melt-free zones; in particular, they expect crossing the plagioclase–spinel boundary to terminate the melting process at least under normal mid-oceanic ridges, so that usually no melting of plagioclase lherzolite will take place. Apart from this, simply the change in dT_s/dp across the phase boundary, which is smaller on the low-p side, can also contribute to the suppression of melting, because the mantle geotherm may intersect $T_s(p)$ again if the mantle is not too hot (Kushiro, 2001). However, experiments indicate that in more complex systems like CMASNF the cusp is not present (Presnall *et al.*, 2002). Phipps Morgan (1997) estimated the temperature drop of material crossing the sp–gt and the pl–sp boundary from below to be 5–15 K and 7 K, respectively. On the other hand, if the Al-phases are exhausted by preceding melting, no

Al is left to form the lower-p mineral at the depth where this transformation would occur; in particular, the amount of Al-spinel is strongly reduced after the first few percent of melting and thus no primary plagioclase lherzolite is likely to appear in the uppermost lithospheric mantle (Kostopoulos and James, 1992), which means that the cooling effect is of no importance. In general, the "out curves" of the Al-phases lie within a few tens of degrees above the solidus.

As already mentioned, another effect of depletion will also introduce a discontinuity in the melting evolution: the exhaustion of cpx takes place at $f \gtrsim 20 \ldots 25\%$ in a normal peridotite and leads to a significant upward shift of the solidus of the residue; thus, only weak, if any, further melting will happen beyond this point, and numerical models (Cordery and Phipps Morgan, 1993) show that a harzburgitic layer without melting can form above the melting zone. Nonetheless, it was demonstrated theoretically (Hirschmann *et al.*, 1999) and experimentally (Pickering-Witter and Johnston, 2000; Schwab and Johnston, 2001) that $(\partial f/\partial T)_p$ is *lower* for systems enriched in incompatible components than for depleted ones, but that they will produce more total melt even so, because they have a lower solidus. – On the other hand, Kostopoulos (1991) stated that in spinel lherzolite clinopyroxene disappears from the parental rock at $f \approx 0.42$, which is about twice as high as the cpx-out values suggested by experiments; he explained this large discrepancy by the fact that in the experiments batch melting prevails, while in nature (near-)fractional melting is the rule and was shown to cause clinopyroxene to be stable in the residue to higher temperatures (Falloon *et al.*, 1988, pp. 1273ff.). Late disappearance of clinopyroxene is corroborated by the finding that even melting of abyssal peridotites comprises four phases (Dick *et al.*, 1984).

Generally speaking, melting cannot be described by a single set of melting proportions, but rather by a set thereof for different compositional regimes and pressures (Kostopoulos, 1991; Gudfinnsson and Presnall, 2000; Walter, 1998). Cpx and to a lesser extent opx are the main components of the melt phase, whereas olivine and spinel only contribute a small fraction or even crystallize, although the olivine content of melts will increase at higher pressures, while opx crystallizes instead (Green *et al.*, 2001; Kinzler, 1997); thus, more depleted abyssal peridotites typical *e.g.* for hotspots contain less pyroxene than the less depleted MORB source (Dick *et al.*, 1984; Johnson *et al.*, 1990).

For plume melting, consideration of the spinel stability field is not sufficient, but even for normal MORB generation many workers assume that the mantle geotherm crosses the dry solidus in the neighbourhood of the sp–gt transition and that melting actually begins in the garnet stability field (see section II.2.5). Thus, it is clear that an ascending hot thermal anomaly will begin to melt at greater depths, definitively in the garnet stability field, and melting of garnet will strongly reduce, if not eliminate, the temperature effect of the gt–sp transition (see section V.1.6).

If these considerations are extended to hydrous melting, it is clear that the initial stage of melting takes place in the garnet stability field, because the presence of water strongly lowers the solidus of peridotite and allows for the existence of OH-bearing minerals (*e.g.* Thompson, 1992). This is especially true in the case of plumes, as it has been suggested that hotspots are also wetspots, *i.e.* mantle regions with elevated water concentrations; examples include Iceland (Jamtveit *et al.*, 2001; Nichols *et al.*, 2002, see section IV.2.2), the Azores (Schilling *et al.*, 1980) and Hawaii (Wallace, 1998), for which estimates range from 300 to as much as 920 ppm. The inclusion of water in melting models then also has important implications for the excess temperature of plumes: Asimow and Langmuir (2003) concluded from chemical modelling that a wet plume produces a given amount of melt resp. a crust of given thickness at lower temperature than a dry one; they did not account for the dynamic effect of dehydration on active upwelling, though. However, in real suboceanic mantle, the

Figure II.4: Simple qualitative
scheme of the main (dry) and hydrous melting regions beneath a
normal MOR and in a ridge-centered
plume; phase transitions from garnet to spinel and from spinel to plagioclase are also shown.   Not to
scale.

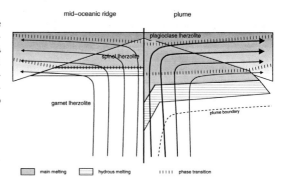

degree reached by hydrous melting will not exceed a few percent at most, because the water
content of 140–200 ppm (Wood, 1995; Saal *et al.*, 2002) is far below the maximum solubility
of water under upper-mantle conditions, while a large water activity is required for melting
near the wet solidus (*e.g.* Green, 1973; Hirschmann *et al.*, 1999). This is also essentially
confirmed by hydrous melting models of this study (see section V.1.4).

Yet a further complication is the possibility of a petrologically heterogeneous mantle
containing streaks of eclogite, pyroxenite or a similar material with a lower solidus than
peridotite (*e.g.* Hofmann and White, 1982; Allègre and Turcotte, 1986; Phipps Morgan and
Morgan, 1999); global patterns of MORB major and trace element composition cannot be
explained without some heterogeneity of the mantle source, especially near hotspots (*e.g.*
Shen and Forsyth, 1995; Salters and Dick, 2002; le Roux *et al.*, 2002). Hirschmann and
Stolper (1996) estimate the fraction of this mantle component to be about 5 % but predict
that it will contribute 15–20 % to the total MORB melt, because pyroxenite has a higher
productivity than peridotite due to its smaller solidus–liquidus interval. As far as plume
melting is concerned, the elevated water contents ascribed to plumes could be due to a
recycled pyroxenite or eclogite source member in the plume, because the pyroxenes, which
are the principal phases of those rocks, can dissolve more water than the olivine which
dominates peridotite (*e.g.* Wallace, 1998). The high content of incompatible elements in
hotspot melts could also be explained by a combination of the effects of low-degree melting
under hydrous conditions and the heterogeneity of the source mantle (Asimow and Langmuir,
2003).

Mantle melting is not only characterized by the conditions of initial melting and the
composition of the main melting domain, though, but it is also influenced by the depth of
final melting. Most of the melt which forms the MORB of the oceanic crust is generated at
shallow depths, *i.e.* in the uppermost few dozen kilometers of the mantle (*e.g.* McKenzie
and Bickle, 1988), but probably not anymore within the plagioclase stability field, judging
from the absence of magmas with eruption temperatures of less than 1150 °C (Kostopoulos
and James, 1992). The thickness of the lithosphere sets an upper boundary to the melting region, especially in the initial stages of a spreading process before it has experienced
significant stretching, so that less melt will be generated under a thick lithosphere, and its
average melting pressure and Fe content will be higher (Fram and Lesher, 1993; Cella and
Rapolla, 1997); a similar effect is expected at ridges with very slow spreading rates where
conductive cooling is important (White *et al.*, 2001) or for off-axis lavas from mantle below
older lithosphere (Haase, 1996).

Differences between fractional and batch melting — In connection with the chemical signature of erupted melts it must be emphasized that the chemical characteristics of the two endmember melting models, batch melting and fractional melting, differ significantly; as Asimow (1999) remarks, trace element analyses from abyssal peridotites require a near-fractional melting model, whereas major elements and modes can be explained better by batch melting. The following equations have been established for the concentration of a trace element (or component) in batch melt:

$$C_f = \begin{cases} \dfrac{C_0}{D + f(1 - D)} & \text{(modal)} \\[2ex] \dfrac{C_0}{D_0 + f(1 - \mathcal{M})} & \text{(non-modal)}, \end{cases} \tag{II.24a}$$

and in a fractional melt increment:

$$C_f = \begin{cases} \dfrac{C_0}{D}(1 - f)^{\frac{1}{D} - 1} & \text{(modal)} \\[2ex] \dfrac{C_0}{D_0}\left(1 - \dfrac{\mathcal{M}f}{D_0}\right)^{\frac{1}{\mathcal{M}} - 1} & \text{(non-modal)}; \end{cases} \tag{II.24b}$$

the concentration in a pooled liquid follows by averaging of eqs. II.24b over f:

$$\mathcal{C}_f = \begin{cases} \dfrac{C_0}{f}\left[1 - (1 - f)^{\frac{1}{D}}\right] & \text{(modal)} \\[2ex] \dfrac{C_0}{f}\left[1 - \left(1 - \dfrac{\mathcal{M}f}{D_0}\right)^{\frac{1}{\mathcal{M}}}\right] & \text{(non-modal)}, \end{cases} \tag{II.24c}$$

(Shaw, 1970; Hertogen and Gijbels, 1976). The corresponding concentration in the solid is $C_s = DC_f$ according to eq. II.17, in particular in the case of non-modal fractional melting:

$$C_s = DC_f = \frac{C_0}{1 - f}\left(1 - \frac{\mathcal{M}f}{D_0}\right)^{\frac{1}{\mathcal{M}}}, \qquad D = \frac{D_0 - \mathcal{M}f}{1 - f} \tag{II.24d}$$

(Shaw, 1970). In these equations, $\mathcal{M} = \sum_i \mathcal{M}_i D_i$, whereby $\mathcal{M}_i$ is the melting coefficient of phase i, which is different from its proportion in the solid in non-modal melting. For a melting phase, the mass balance condition eq. II.22 must be fulfilled. Strictly speaking, eqs. II.24 are only valid for constant D and $\mathcal{M}$, and only until one phase in exhausted at $f = x_{i0}/\mathcal{M}_i$, but they can be extended for variable D and $\mathcal{M}$ and for accounting of phase exhaustion, although fractional melting is not easily cast in a general, straightforwardly usable form anymore (Hertogen and Gijbels, 1976). For small enough increments and proper interfacing of discontinuities in the evolution of the melting system, however, one can use eqs. II.24. Examples are plotted in figure II.5.

The consequences of fractional melting compared to batch melting are expected to be a roughly twice as high mean pressure of melt generation, a smaller degree of melting, resulting in a thinner crust, and a greater enrichment of incompatible elements in the melt (Ribe, 1988; Langmuir *et al.*, 1992; Kushiro, 2001). The composition of both the incremental melts and the residue differ significantly for fractional and for batch melting (Presnall, 1969), although the compositions of pooled melts from polybaric fractional melting and batch melting at a certain intermediate pressure are quite similar (Shaw, 1970), as confirmed by experiments (Hirose and Kawamura, 1994). $(\partial f / \partial T)_p$ is smaller for fractional melting than for batch

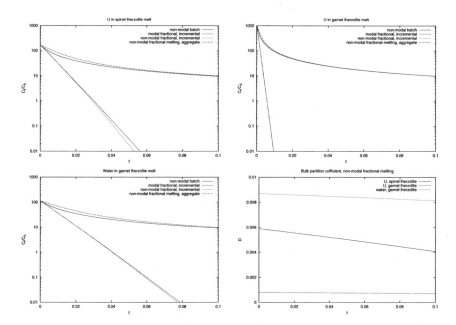

Figure II.5: Plots of eqs. II.24. Normalized concentrations of U in spinel and garnet lherzolite melt and of water in garnet lherzolite melt; partition coefficients of U and water in lherzolite. Note the similarity between the concentrations in batch melts and accumulated fractional melts. D_i and $\mathcal{M}_i$ data are from Hirth and Kohlstedt (1996), McDonough and Rudnick (1998), Johnson *et al.* (1990) and Walter (1998).

melting near the solidus due to the faster initial change of fractional melt compositions, and possibly also for high f, but higher in between (Hirschmann *et al.*, 1999). Together with the fact that fractional melting requires temperatures higher by some tens of kelvins to result in a given melting degree, this explains why nonetheless batch fusion generates a larger amount of melt.

None of both models is likely to match reality; instead, it is usually assumed that the melt resides in the matrix only for very low melt fractions, whereas interconnection between the melt bodies is established as φ exceeds a few percent or tenths of a percent (see section II.1.5) and leads to segregation (Wilson, 1989). Thus, an intermediate form characterized by batch resp. equilibrium melting up to a critical porosity value and continuous extraction beyond it is expected to occur (McKenzie, 1985a; Johnson *et al.*, 1990; Langmuir *et al.*, 1992; Iwamori *et al.*, 1995; Asimow, 1999; Lundstrom *et al.*, 2000). In such an intermediate-type melting model, eqs. II.24 are not applicable in a strict sense, but one can handle this case by treating the melt as a mantle component with $\mathcal{M}_i = 0$ and $D_i = 1$ in eq. II.24b (Fitton *et al.*, 1997; Salters and Longhi, 1999). In the hydrous melting models of this study, this procedure is adopted to account for the effect of the prescribed retained melt fraction (see sections III.4.2, V.1.4).

II.3.3   *Parameterizations of solidus, liquidus, melt fraction, and composition*

For use in numerical calculations it is useful to describe the melting behaviour of the material with closed-form analytical expressions. The alternative would be rigorous thermodynamical modelling of the chemical and petrological composition for every grid point, which is too computationally expensive to be carried out in the framework of a three-dimensional mantle-dynamical model, although it is to be expected that this limitation can be overcome in a few years with even more powerful processors; apart from the computational aspect, thermodynamical models such as the MELTS algorithm (Ghiorso and Sack, 1995) need information about poorly constrained variables and are known to have certain difficulties in reproducing experimental results, especially for low melting degrees (see Hirschmann *et al.*, 1998, for a review). Therefore, parameterizations of relevant quantities are used in this study, and some expressions for $T_s(p)$, $T_l(p)$, φ, and X of interest are discussed in this section. An own compilation of data and parameterization of $T_s(p)$ and $T_l(p)$, including newer experiments and restricted to a single rock type (KLB-1) and to forsterite, respectively, was also made and is compared to those of other authors. Some special attention will be paid to the influence of water, although much less data are available here. The data used and further details are summarized in appendix C. At this point it must be emphasized that melting experiments are isobaric and performed under equilibrium conditions in most cases, while melting in the mantle is an approximately adiabatic non-equilibrium process, as has been discussed in sections II.3.1 and II.3.2; thus, the results of these experiments have to be used with care and are likely to deviate from mantle melting *e.g.* with respect to chemical compositions of melts (McKenzie and Bickle, 1988; Kostopoulos, 1991). Besides, the experiments were carried out with rather fertile peridotites in many cases, thus yielding solidus and liquidus temperatures for undepleted mantle; in reality and under the assumption that the melting process is mostly fractional, T_s and T_l depend on the melting degree and are shifted upward for increasingly depleted compositions, so that the effective melting degree and melting pressure are 5–15 % lower and the potential temperature necessary to produce *e.g.* the normal oceanic crust is by some tens of kelvins higher than in the case of batch melting (Iwamori *et al.*, 1995). Another aspect not paid much attention to yet is the influence of alkali metals on the solidus: Hirschmann (2000) showed that small concentrations of incompatible alkalis can lower T_s by some tens of kelvins, depending on pressure, and may be more significant than variations in the Mg number.

Solidus temperature — Many experiments have been performed to determine the anhydrous solidus of spinel resp. garnet peridotite over a wide p–T range. McKenzie and Bickle (1988) compiled data from a large number of anhydrous melting experiments and chose

$$p = \frac{T_s - 1100}{136} + 4.968 \cdot 10^{-4}\, e^{0.012(T_s - 1100)} \tag{II.25a}$$

with p in GPa and T_s in °C as a fitting function. This equation has the disadvantage that it is quite complicated and gives $p(T_s)$ instead of $T_s(p)$, which would be a more useful form for a modelling program; thus, it is a bit inconvenient. Langmuir *et al.* (1992) derived the linear function

$$T_s = 1118 + 127p \tag{II.25b}$$

from a subset of the same data. Although eq. II.25b may serve for some general considerations, it is too simple to describe the mantle solidus adequately, as it does not take phase changes into account and is not applicable at greater depths; it also neglects the curvature of the solidus, which is certainly visible in the data at least for the spinel and garnet stability

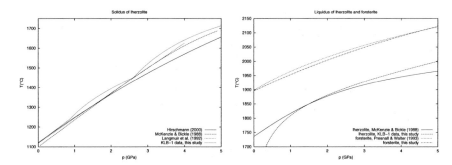

Figure II.6: Some parameterizations of the solidus of lherzolite (left) and the liquidi of lherzolite and forsterite (right) in the pressure range from 0 to 5 GPa.

fields. Hirschmann (2000, Tab. 2, and pers. comm.) recommended the parameterization

$$T_{\rm s} = 1120.66061 + 132.899012p - 5.1404654p^2, \qquad\qquad \text{(II.25c)}$$

which is based on newer data and at least accounts for the curvature of the solidus, although not for phase boundaries. On the other hand, the cusps in the solidus of a real rock are not as sharp as in pure systems like CMAS due to solid-solution effects, and the phase transition actually is not a step-like boundary, but rather smeared over a certain p resp. z interval *e.g.* of about 0.1–0.2 GPa for the sp–gt transition (Robinson and Wood, 1998); this justifies the use of a smooth curve.

An additional function of the form $T_{\rm s} = c_0 + c_1 p + c_2 p^2 + c_3 p^3 + c_4 \ln p + c_5 \sqrt{p}$ with fitting constants c_i has been derived from the compilation detailed in appendix C. It uses only data points for samples of KLB-1, because this peridotite is said to be quite representative for the upper mantle but has a higher solidus than more alkali-rich rocks like PHN 1611, as confirmed by Hirschmann (2000). Comparison of solidus curves derived from different datasets (figure II.6) reveals a variability of the order of some tens of kelvins due to compositional variations (*e.g.* Na content) which could be regarded as minor at first sight (Walter and Presnall, 1994; Hirschmann, 2000). However, the influence of rock composition, especially Mg# and the content in some minor and incompatible elements such as alkali metals, should be obvious after the onset of (fractional) melting, where the solidus continually shifts upward with increasing depletion; for instance, Robinson and Wood (1998) estimated an increase of $7\,^{\rm K}/\%\,{\rm melt}$. The effect can be pressure-dependent due to changes in the partition coefficient, though: Hirschmann (2000) points out that the effect of Na is smaller at higher pressures, because the increasing clinopyroxene mode enhances Na compatibility. In general, elevated Na_2O, K_2O and FeO lower $T_{\rm s}$, Cr_2O_3 raises it (Kinzler, 1997). The effect of different solidi on melting will be investigated in section V.1.7.

It should be noted that the above parameterizations are for peridotite. A general solidus parameterization for eclogite or pyroxenite is not available, neither is one for mixed materials, which would probably represent a heterogeneous mantle more closely than pure peridotite. Hirschmann and Stolper (1996) estimated that eclogite and pyroxenite solidi are 50–150 K lower than lherzolite solidi, but have roughly similar slopes; this seems to be in agreement with experiments on lherzolite–basalt mixtures with varying proportions by Yaxley (2000), who estimated a decrease by 25–50 K for a mixture with 25 % basalt at 3.5 GPa, whereas Kogiso and Hirschmann (2001) located the solidus of clinopyroxenite between 1 and 2 GPa within the range of common peridotite solidi, or only slightly below. If melting of a heterogeneous mantle were considered in this study, it would hence be the most pragmatic approach to approximate the solidus of pyroxenite by a formula like eq. II.25c, lowered by some dozen kelvins, but this path will not be followed further.

Liquidus temperature — Data for the liquidus temperature of a peridotite are quite scarce; McKenzie and Bickle (1988) could only use some data points from Takahashi (1986), from which they derived the expression

$$T_l = 1736.2 + 4.343p + 180 \arctan \frac{p}{2.2169}. \tag{II.26}$$

It is, however, questionable if using the liquidus of peridotite is appropriate at all: as the melting process is probably rather fractional than batch-like, the liquidus of the last residual phase, usually forsterite, is likely to be more realistic (Iwamori *et al.*, 1995). Therefore, the liquidus function

$$T = 1895.5 + 60.9543p - 3.52246p^2 + 0.0801926p^3 \tag{II.27}$$

has been derived by fitting experimental data on melting of forsterite up to 16.7 GPa (see appendix C.3.2).

Melt fraction/degree of melting — Using the homologous temperature $\tilde{T}$ defined in eq. II.23 and keeping in mind that $0 \leq \tilde{\varphi} \leq 1$, McKenzie and Bickle (1988) derived the widely used formula

$$\tilde{\varphi} = \tilde{T} + (\tilde{T}^2 - 0.25)(0.4256 + 2.988\tilde{T}) + 0.5. \tag{II.28a}$$

To take into account the observation that mantle rock was found to behave similar to a eutectic material near the solidus (*e.g.* Green *et al.*, 2001), McKenzie and Bickle (1990) modified the fit to the dataset underlying eq. II.28a as follows:

$$\tilde{\varphi} = \begin{cases} 16.14(\tilde{T} + 0.5) & -0.5 \leq \tilde{T} \leq -0.49 \\ 0.8386[\tilde{T} + (\tilde{T}^2 - 0.25)(0.785 + 2.2867\tilde{T} + 5.182\tilde{T}^2)] + 0.5807 & -0.49 \leq \tilde{T} \leq 0.5; \end{cases} \tag{II.28b}$$

this causes a considerable initial increase in $\tilde{\varphi}$ up to $\tilde{\varphi}_e = 0.1614$ at almost constant homologous temperature close to the solidus, as it would be expected for eutectic melting. As the temperature of the melting rock stays at the solidus until φ_e is reached, the p–T path of a rock parcel follows the solidus over a short interval before it enters the supersolidus domain. McKenzie and Bickle (1990) reported that this leads to a slightly larger melt production than a calculation without eutectic behaviour as in their earlier paper. On the other hand, different shapes of melting functions have been observed in experiments, some of which are in blatant contrast to the parameterizations of McKenzie and Bickle (1988, 1990), as mentioned in section II.3.1. A possible explanation apart from compositional differences is that the very low melt fractions of incipient melting have been overlooked in experiments, which resulted in too low f values respectively too high estimates for T_s. If this is the case, a slowly increasing low-f tail would have to be prepended to the cubic parabola in the McKenzie and Bickle (1988) parameterization. A parameterization of this type has been developped by Katz *et al.* (2003) for anhydrous melting as a preparatory step for their parameterization of hydrous melting (see below):

$$f = \begin{cases} \left(\frac{T - T_s}{T_{l,\mathrm{mod}} - T_s}\right)^{\beta_f}, & f \leq f_{\mathrm{cpx-out}} \\ f_{\mathrm{cpx-out}} + (1 - f_{\mathrm{cpx-out}}) \left(\frac{T - T_{\mathrm{cpx-out}}}{T_l - T_{\mathrm{cpx-out}}}\right)^{\beta_f} & f > f_{\mathrm{cpx-out}} \end{cases} \tag{II.29}$$

with

$$T_{\text{cpx-out}} = f_{\text{cpx-out}}^{1/\beta_{\text{f}}}(T_{\text{l,mod}} - T_{\text{s}}) + T_{\text{s}};\qquad(\text{II.29a})$$

the two cases are for distinguishing one melting interval with cpx present and one beyond cpx-out. $T_{\text{l,mod}}$ is a hypothetical "modal" liquidus for modally melting peridotite, for which a similar form as for the solidus eq. II.25c can be assumed.

Iwamori *et al.* (1995) stated that the degree of melting f in a near-fractional melting model should rather be expressed as a function of pressure and homologous temperature than as a function of $\tilde{T}$ alone as in eqs. II.28. As most of the melting experiments produce batch melts, an $f(\tilde{T}, p)$ relation for fractional melting is estimated by stretching its batch melting analogue between the (common) solidus and the liquidus of the fractional residue in order to work around the lack of values for $(\partial T/\partial p)_f$ and $(\partial T/\partial f)_p$; Iwamori *et al.* (1995) claimed that this solution gives a reasonable fit to incremental batch melting data from Hirose and Kawamura (1994), although the existence of temperature intervals without melt generation due to exhaustion of certain solid phases is not explicitly accounted for.

They also attempted to derive relations for water-undersaturated melting by stretching the relation for the dry system between the wet solidus and the dry liquidus; it is questionable, however, if simple stretching is adequate, because it is likely that the water will be absorbed by the melt after the first few percent of melting, and the low-degree melting tail is not reproduced by this procedure. It seems more appropriate to consider the T_{s}-lowering effect of water (eq. II.20a) and treat water as an incompatible trace element according to eqs. II.24 (Katz *et al.*, 2003). For a two-component system consisting of a nearly pure solid A and a small amount of an incompatible component B, Hirschmann *et al.* (1999) quoted the relation

$$b_{T_{\text{s}}} = \frac{RT_{\text{s,A}}^2}{L_{\text{m,A}}}\qquad(\text{II.30})$$

for the proportionality constant $b_{T_{\text{s}}}$ between the mole fraction of B and the solidus depression of A in eq. II.20a. Setting $L_{\text{m,A}} = T_{\text{s,A}}S_{\text{m,A}}$ and combining eqs. II.20a, II.30 and the non-modal version of eq. II.24b and assuming that A is lherzolite rather than a chemically pure solid yields the expression

$$T_{\text{s,w}}(f) = T_{\text{s}}\left[1 - \frac{RC_{\text{B},0}}{S_{\text{m}}D_0}\left(1 - \frac{\mathcal{M}f}{D_0}\right)^{\frac{1}{\mathcal{M}}-1}\right]\qquad(\text{II.31a})$$

for the solidus of A+B, where B could be water, but also an oxide of an incompatible element, *e.g.* Na$_2$O, with a bulk concentration $C_{\text{B},0}$. With a constant $S_{\text{m,A}} = 300\,\text{J}/\text{kg\,K}$ (Hirschmann *et al.*, 1999) and $T_{\text{s,A}} = 1723\,\text{K}$, this would yield $b_{T_{\text{s}}} \approx 276\,\text{K}$, or about $26.5\,\text{K}/\text{wt\,\%}$; note that $b_{T_{\text{s}}}$ is not constant, but is a function of the solidus temperature and hence depth-dependent.

Alternatively, Katz *et al.* (2003) used eq. II.20b to derive hydrous solidi. Assuming non-modal fractional melting and taking into account water saturation, this would yield

$$T_{\text{s,w}}(f,p) = \begin{cases} T_{\text{s}} - \tilde{b}_{T_{\text{s}}}\left[\frac{X_{\text{H}_2\text{O},0}}{D_0}\left(1 - \frac{\mathcal{M}f}{D_0}\right)^{\frac{1}{\mathcal{M}}-1}\right]^{\nu_{\text{s}}} & X_{\text{H}_2\text{O}} < X_{\text{H}_2\text{O,sat}}, \\ T_{\text{s}} - \tilde{b}_{T_{\text{s}}}(c_{\text{sat}}p^{\nu_{\text{sat}}} + p)^{\nu_{\text{s}}} & X_{\text{H}_2\text{O}} \geq X_{\text{H}_2\text{O,sat}}, \end{cases}\qquad(\text{II.31b})$$

where the term in parentheses in the case $X_{\text{H}_2\text{O}} \geq X_{\text{H}_2\text{O,sat}}$ is a pressure parameterization of water saturation $X_{\text{H}_2\text{O,sat}}$ in the melt, and $\nu_{\text{s}} = 0.75$, $\nu_{\text{sat}} = 0.6$, $c_{\text{sat}} = 12\,\text{wt\,\%}/\text{GPa}^{\nu_{\text{sat}}}$, and $\tilde{b}_{T_{\text{s}}} = 43\,\text{K}/\text{wt\,\%}$ (Katz *et al.*, 2003). The example in figure II.7 suggests that both expressions yield quite similar results in the probably most important pressure range at reasonable water contents for normal MOR; the difference for a plume is a bit stronger,

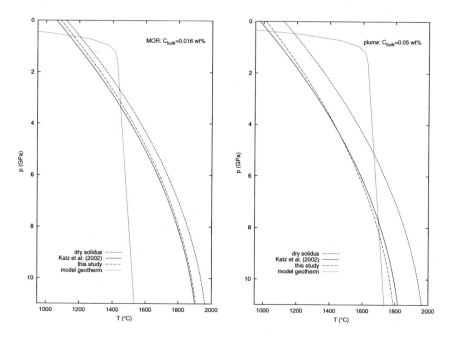

Figure II.7: Depression of the peridotite solidus eq. II.25c (Hirschmann, 2000) caused by 0.016 wt. % and 0.05 wt. % bulk water, respectively, as calculated by the parameterization of Katz *et al.* (2003) and by eq. II.31a, with $D_0 = 0.01$ (Hirth and Kohlstedt, 1996) and $S_m = 300\,\text{J/kg}\,\text{K}$ (Hirschmann *et al.*, 1999); saturation was not considered. For the geotherm, a potential temperature of 1410 °C is assumed. Even the rather small amount of water in normal MORB-source mantle shifts the onset of melting about 500 MPa to higher pressure, which corresponds to nearly 20 km. In a plume with a higher water content of 0.05 wt. % and a temperature excess of 200 K, the onset of melting would occur at nearly 1.5 times as high pressures than in the anhydrous case.

but it seems that the assumptions in the derivation of eq. II.31a are justified. In view of the scarcity of experimental data on hydrous melting and the uncertainty in the other parameters, *e.g.* S_m, it is difficult to decide which method is better, but the agreement is encouraging. As can also be seen from the form of both of these equations, the solidus depression is essentially removed after a few percent melting, although mathematically T_s is not reached until $f = {}^{D_0}\!/_{\mathcal{M}}$, which corresponds to several tens of percents for mantle rock. One of these equations could be plugged into the low-degree f formula

$$f = \left(\frac{T - T_{s,w}}{T_{l,\mathrm{mod}} - T_s} \right)^{\beta_f}, \quad \beta_f = 1.5 \tag{II.32}$$

proposed by Katz *et al.* (2003, and pers. comm.) and be solved for f using a root-finder. The differences between the approach in this study and the one by Katz *et al.* (2003) are that they use the modal version of eq. II.24a and have shifted eq. II.25c by about 35 K downwards, which is not generally done here because assessing an appropriate value for the shift seemed to be difficult, although Katz *et al.* (2003) apparently indeed achieved a closer fit to data points near the solidus; furthermore, in this study a "fractional modal liquidus" is constructed by mapping the modal liquidus of Katz *et al.* (2003) from the solidus–liquidus interval of peridotite into the interval between the forsterite liquidus and the peridotite liquidus, in analogy to the procedure of Iwamori *et al.* (1995) for anhydrous melting.

Modal and chemical composition — For the calculation of physical properties and the resulting dynamic behaviour of the melt and the residual mantle rock as well as the composition of the generated crust, it is of great interest to know the chemical or rather the petrological composition of the source material as a function of pressure and temperature. Several attempts have been made to develop numerical models of the chemical evolution of melting rock, but all of them still seem to have difficulties in reproducing experimental data correctly (Falloon *et al.*, 1999), especially for low-degree melts; on the other hand, in spite of significant efforts (*e.g.* Presnall, 1986), no sufficiently comprehensive procedure has yet been developped to solve the particular problem of the modal composition, because it is a very difficult task and demands a large experimental database to describe the compositional changes of a melting rock consisting of four or five different solid solutions consisting of more than six oxide components (Dick *et al.*, 1984) for a wide range of $p–T$ conditions and additionally undergoing transitions of the Al-bearing phase, which change the composition and melting behaviour. However, it is very much worth the effort, because this would make possible an integrated physical and chemical model of the mantle. One promising attempt of a combined dynamical–chemical model had been made by Cordery and Phipps Morgan (1993).

One approach is the investigation of the modal composition of certain simplified systems, which nonetheless contain the most important oxides, *e.g.* CMASF, CMASN, and CMASNF (Gudfinnsson and Presnall, 2000; Walter and Presnall, 1994; Presnall *et al.*, 2002). Kostopoulos (1991) has examined the melting relations of the system CMASF+Cr and deduced the development of modal compositions of an initially fertile spinel lherzolite (see table D.1) along a dry fractional melting path. To calculate the modal composition of the mantle residue, he solved eq. II.22 for f and X_f and derived melting proportions $\mathfrak{M}_i$ for four intervals of different depletion in the spinel facies (see table C.6); the melting intervals are mostly defined by exhaustion of a phase, where cpx exhaustion is the most important change. Kostopoulos and James (1992) extended these relations to the garnet facies by considering the mineral transformations at the spinel–garnet transition.

For the calculation of the composition of the residue, one can transform eq. II.22 under the assumption that X_0 is reset to 1 after each increment of melting and recrystallization into

$$\frac{\mathrm{d}X_{s_i}}{\mathrm{d}f} = c_\varphi X_{s_i} - \mathfrak{M}_i, \tag{II.33a}$$

which has the solution

$$X_{s_i}(f) = \frac{(c_\varphi X_{s0_i} - \mathfrak{M}_i)e^{c_\varphi(f-f_0)} + \mathfrak{M}_i}{c_\varphi}; \tag{II.33b}$$

(see appendix C.4); c_φ is the fraction of the reacting material which ends up as melt and does not crystallize as another solid phase. Kostopoulos and James (1992) have provided an overall polynomial fit to the solution of eq. II.33a for the spinel facies describing the modal composition as a function of f for fractional melting; alternatively, one can solve the equation for each melting interval and use the analytic solution eq. II.33b (figure II.8). The melting proportions for the spinel and the garnet stability field and $X_{s_i}(f)$ functions are given in appendix C.4. There have been other attempts to parameterize the modal composition of mantle rocks residues for the whole melting region (*e.g.* Kinzler, 1997; Niu, 1997; Ryabchikov,

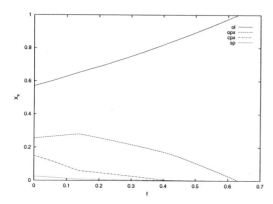

Figure II.8: $X_s(f)$ parameterization of fractional melting of spinel lherzolite with the analytical solution eq. II.33b for the KLB-1-like starting modal composition of Kostopoulos (1991) listed in table D.1 and the melting proportions of table C.6; in this case, $c_\varphi = 1$.

1998; Walter, 1998; Longhi, 2002), but these attempts are somewhat restricted, apply only to batch melting experiments or do not lend themselves so straightforwardly to incorporation in a dynamical melting program; moreover, the one by Niu (1997) has been subject to some controversy (*e.g.* Baker and Beckett, 1999).

Individual chemical components can also be expressed in terms of f. For instance, Hellebrand *et al.* (2001) investigated the correlation between the concentration of melting indicators such as REE and major elements in clinopyroxene and spinel and derived the relation

$$f = 10 \ln(\mathrm{Cr}\#) + 24 \qquad (\mathrm{II.34})$$

for the melting degree (in percent); the formula is valid for $0.1 \leq \mathrm{Cr}\# \leq 0.6$ as long as clinopyroxene is present in the rock and is thus expected to be applicable for MORB, while in plume basalts higher $\mathrm{Cr}\#$ and maybe exhaustion of clinopyroxene are to be expected (*e.g.* Green *et al.*, 2001). Hellebrand *et al.* (2001) stated that eq. II.34 is not sensitive to the melting mechanism; for their calculations they used a cpx content of 0.14, which is similar to that of KLB-1 and the one used by Kostopoulos and James (1992). Solving eq. II.34 for $\mathrm{Cr}\#$ would allow to predict a geochemical standard indicator from numerical convection–melting models. – Furthermore, one can reformulate geothermobarometers or geothermometers such as those by Putirka *et al.* (1996) or Gudfinnsson and Presnall (2001) and combine them with additional assumptions or information *e.g.* on element partitioning to derive the composition of certain components of the residue.

II.4   MELT FLOW IN A POROUS MATRIX

As already pointed out in sections I.1 and II.1.5, a certain fraction of the melt does not remain in its source region, but segregates. The flow of the melt through the porous matrix resp. the relative motion of melt and matrix can be described as a two-phase flow with mass exchange between the phases. The underlying theory has been developped in a very consistent and comprehensive form by McKenzie (1984); recently, Bercovici *et al.* (2001) have introduced a different approach, but inclusion and comparison of this method with McKenzie's is beyond the scope of this study, as is melt transport within the lithosphere and the corresponding extension of porous flow to viscoelastic and viscoplastic rheologies (Connolly and Podlachikov, 1998; Vasilyev *et al.*, 1998).

II.4.1   Basic equations

Melt generation and migration through the partially molten mantle is described as a two-phase flow, where the melt is moving with a velocity $\vec{v}_f$ different from the rock matrix velocity $\vec{v}_s$ (McKenzie, 1984). For a proper description of melt generation, the equilibrium conditions for the various phases of mantle rock have to be combined with the energy conservation law, which is quite complicated for real rock compositions.

Conservation of mass — The conservation of mass cannot be formulated in such a simple way as in eq. I.1, because melt and matrix are not conserved independently, as mass is exchanged by melting or freezing. Therefore, mass conservation for the melt is

$$\frac{\partial(\varphi\varrho_f)}{\partial t} + \nabla \cdot (\varphi\varrho_f \vec{v}_f) = \Gamma, \tag{II.35a}$$

and mass conservation for the matrix is

$$\frac{\partial}{\partial t}\left[(1-\varphi)\varrho_s\right] + \nabla \cdot \left[(1-\varphi)\varrho_s \vec{v}_s\right] = -\Gamma \tag{II.35b}$$

(McKenzie, 1984); the RHS of the equations express the mass exchange rate between melt and matrix in the matrix reference system. For a multiphase system with N components, these equations can be extended by solving the conservation equation for each component individually subject to the constraint $\sum_{n=1}^{N} X_{n_f} = \sum_{n=1}^{N} X_{n_s} = 1$ (Ribe, 1985b). Extension of melt transport to a combination of porous flow and channel flow *e.g.* in dikes to account for the fact that segregation does not only happen by porous flow (see section II.4.2) requires subtraction of the rate of suction from the porous domain into the channels from the RHS of eq. II.35a (Iwamori, 1993b).

Conservation of momentum — In principle, the momentum equations for the melt and the matrix must include not only the individual momenta, but also the interaction between both phases, especially the control the matrix exerts on the melt by shear and compaction. Assuming that momentum is chiefly transmitted by viscous diffusion and changes very slowly, momentum conservation for the melt can be written by expanding the classical Darcy's law for buoyant two-phase flow,

$$\vec{v}_f' - \vec{v}_s' = -\frac{k_\varphi'}{\eta_f'\varphi}\nabla'(p' - \varrho_f' g z') \tag{II.36}$$

(*e.g.* Turcotte and Schubert, 1982), to a set of equations describing the momentum of the melt as

$$\vec{v}_f' - \vec{v}_s' = -\frac{k_\varphi'}{\eta_f'\varphi}\left[-(1-\varphi)\Delta\varrho_f' g\,\vec{e}_z \right.$$
$$\left. + \nabla' \cdot \left\{\eta_s'\left[\nabla'\vec{v}_s' + (\nabla'\vec{v}_s')^{\mathrm{T}} - \frac{2}{3}(\nabla' \cdot \vec{v}_s')\mathbf{1}\right] + \eta_b'(\nabla' \cdot \vec{v}_s')\mathbf{1}\right\}\right] \tag{II.37a}$$

and the conservation of momentum for the matrix as

$$0 = \left[(1-\varphi)\varrho_s' + \varphi\varrho_f'\right]g\vec{e}_z - \nabla p'$$
$$+ \nabla' \cdot \left[(1-\varphi)\left\{\eta_s'\left[\nabla'\vec{v}_s' + (\nabla'\vec{v}_s')^{\mathrm{T}} - \frac{2}{3}(\nabla' \cdot \vec{v}_s')\mathbf{1}\right] + \eta_b'(\nabla' \cdot \vec{v}_s')\mathbf{1}\right\}\right] \tag{II.37b}$$

(after McKenzie, 1984; Schmeling, 2000). In eq. II.37a, the first term represents the buoyancy of the melt, the second the viscous stress exerted by the matrix and the third the compaction; similarly, in eq. II.37b the first term denotes the buoyancy, the second the lithostatic and dynamic pressure and the third the sum of viscous and compaction stress. η_b is the bulk viscosity of the matrix, which can be defined in different ways, *e.g.* by a porosity-dependent semi-analytical relation like eq. III.17 or a function relating the compression rates $\nabla \cdot \vec{v}_s$ to the pressure difference between fluid and solid (Scott and Stevenson, 1986). McKenzie (1985a, 1987) estimated a bulk viscosity of $3 \cdot 10^{18}$ Pa s for an olivine matrix, based on experimental studies at $\varphi = 0.15$; if η_b is assumed porosity-dependent, as one would expect for a property of the bulk material, this value is though likely to be too low by some orders of magnitude for the low porosities postulated by him to prevail in the mantle. With the permeability–porosity relation eq. II.8 and after subtracting the lithostatic part of the stress tensor, these equations can be non-dimensionalized:

$$\vec{v}_f - \vec{v}_s = \frac{\varphi^{n-1}}{Rt} \frac{\eta'_{f0}}{\eta'_f} \left[Rc_\varphi \frac{\Delta \varrho'_f}{\Delta \varrho'_{f0}} (1 - \varphi) \vec{e}_z - \nabla \cdot \left\{ \eta_s \left[\nabla \vec{v}_s + (\nabla \vec{v}_s)^{\mathrm{T}} \right] + \eta_b (\nabla \cdot \vec{v}_s) \mathbf{1} \right\} \right], \quad \text{(II.38a)}$$

$$0 = \left(Ra\, T + Rc_\varphi \frac{\Delta \varrho'_f}{\Delta \varrho'_{f0}} \varphi - Rc_f f \right) \vec{e}_z - \nabla \tilde{p}$$

$$+ \nabla \cdot \left[(1 - \varphi) \left\{ \eta_s \left[\nabla \vec{v}_s + (\nabla \vec{v}_s)^{\mathrm{T}} \right] + \eta_b (\nabla \cdot \vec{v}_s) \mathbf{1} \right\} \right] \quad \text{(II.38b)}$$

(after Schmeling, 2000), whereby incompressibility of the matrix has been assumed except for compaction (compaction Boussinesq approximation). $Rt = {}^{b\eta'_{f0} z'^2_m}/{\eta'_0 a'^2}$ is the melt retention number, and $\Delta \varrho'_f$ can be computed with eq. II.6; η'_{f0} in the definition of Rt is a reference viscosity, and in contrast to Schmeling (2000), eq. II.38a allows for viscosity variations in the melt. The RHS of eqs. II.38 contain expressions which can be termed "nonhydrostatic pressures" and whose magnitude compared with the relative buoyancy of the melt determines by which material parameters segregation is controlled: if their gradient is larger than the buoyancy, the matrix viscosities dominate, otherwise η_f and ϱ_f are the most important factors (McKenzie, 1985a; Ribe, 1987), which in general is probably the case in the earth. Eq. II.38a then allows to quantify the prediction of section II.1.4 that the decrease of η_f with depth cancels or even overcompensates the decrease of the density contrast between melt and matrix with respect to melt mobility in the lower part of the melting region, because η_f is in the denominator, whereas $\Delta \varrho_f$ is in the numerator. Estimates for the segregation velocity $\vec{v}'_f - \vec{v}'_s = \vec{v}'_{\mathrm{seg}}$ from other studies range from 0.5–1.5 m/a (Tanimoto and Stevenson, 1994; Scott and Stevenson, 1989) to 100 m/a (Su and Buck, 1993).

For lithostatic equilibrium, eq. II.37b yields

$$\frac{\mathrm{d}p}{\mathrm{d}z} - \left[(1 - \varphi)\varrho_s + \varphi \varrho_f \right] g = 0, \quad \text{(II.39)}$$

which means that the pressure compensates the weight of the overlying mantle; if $\mathrm{d}p/\mathrm{d}z$ is greater than the lithostatic pressure gradient, the rock dilates, if it is smaller, it compacts. Considering a layer with an impermeable lower and a free upper boundary, it can be shown that a compacting boundary of initial thickness

$$l_c := \sqrt{\frac{\eta_b + \frac{4}{3}\eta_s}{\eta_f} k_\varphi} = \sqrt{\frac{(\eta_b + \frac{4}{3}\eta_s)a^2}{\eta_f b} \varphi^n} \quad \text{(II.40)}$$

forms (McKenzie, 1984); for the second step, the permeability–porosity relation eq. II.8 has been used. l_c is the compaction length over which the compaction rate decreases to $1/e$ of its maximum value; it is the characteristic length scale of compaction processes. Compaction is driven by the pressure difference between melt and matrix and depends on the rheological properties of both; if the layer is much smaller than l_c, it only depends on the matrix viscosities. – If the upper boundary is impermeable, the upwelling melt will accumulate there and cause negative compaction, *i.e.* dilatation of the matrix, until it decomposes at high melt fractions. This situation is expected to be of importance for the partially molten region of the mantle, which is bounded by the impermeable solid mantle from below and by the equally impermeable cold lithosphere from above; however, it is unlikely that melt fractions become so large that the matrix disintegrates.

It is of interest to estimate if compaction is an important factor in the geodynamical framework of the melting regions in the upper mantle. In 1-D and 2-D models of mantle upwellings, the compaction length was found to be usually three orders of magnitude smaller than the melting region in the mantle, which covers a depth range of $10\ldots100\,\mathrm{km}$ (Ribe, 1985a). In this limiting case, the contribution of compaction to melt flow is restricted to small distances and is negligible compared with the effect of buoyancy and shear deformation; therefore it is concluded that Darcy's law (eq. II.36) provides an adequate description of melt segregation on a large scale (Ribe, 1985a, 1987). On the other hand, Ghods and Arkani-Hamed (2002) observed in models of both melting and non-melting diapirs of a scale smaller than plumes but larger than l_c the development of porosity variations related to compaction of the matrix, but their models are not all directly applicable to the mantle *e.g.* with respect to the permeabilities and retained melt fractions chosen. – Apart from this large-scale behaviour, it is quite probable that compaction is of importance in small-scale segregation in the neighbourhood of (short-lived) melt channels or dikes (section II.4.2), because in this case the pressure gradient, which is inversely proportional to the vein size, is much larger and might well be of a similar magnitude as the buoyancy-related pressure gradient (Sleep, 1988).

A further issue with eq. II.40 is the assumption of constant viscosity. Connolly and Podlachikov (1998) remark that constant viscosity models can only be applied on length scales of a few e-fold lengths of decrease of η, which for a thermally activated viscosity model is

$$l_\eta = \frac{R\underline{T}^2}{\mathcal{E}\frac{\mathrm{d}T}{\mathrm{d}z}}, \tag{II.41}$$

and only if $l_c \ll l_\eta$; only then will l_c be the length scale of compaction. While this condition is probably not fulfilled in the lithosphere (Connolly and Podlachikov, 1998), it should hold in the melting region of the mantle due to the high temperatures and flat geothermal gradients; $\underline{T} \approx 1700\,\mathrm{K}$ and $\mathcal{E} = 610\,\mathrm{kJ/mol}$ (Kohlstedt and Zimmerman, 1996) yield $l_\eta \approx 130\,\mathrm{km}$, *i.e.* 1–2 orders of magnitude more than common estimates for l_c. – A similar problem could arise if the viscosity changes due to dehydration.

Conservation of energy — In contrast to momentum, energy is not individually conserved in the melt and the matrix, but only in the whole system, because mass is exchanged and the heat can diffuse from one phase to the other. The general heat equation includes several contributions such as advection and conduction of heat, dissipation, heat of fusion, internal heat production etc. It also has to account for the effects of relative motion between melt and matrix, which mainly are transport of heat due to the smaller heat loss of ascending segregated melt and the release of gravitational energy which leads to dissipation. Both factors increase the production of melt, although they were not considered in section II.3.1; in particular, the migration of melt in a melting plume head has been shown to homogenize the T field (Ceuleneer *et al.*, 1993), but some effects, *e.g.* dissipation, are not significant in terms of melt production in most or all parts of the melting domain (Asimow, 2002). The heat equation is an extension to eq. I.3b:

$$(1 - \varphi)\varrho_s \left[c_{p_s} \left(\frac{\partial}{\partial t'} + \vec{v}_s' \cdot \nabla' \right) T' - \frac{\alpha_s T'}{\varrho_s} \left(\frac{\partial}{\partial t'} + \vec{v}_s' \cdot \nabla' \right) p' \right]$$

$$+ \varphi \varrho_f \left[c_{p_f} \left(\frac{\partial}{\partial t'} + \vec{v}_f' \cdot \nabla' \right) T' - \frac{\alpha_f T'}{\varrho_f} \left(\frac{\partial}{\partial t'} + \vec{v}_f' \cdot \nabla' \right) p' \right]$$

$$= \nabla' \cdot (k_T \nabla' T') + H' - T' \Delta S'_m \Gamma'$$

$$+ \frac{\eta_f' \varphi^2}{k_\varphi} (\vec{v}_f' - \vec{v}_s')^2 + \left(\eta_b' - \frac{2}{3}\eta_s' \right) (\nabla' \cdot \vec{v}_s')^2 + \frac{\eta_s'}{2} [\nabla' \vec{v}_s' + (\nabla' \vec{v}_s')^T] \cdot [\nabla' \vec{v}_s' + (\nabla' \vec{v}_s')^T] \quad \text{(II.42)}$$

(McKenzie, 1984; Ribe, 1987), and it is challenging to solve it in this form. Therefore, it is simplified by casting it in the form already shown in eq. I.3b using the matrix velocity in order to make it numerically tractable more easily and less expensively; it should be noted that thermal equilibrium between matrix and melt is assumed here, although this is not strictly fulfilled in reality. It must be emphasized that this formulation of the conservation of energy still does not take into account the influence of the chemical/petrological composition of the rock.

Conservation of composition — With respect to the considerations about trace element composition made in sections II.3.1 and II.3.2, it is worth mentioning a fourth conservation law relevant to melting, namely the conservation of chemical composition. If chemical equilibrium between matrix and melt prevails, it can be shown that the effective transport velocity of a given element travelling as a solute in the melt is *not* equal to $\vec{v}_f$, but is smaller and depends on the partition coefficient and the melt fraction:

$$\vec{v}_{C,\text{eff}} = \frac{\varrho_f \varphi \vec{v}_f + \varrho_s (1 - \varphi) D \vec{v}_s}{\varrho_f \varphi + \varrho_s (1 - \varphi) D} \quad \text{(II.43)}$$

(Spiegelman and Elliott, 1993; McKenzie, 1984; Navon and Stolper, 1987); this phenomenon is known as "chromatographic" melt migration. The results summarized in sections II.1.5 and II.4.2 suggest though that chemical equilibrium between melt and matrix is not likely to be achieved in many situations, especially at higher segregation velocities. Iwamori (1993a,b) presented a one-dimensional model of segregation through a matrix of spherical grains taking into account chemical disequilibrium which describes the transition of melt from the porous flow domain to the channels by a suction term. Comparison of his numerical results with data from peridotites indicates that for equilibrium melting a substantial part of the melt must be retained in the matrix, while efficient melt draining is necessary to explain the peridotite chemistry in a disequilibrium context; as the results from permeability investigations (see section II.1.5) suggest that melting is near-fractional, the numerical models of Iwamori (1993b) support the statement that the melting process is associated with chemical disequilibrium and that at least 80 % of the melt is transported through channels. The degree of equilibration of segregating melt depends subtly on segregation velocity and channel spacing and can vary between different groups of chemical indicators, at least if only diffusive processes are taken into account (Kenyon, 1998). Recent geochemical studies on radioisotope disequilibria of MORB indicated that the parent magma was a mixture of disequilibrium melt from channels and equilibrated melt from porous flow (Sims *et al.*, 2002), but also admitted a suction parameter of at least *ca.* 0.4–0.5 (Lundstrom, 2000; Jull *et al.*, 2002).

Magmons — For completeness, some outlines of certain special solutions of the equations of porous flow are presented in the following, although they will not be examined further: in addition to the normal percolation of melt, the equations of porous flow II.35 and II.37a have another class of solutions which describe nonlinear porosity waves (Scott and Stevenson, 1984; Richter and McKenzie, 1984; Spiegelman, 1993b). These so-called magmons are regions of high porosity which arise from strongly molten boundary layers between an impermeable bottom margin and a low-porosity region (Scott and Stevenson, 1984) and travel as solitary waves with porosity-dependent speed; the waveform travels faster than both the background and the magmon liquid. Scott and Stevenson (1984, 1986) assessed that they could reach a vertical extension of a few kilometers and velocities of some centimeters or decimeters per year and showed them to almost conserve their shapes in collisions in many cases, but to be unstable to perturbations of higher dimension than their own. It is not clear whether magmons exist in the earth, but Scott and Stevenson (1986) estimated that they are likely to be unimportant in regions of broad upwelling resp. considerable variations in background porosity such as spreading centers, where the migrating melt is entirely supplied by the region in which magmons would propagate, because they would degrade in such an environment (Scott, 1988). They might be important in a setting with a stable partially molten domain into which melt is supplied from outside, *e.g.* regions of low permeability as at the freezing front under cold oceanic lithosphere, and develop into shock waves whose steep gradients in porosity and viscosity can result in significant compaction even if l_c is small (Spiegelman, 1993a).

With respect to this study, the possible formation and propagation of magmons will not be considered further, and is, in fact, not expected to be observable at all. The abovementioned investigations suggest that magmons are rather small-scale structures compared *e.g.* with the lengthscale of convection cells, and lie below the resolution of the numerical grids used in the models of this study; a higher resolution would be prohibitively costly. Apart from this technical reason, the method of melt extraction applied here includes an imposed cut-off of porosity peaks at each timestep, which suppresses the formation of more pronounced instabilities and thus also the evolution of magmons; details of the melt-related algorithm are given in section III.2.1.

II.4.2 Structural control of melt segregation, focussing, and extraction

The large number of different factors influencing the generation and segregation of melt and the compaction of the matrix has stimulated a variety of proposals for mechanisms which would both make the observed efficient transport of melt possible and explain the focussing of melt from a supposedly broad volume of generation into a relatively narrow eruption zone; these mechanisms must also explain the fact that MORB are not in chemical equilibrium with residual mantle rocks: as discussed in section II.3.1, MORB are likely to be formed as a mixture of melts from different depth/pressure levels which are partly much deeper than the base of the crust, as indicated *e.g.* by their undersaturation in orthopyroxene and their REE signature (Kelemen *et al.*, 1997) as well as their remarkable chemical homogeneity with regard to several characteristics.

Many of these models include the replacement of the scalar permeability by a tensor, either *a priori* (Phipps Morgan, 1987) or by deriving mechanisms for the formation of high-porosity channels or dikes with a uniform orientation which result in bulk anisotropy (Stevenson, 1989; Ceuleneer and Rabinowicz, 1992; Aharonov *et al.*, 1995). In any case, it seems to be necessary that melt be transported by some kind of flow along oriented paths instead of simple porous flow in a substantial part of the mantle beneath a MOR (Kelemen *et al.*, 1997), but apparently also partially in plumes (Hauri and Hart, 1994); these channels are likely to have a spacing of less than 1 km, but more than 0.1 m, because the melt is expected to be in thermal, but not in chemical equilibrium with the rock matrix (Iwamori, 1993a,b; Iwamori *et al.*, 1995), at least not beyond a certain distance of its source region. Magnetotelluric measurements from the RAMESSES experiment also indicated that down to a depth of 50 km, the electrical resistivity is too high to be reconciled with an interconnected melt network, while below this layer, low resistivities are in agreement with a small, interconnected melt fraction (Sinha *et al.*, 1998; Heinson *et al.*, 2000).

Richardson *et al.* (1996) considered the existence of cylindrical melt conduits of whatever kind in a porous matrix and showed that a channel of width b_{ch} exists as a distinct structure if

$$\frac{b_{ch}}{l_c} > \sqrt[4]{\frac{\eta_f}{\eta}}, \qquad (II.44)$$

and that there are three domains of melt flow in the channel–matrix system: In the inner parts of the conduit, melt ascends rapidly and is isolated from the matrix; within the matrix, the melt flows very slowly and does not enter the channel; across the boundary, melt is cycled from the channel margins into the adjacent matrix and back in a closed flow system. The formation of such a channel is though not covered in their model.

If the existence of a drainage system consisting of some kind of oriented channels is assumed, a link between the grain-size and the dike-size level of it has to be established. Therefore, Hart (1993) introduced the concept of a fractal tree for melt drainage. The dimensions of the channels are unconstrained so far, but could be assessed by combining data from dunite channels in ophiolites[4] and measurements in laboratory samples with estimates on the depths of melt generation. He estimated the velocity increase to roughly one order of magnitude every 10 generations, so that chemical equilibrium between matrix and melt will basically have ceased after a few kilometers along the melt path. Inserting rough estimates of $\eta_f = 1\,Pa\,s$, $\eta = 10^{19}\,Pa\,s$ in eq. II.44 yields $O(b_{ch}/l_c) > 10^{-5}$, which can be achieved for b_{ch} of a few millimeters or larger and l_c on the kilometer scale or less. Thus, the existence of such cylindrical channels is quite possible in the mantle beyond a certain distance of the source region, but one should bear in mind that tabular channels are preferred over tubes at least on larger lengthscales beneath MOR.

Isotropic corner flow and related models — Corner flow models (Batchelor, 1967) can be used as a simple approximation to the movement of the mantle *e.g.* below a spreading ridge and permit the calculation of an analytical solution, although in the real mantle one would expect a certain departure from corner flow. In the model of Spiegelman and McKenzie (1987), the flow of the partially molten mantle is imposed by the divergence of the lithosphere at the ridge, and the resulting nonhydrostatic pressure gradient focusses the melt towards the ridge, especially in the parts near the spreading center, where the gradient is high, while in the outer parts of the upwelling column, it is rather accumulated at the bottom of the lithosphere and moved with the plate. Hence, focussing is more efficient

---

[4]Ophiolites are tectonic key structures for the verification of melt extraction mechanisms. Several authors have reported the existence of discordant dunite bodies in the mantle segments of different ophiolites and believe that they have been generated in upwelling mantle by one or more of the mechanisms mentioned in this section – in particular, by reactive flow and/or cracks – and represent conduits for focussed flow; this is supported by evidence for chemical equilibrium between them and MORB (Kelemen *et al.*, 1997). Their estimate for the proportion of dunite conduits in the mantle is 0.5...1.5 %, depending on the fraction of large dikes which flush the largest part of the melt; the channels are not evenly distributed in an ophiolite though, but tend to cluster while being totally absent in other areas (Kelemen *et al.*, 2000), and the timescale of their formation seems to be short (Suhr, 1999).

at high matrix shear viscosities (Ribe, 1987); Spiegelman and McKenzie (1987) estimated that $\eta_s = 10^{20} \ldots 10^{21}$ Pa s is necessary to achieve the observed oceanic crust thicknesses for $\varphi = 2 \ldots 4\%$. This viscosity is much higher than some independent estimates which indicate $\eta_s = 10^{19}$ Pa s or less for the asthenosphere. Hence, this mechanism may contribute, but cannot explain alone the concentration of the melt in the ridge region (*e.g.* Phipps Morgan, 1987; Scott and Stevenson, 1989) unless higher porosities are admitted. Alternatively, and more likely in view of the results on permeability (see section II.1.5), a dehydrated restite layer as the lower part of the "compositional lithosphere" (Phipps Morgan, 1997, see sect. I.3.2) could also provide sufficiently high viscosities to make focussing by viscous stress gradients a possible mechanism for explaining the narrowness of the crust production zone; in fact, petrophysical considerations (Hirth and Kohlstedt, 1996) and numerical experiments (Braun *et al.*, 2000; Choblet and Parmentier, 2001) suggested viscosities of $10^{20} \ldots 10^{21}$ Pa s for the main melting region. – In contrast, melt migration in corner flow models for a plume beneath a stationary plate is buoyancy-dominated and can be divergent (Ribe and Smooke, 1987), although strong upwelling or large porosity gradients can also enable focussing in this case (Li and Spohn, 1991).

As an alternative, especially beneath slow-spreading ridges, several numerical convection models derived from a corner-flow-like situation led to the proposition that the downgoing currents of small-scaled convection rolls caused by lateral density inhomogeneities due to melting of the mantle constrain the upwelling to a narrower region (Scott and Stevenson, 1989; Sotin and Parmentier, 1989; Su and Buck, 1993; Cordery and Phipps Morgan, 1993, see sect. I.2.1); similarly, viscosity variations due to dehydration and melting-related creep mechanism changes also have been shown to result in flow focussing (Braun *et al.*, 2000; Choblet and Parmentier, 2001).

Magmon channelling — As mentioned in section II.4.1, large gradients in porosity may lead to the formation of solitary waves. Hence, it has been suggested that the melt flowing upwards through the porous mantle matrix and encountering the impermeable boundary of the lithosphere may develop two-dimensional porosity waves parallel to the slope of the freezing front acting as channels oriented towards the spreading center (Sparks and Parmentier, 1991, 1994; Spiegelman, 1993b; Ghods and Arkani-Hamed, 2000). A precondition for channel formation parallel to the freezing front to take place instead of simple freezing of melt at the base of the lithosphere is that the freezing zone is constrained to a thickness of not much more than the compaction length; otherwise the compaction due to the viscous reaction of the matrix would be negligible.

Solution-induced channels — The concept that chemical reactions of a fluid migrating through a porous matrix result in the formation of finger-like dissolution instabilities at the reaction front (*e.g.* Ortoleva *et al.*, 1987) can be applied to melt undersaturated in certain components (especially opx) flowing through a rigid or a viscously compacting porous peridotite matrix with a solubility gradient (Aharonov *et al.*, 1995; Spiegelman *et al.*, 2001) to explain the chemical disequilibrium of MORB with upper mantle peridotites and the existence of so-called replacive dunite in ophiolites: if the melt is *e.g.* undersaturated in orthopyroxene, which is the case with MORB melts at low pressure, it would dissolve opx from the matrix in the shallow mantle near the crust–mantle transition during porous flow and thus increase the porosity along its path, which would in turn cause more melt to pass through this region and reinforce the reaction. By linear analysis and numerical models it could be shown that the combination of dissolution, diffusion, compaction, and possibly even melting leads to the formation of channels with a spacing smaller than l_c, if l_c is much greater than the reaction length (Aharonov *et al.*, 1995; Spiegelman *et al.*, 2001); the dominating spacing ranged from approximately 10 cm and 1 km in these models.

Field investigations on the Ingalls and Oman ophiolites by Kelemen *et al.* (2000) and Braun and Kelemen (2002) support the solution model through evidence for melt transport through dunite channels in at least the upper 30 km of the mantle. In the latter, more comprehensive study including channels with b_{ch} ranging from millimeters to ~ 100 m, a cumulative distribution[5]

$$\mathfrak{N}_{ch} = \int\limits_{b_{ch,min}}^{b_{ch,max}} \frac{B_{dun}\Lambda}{b_{ch}^{\Lambda+1}}\, db_{ch} = \frac{B_{dun}}{b_{ch,min}^{\Lambda}} - \frac{B_{dun}}{b_{ch,max}^{\Lambda}} \approx \frac{B_{dun}}{b_{ch,min}^{\Lambda}} \qquad (\text{II.45})$$

with an average $\Lambda = 1.11$ and $B_{dun} = 0.088\,\text{m}^{\Lambda-1}$ was found[6]; the latter quantity is related to the dunite fraction, which ranges from 40 to 60 % in most samples of their study. However, Kelemen *et al.* (2000) warned that their relations were derived from measurements in ophiolite areas were dunite channels were actually present, which is not the case throughout a whole ophiolite complex. They remarked that the channels can be initiated by different processes, *e.g.* a small crack (see Suhr, 1999) or an initial chemical heterogeneity; however, Suhr *et al.* (2003) questioned on the basis of detailed chemical analyses of such dunite bodies that the largest of them are generally related to melt drainage. In laboratory experiments with melts saturated only in olivine and synthetic peridotites with varying pyroxene contents, finger-like high-porosity structures also developped in the rock matrix after some hours of disequilibrium porous flow (Daines and Kohlstedt, 1994).

Stress-induced channels — McKenzie (1985a) predicted that instabilities in melt percolation would result in the formation of high-porosity channels, in which the melt would be flushed to the surface at very high rates and isolate them chemically from the matrix. Stevenson (1989) then showed in a two-dimensional model that a partially molten viscous medium in a large-scale deformation field develops small-scale instabilities of the melt distribution if η is a function of φ; these instabilities form by melt migration parallel to the axis of minimum compressional stress and its accumulation in channels perpendicular to this direction and grow at most proportionally to $-\dot{\varepsilon}_{0_{max}} d\ln\eta/d\varphi$, which is estimated to be fast on a geological timescale, so that melt flow in these veins will dominate over ordinary porous flow. This observation was confirmed by Richardson (1998) in a numerical study of flow under deviatoric strain. The physical reason for the instability is the fact that high η correspond to low φ and that melt is pressed into regions of low p_l by the difference between ambient fluid pressure and the least compressive stress (*e.g.* Sleep, 1988; Rubin, 1998) which correspond to high φ anyway. Stevenson (1989) showed that under favorable conditions an only slightly perturbed initial melt distribution evolves into a pattern of melt-soaked veins with random spacing on the order of meters on average, separated by domains with a porosity several orders of magnitude smaller than that of the channels; the evolution of the system is very sensitive to the choice of the parameters, though. He proposed that the tips of these veins may act as nucleation points for fractures, because the stress concentrates there, especially for veins whose length is about l_c. Sleep (1988) showed that veins grow and become more closely spaced until their melt drainage domains overlap; the merging of veins then leads to further concentration of melt, because the large ones have a lower fluid pressure and suck melt from

---

[5]defined as the number of channels per meter larger than a certain minimum width

[6]Having in mind that a pyroxenite streak could initiate a channel (Lundstrom *et al.*, 2000), it is noted that Allègre and Turcotte (1986) found a power-law relation between $\mathfrak{N}$ and b for pyroxenite layers in the Beni Bousera ophiolite; their Λ of 0.85. . .0.9 lies within the scatter of the values in the study of Braun and Kelemen (2002).

the smaller ones. However, a certain total strain has to be accumulated before channels can form (Richardson, 1998).

Dikes — In comparison with the velocities in the deep source regions in the mantle, the final step of melt migration happens much more rapidly, thus requiring an efficient flushing mechanism. Taking mantle viscosity as the crucial material parameter, rheological and geological considerations suggest that dikes form in the uppermost part of the mantle and in the crust, which are sufficiently stiff due to their relatively low temperature to show brittle reaction to stresses (Nicolas, 1990). Dikes can possibly form within the replacive dunite bodies due to the high porosity and unusually large grains by hydrofracture, but only if the viscosity is high enough and the compaction length is on the order of 10 km, which is only to be expected near the lithospheric base or another region of potentially sharp porosity gradients (Kelemen *et al.*, 1997); on the other hand, Kelemen *et al.* (2000) suggested that small cracks can also be nucleation points for the formation of dunite channels. In contrast, melt transport by dikes probably plays only a minor role in anomalously hot mantle, *e.g.* in hotspots, where the viscosity is reduced by at least one order of magnitude (Sleep, 1988); on the other hand, it has been suggested that at least for the formation and ascent of larger dikes, partially molten mantle rock can be treated as a poroelastic/brittle rather than as a viscous material with respect to fracturing (Rubin, 1998).

The prevailing stress field controls the onset of fracturing. Partially molten rock may exhibit ductile fracture by viscous deformation of grains accommodating ruptures initiated by melt which propagates along their boundaries (Fowler, 1990) or along planes of weakness such as boundaries between harzburgite and pyroxenite (Suhr, 1999). This will happen if the fluid overpressure $\Delta p = g \Delta \varrho \, \Delta z_d$ in an interconnected melt column of height Δz_d exceeds the peridotite yield stress, which was estimated as 40 or 50 MPa (Nicolas, 1990) but is possibly as low as 1 MPa in the melting region or 10–15 MPa beneath a freezing boundary (Rubin, 1998; Maaløe, 1998); *e.g.* for a dike at the base of the lithosphere, the critical value for Δz_d is only a few kilometers. The fracture criterion is probably fulfilled in the more viscous resp. poroelastic thermal boundary above the partially molten zone. Thus, fractures are expected to appear in this shallowest part of the mantle and the lithosphere and crust and provide an efficient means of melt drainage (Fowler, 1990; Turcotte and Phipps Morgan, 1992), in accordance with observations in ophiolites (Ceuleneer and Rabinowicz, 1992); however, porous flow from the surrounding matrix into the growing dike was shown to enable downward growth of a dike nucleating at the base of the lithosphere into the partially molten region, in addition to its principal upward propagation (Fowler and Scott, 1996).

The stress field resulting from corner flow beneath MOR does though not generate extensional dikes whose orientation would focus melt to the spreading center for $z < x$ in an isoviscous mantle (Sleep, 1984; Phipps Morgan, 1987); in the deeper parts of a plume melting zone, where the upwelling is narrowing, σ_1 is even subhorizontal, which will hinder straight melt ascent or even lead to melt trapping (Ceuleneer *et al.*, 1993). On the other hand, the stress-controlled dike orientation could be more favourable, although not ideal, if the viscosity distribution leads to the formation of small convection rolls near the ridge which would deflect the mantle flow toward the spreading center (Ceuleneer and Rabinowicz, 1992). As an alternative, Phipps Morgan (1987) proposed that the permeability tensor is related to a mantle fabric induced by finite strain accumulation in such a way that the principal permeabilities coincide with the major axes $\mathbf{e}_i$ of the finite strain ellipsoid; in a two-dimensional corner flow setting, for which McKenzie (1979) has calculated finite strain solutions, a mechanism relating $\mathbf{k}_\varphi$ and $\mathbf{e}$ through $k_{\varphi\text{max}}/k_{\varphi\text{min}} = e_{\text{max}}/e_{\text{min}}$ would indeed result in melt transport toward the spreading center, provided that no other processes suppress the formation of the fabric. – The orientation and propagation path of ascending fluid-filled fractures does though not depend solely on the large-scale principal stresses, but is also influenced by stress gradients (Dahm, 2000; Muller *et al.*, 2001) and the stress fields of other dikes (Ito and Martel, 2002).

As melt transport through dikes is much faster than porous flow, an extraction event will leave the corresponding part of the mantle free of mobile melt. Therefore, the extraction of melt through a dike must be an episodic process: a cycle, estimated to last *ca.* 50 a at a slow-spreading ridge, begins with the growth of a vein dominated by porous flow from its surroundings, which gradually transforms into a rising dike scavenging from smaller veins on its path, and is terminated over a few weeks by flushing and closure of the old dike and formation of a new vein (Sleep, 1988; Nicolas, 1986, 1990).

II.4.3   Flow in an anisotropic matrix

As discussed in the previous section, the permeability is probably macroscopically anisotropic beneath spreading ridges and in plumes, at least in certain parts of the partially molten region; more precisely, it seems plausible to assume that it has an anisotropic component superimposed on an isotropic background. As an idealization, such a medium can be described by splitting it into an "background" domain 1 with an isotropic permeability $k_{\varphi 1}$ and a porosity φ_1 and a "channel" domain 2 with a different intrinsic isotropic permeability $k_{\varphi 2}$ and a porosity φ_2 (figure II.9); the total volume occupied by domain 2 is assumed to be considerably smaller than that of domain 1, but its porosity is larger. If the porosity of a given volume could be treated as homogeneous in an isotropic porous medium, the bulk anisotropic mantle considered here can only be described by an average porosity,

$$\overline{\varphi} = \frac{1}{V_{\text{tot}}} \int\limits_{V_{\text{tot}}} \varphi(x, y, z)\, \mathrm{d}V = (1 - \mathcal{B})\varphi_1 + \mathcal{B}\varphi_2, \qquad \mathcal{B} = \frac{b_{\text{ch}}}{a_{\text{ch}}}, \qquad (\text{II.46})$$

whereby the values of φ_1 and φ_2 can be determined using some suction parameter describing the distribution of total melt on the two domains, as suggested by Iwamori (1993b) (see p. 52), and $\mathcal{B}$ is the volume fraction of domain 2 in the whole region of interest. The fundamental idealization made now is that the suction parameter accounts for the entire exchange of melt between the domains, and that flow in one domain is independent from the other; if flow through the conduit walls can be treated this way, the model of Richardson *et al.* (1996, see p. 54) suggests that this simplification is justified. The background flow in domain 1 can then be described by the equations summarized in section II.4.1, with an appropriately determined $(1 - \mathcal{B})\varphi_1$, *i.e.* the medium is regarded as being free of any structural features influencing the flow, which implies $b_{\text{ch}} \ll a_{\text{ch}}$ in some sense. In contrast, the geometrical structure of the more or less narrow channels is substantial for the channelized flow, so this domain is represented as a medium with conduits with porosity φ_2 and width b_{ch}, separated by melt-free, impermeable rock units, *i.e.* the underlying assumption is $k_{\varphi 2} \gg k_{\varphi 1}$; the resulting permeability model would hence be a transversely isotropic medium with the direction of lowest permeability parallel to its symmetry axis. The volume fraction $\mathcal{B}$ occupied by the channel domain should be derived from field observations; ideally, one could use distribution laws such as eq. II.45, although in this particular case, problems with chosing a proper normalization prevented straightforward use. Hence, $\mathcal{B}$ is a free parameter as in the model of Jull *et al.* (2002); however, it is in fact constrained by the suction parameter and by φ_2.

The medium is treated as a two-dimensional one, in the sense that the channels are tabular, *i.e.* their extent in the direction perpendicular to the plane of figure II.9 is much

Figure II.9: Scheme of oriented porous channels with the intrinsic φ_2 in a matrix with a different porosity φ_1.

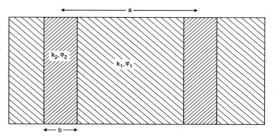

larger than b_{ch}; for the representation in a numerical grid, this would mean that the plane normal to $\vec{b}_{\mathrm{ch}}$ at a grid point extends as far as the corresponding cell volume, which should be of the order of a few kilometers and seems to be geologically reasonable. To describe the flow in domain 2 with the established equations resp. in the simplest form with Darcy's law eq. II.36, the permeability must be written as a second-rank tensor to include the influence of channel orientation:

$$\vec{v}_{\mathrm{f}} - \vec{v}_{\mathrm{s}} = -\frac{\mathbf{k}_{\varphi\mathrm{ch}}}{\eta_{\mathrm{f}}\mathcal{B}\varphi_2}[\nabla\tilde{p} - (1 - \mathcal{B}\varphi_2)\Delta\varrho_{\mathrm{f}}g\vec{e}_z], \tag{II.47}$$

with primes being dropped. To derive its explicit form for a given permeability model, it is most convenient to consider vertically oriented dikes, because in this case, the only nonzero component of $\mathbf{k}_{\varphi\mathrm{ch}}$ is $k_{\varphi\mathrm{ch}_{zz}}$. For application of the non-dimensionalized version of the equation of melt momentum, eq. II.38a, the melt retention number Rt has to be adapted, but given the simple geometrical model, it is possible to split the tensor into a scalar reference permeability $k_{\varphi\mathrm{ch}}^0$ or retention number Rt_{ch}^0 and a dimensionless scaling tensor $\mathbf{k}_\varphi$ containing the information about orientation.

Reference permeability models for the channel domain — Geological observations and the mechanisms of channel and dike formation suggest that the most obvious geometric model for the derivation of $k_{\varphi\mathrm{ch}}^0$ is a set of parallel tabular veins or dikes of width b_{ch} and center-to-center distance a_{ch} as in figure II.9. For the case of cracks, domain 2 consists entirely of melt, *i.e.* $\varphi_2 = 1$, and assuming melt transport by Poiseuille flow, Darcy's law can be written for parallel dikes as

$$\mathcal{B}\vec{v}_{\mathrm{seg}} = -\frac{b_{\mathrm{ch}}^2}{12\eta_{\mathrm{f}}}\nabla p = -\frac{a_{\mathrm{ch}}^2}{12\eta_{\mathrm{f}}}\mathcal{B}^2\nabla p, \tag{II.48a}$$

so that it is possible to derive the permeability in the dike plane, the anisotropic reference permeability, and a melt retention number for channel flow as

$$k_{\varphi\mathrm{ch}}^0 = \frac{b_{\mathrm{ch}}^2}{12} = \frac{a_{\mathrm{ch}}^2}{12}\mathcal{B}^2 \tag{II.48b}$$

$$Rt_{\mathrm{ch}}^0 = \frac{12\eta_{\mathrm{f}0}z_{\mathrm{m}}^2}{\eta_0 b_{\mathrm{ch}}^2}; \tag{II.48c}$$

a similar model was also proposed by Maaløe (1998). The spacing $a_{\mathrm{ch}} - b_{\mathrm{ch}}$ has the same order of magnitude as the compaction length l_{c} in eq. II.40 (Sleep, 1988; Ribe, 1986); more precisely, it is proportional to $\sqrt{l_{\mathrm{c}}a}$ (Stevenson, 1989). The width of the channels in Maaløe's (1998) layer model is estimated to be of the order of 5–50 cm; velocities in this kind of channels could be as high as some millimeters per second. Similarly large values were estimated by McKenzie (2000) considering the transport of millimeter-to-centimeter-sized xenolith nodules.

As an alternative, one could assume that the conduits form an array of parallel cylindrical channels; in his fractal tree model, Hart (1993) uses a network of cylindrical tubes. For the simpler case of parallel tubules of diameter b_{tb} arranged in a pattern of hexagonal symmetry with center-to-center distance a_{tb}, the porosity and the permeability would be:

$$\mathcal{B} = \frac{A_{tb}}{A_{hex}} = \frac{\pi}{6\sqrt{3}} \left(\frac{b_{tb}}{a_{tb}}\right)^2 \tag{II.49a}$$

$$k_{\varphi ch}^0 = \frac{3\sqrt{3}a_{tb}^2}{16\pi}\mathcal{B}^2 = \frac{\pi}{192\sqrt{3}}\frac{b_{tb}^4}{a_{tb}^2}. \tag{II.49b}$$

As $\varphi_2 = 1$ for the above models, one can determine φ_1 directly from eq. II.46 if the average melt content in a region is known.

The two-porosity model of Lundstrom (2000) is conceptually a variant of this; however, he assumes that in both the background matrix and in the dunite channel, isotropic porous flow takes place and that the difference between both is due to different parameters to be used in the Kozeny–Carman equation II.8. This case has been treated in connection with the flow in porous tabular dunite channels of different size b_{ch} and a cumulative-number–size distribution of the type of eq. II.45 by Braun and Kelemen (2002). From their eq. 12 for the total channelized melt flux in a region with length scale l, channel widths between $b_{ch,max}$ and $b_{ch,min}$ and a maximum channel porosity φ_{max2},

$$q_{tot} = \frac{a^2\varphi_{max2}^n B_{dun}\Lambda l\Delta\varrho g}{b\eta_f b_{ch,max}^\Lambda}(b_{ch,max} - b_{ch,min}), \tag{II.50}$$

an effective permeability in the direction of the dunite channel flow can be derived by dividing q_{tot} by $-l\varphi_2$ and comparing with Darcy's law eq. II.36:

$$k_{\varphi ch}^0 = \frac{a^2\varphi_{max2}^n B_{dun}\Lambda(b_{ch,max} - b_{ch,min})}{b\,b_{ch,max}^\Lambda}. \tag{II.51}$$

Orientation of the permeability tensor — In the case of dikes and stress-induced channels, the orientation of the symmetry axis follows, at least to a first approximation, from the stress field resulting from the mantle current velocity. Thus, it is possible to calculate the orientation for an arbitrary convection model by solving the eigenvalue problem for the stress tensor; besides, analytical solutions can also be achieved if the current is known in closed form.

The viscous deviatoric stress tensor in a viscous medium,

$$\sigma_{ij} = 2\eta\dot{\varepsilon}_{ij} = \eta\left(\frac{\partial v_i}{\partial x_j} + \frac{\partial v_j}{\partial x_i}\right), \tag{II.52}$$

is given by

$$\Leftrightarrow \boldsymbol{\sigma}(x,z) = \begin{pmatrix} \sigma_{xx} & \sigma_{xz} \\ \sigma_{xz} & \sigma_{zz} \end{pmatrix} = 2\eta\begin{pmatrix} \frac{\partial v_x}{\partial x} & \frac{1}{2}\left(\frac{\partial v_x}{\partial z} + \frac{\partial v_z}{\partial x}\right) \\ \frac{1}{2}\left(\frac{\partial v_x}{\partial z} + \frac{\partial v_z}{\partial x}\right) & \frac{\partial v_z}{\partial z} \end{pmatrix}. \tag{II.52a}$$

for a two-dimensional (x, z) model. To compute the principal stresses and the smallest and largest compressive stress, a principal axis transformation is performed. The solution of the corresponding eigenvalue problem results in the eigenvalues

$$\sigma_{1,3} = \eta \left[\left(\frac{\partial v_x}{\partial x} + \frac{\partial v_z}{\partial z} \right) \pm \sqrt{ \left(\frac{\partial v_x}{\partial x} - \frac{\partial v_z}{\partial z} \right)^2 + \left(\frac{\partial v_x}{\partial z} + \frac{\partial v_z}{\partial x} \right)^2 } \right] \qquad \text{(II.53a)}$$

and in the eigenvector matrix

$$\mathbf{R} = \begin{pmatrix} -\frac{\partial v_x}{\partial z} - \frac{\partial v_z}{\partial x} & -\frac{\partial v_x}{\partial z} - \frac{\partial v_z}{\partial x} \\ \frac{\partial v_x}{\partial x} - \frac{\partial v_z}{\partial z} - \sqrt{\left(\frac{\partial v_x}{\partial x} - \frac{\partial v_z}{\partial z} \right)^2 + \left(\frac{\partial v_x}{\partial z} + \frac{\partial v_z}{\partial x} \right)^2} & \frac{\partial v_x}{\partial x} - \frac{\partial v_z}{\partial z} + \sqrt{\left(\frac{\partial v_x}{\partial x} - \frac{\partial v_z}{\partial z} \right)^2 + \left(\frac{\partial v_x}{\partial z} + \frac{\partial v_z}{\partial x} \right)^2} \end{pmatrix}$$
$$\text{(II.53b)}$$

with which the stress tensor can be turned into

$$\tilde{\boldsymbol{\sigma}} = \mathbf{R}^{\mathrm{T}} \boldsymbol{\sigma} \mathbf{R} = \begin{pmatrix} \sigma_1 & 0 \\ 0 & \sigma_3 \end{pmatrix}; \qquad \text{(II.54)}$$

here the relation $\mathbf{R}^{\mathrm{T}} = \mathbf{R}^{-1}$ was applied, because $\boldsymbol{\sigma}$ is symmetric and $\mathbf{R}$ thus orthogonal, and the first component of the eigenvectors, which is arbitrary because the eigenspaces of both eigenvalues are one-dimensional, was chosen to be $-\left(\frac{\partial v_x}{\partial z} + \frac{\partial v_z}{\partial x} \right)$. More important for the task considered here is the fact that $\mathbf{R}$ can also be used to rotate the permeability tensor, which is symmetric and positive definite (Dagan, 1989) and thus reducible to its main permeabilities, from the natural coordinate system into the principal axis system of the stress tensor and vice versa:

$$\mathbf{k}_\varphi = \mathbf{R} \mathbf{k}_\varphi^0 \mathbf{R}^{\mathrm{T}} = \begin{pmatrix} k_{\varphi xx} & k_{\varphi xz} \\ k_{\varphi xz} & k_{\varphi zz} \end{pmatrix}; \qquad \text{(II.55)}$$

the initial $\mathbf{k}_\varphi^0$ is assumed diagonal and has to be rotated in the contrary direction to yield $\mathbf{k}_\varphi$ in the natural coordinate system. – The general three-dimensional case can also in principle be solved analytically in terms of velocity derivatives and using the Cardano formula, but the expressions become extremely complicated, so that the numerical solution of the eigenvalue problem seems to be more reasonable. It can be achieved easily with standard mathematical software (*e.g.* LAPACK, Anderson *et al.*, 1999).

In the following, the solution for the classical two-dimensional corner flow model applicable to normal mantle beneath a spreading ridge is given as an example. Consider the streamfunction

$$\Psi(x, z) = Ax - Bz \arctan \left(\frac{x}{z} \right) \qquad \text{(II.56)}$$

(after Spiegelman and McKenzie, 1987; Batchelor, 1967); with the boundary conditions

$$v_x(x, 0) = v_{\mathrm{r}} \qquad v_z(x, 0) = 0 \qquad v_x(0, z) = 0$$

$$\sigma_{xz}(0, z) = 0 \Leftrightarrow \frac{\partial v_x}{\partial z} + \frac{\partial v_z}{\partial x} = 0, \qquad \text{(II.57)}$$

the coefficients in eq. II.56 become $A = 0$ and $B = \frac{2v_{\mathrm{r}}}{\pi}$. This leads to the velocity components

$$v_x(x, z) = -\frac{\partial \Psi}{\partial z} = \frac{2v_{\mathrm{r}}}{\pi} \left(\arctan \frac{x}{z} - \frac{xz}{x^2 + z^2} \right) \qquad \text{(II.58a)}$$

$$v_z(x, z) = \frac{\partial \Psi}{\partial x} = -\frac{2v_{\mathrm{r}}}{\pi} \frac{z^2}{x^2 + z^2} \qquad \text{(II.58b)}$$

(see Phipps Morgan *et al.*, 1987) and with eq. II.52a to the viscous deviatoric stress tensor

$$\boldsymbol{\sigma}(x,z) = \frac{8\eta v_{\mathrm{r}} x}{\pi \left(x^2 + z^2\right)^2} \begin{pmatrix} xz & \frac{1}{2}\left(z^2 - x^2\right) \\ \frac{1}{2}\left(z^2 - x^2\right) & -xz \end{pmatrix} \tag{II.59}$$

with the eigenvalues

$$\sigma_{1,3} = \pm \frac{4\eta v_{\mathrm{r}} x}{\pi \left(x^2 + z^2\right)}. \tag{II.60a}$$

With normalization, the rotation matrix is then

$$\mathbf{R}(x,z) = \frac{1}{\sqrt{2(x^2 + z^2)}} \begin{pmatrix} x+z & x-z \\ z-x & x+z \end{pmatrix}; \tag{II.60b}$$

another usable form could have been calculated from eq. II.53b. The diagonal form of the stress tensor finally comes out as

$$\tilde{\boldsymbol{\sigma}}(x,z) = \begin{pmatrix} \sigma_1 & 0 \\ 0 & \sigma_3 \end{pmatrix} = \mathbf{R}^{\mathrm{T}} \boldsymbol{\sigma} \mathbf{R} = \frac{4\eta v_{\mathrm{r}} x}{\pi \left(x^2 + z^2\right)} \begin{pmatrix} 1 & 0 \\ 0 & -1 \end{pmatrix}, \tag{II.61}$$

which can already be seen from eq. II.60a. Inserting into eq. II.55 a permeability tensor for a vertical-channel model with $k_{\varphi_{zz}}^0$ as determined from one of the permeability models from the previous paragraph as the only non-zero element, and using the rotation matrix eq. II.60, the permeability tensor in the model is

$$\mathbf{k}_{\varphi\mathrm{ch}} = \frac{k_{\varphi_{zz}}^0}{2(x^2 + z^2)} \begin{pmatrix} (x-z)^2 & x^2 - z^2 \\ x^2 - z^2 & (x+z)^2 \end{pmatrix}. \tag{II.62}$$

If the melt is driven only by buoyancy through a system of dikes or cylindrical tubules, its ascent velocity relative to the matrix is

$$\begin{aligned} \vec{v}_{\mathrm{seg}} &= -\frac{k_{\varphi_{zz}}^0 (1-\mathcal{B}) \Delta \varrho_{\mathrm{f}}}{2\eta \mathcal{B}(x^2 + z^2)} \begin{pmatrix} (x-z)^2 & x^2 - z^2 \\ x^2 - z^2 & (x+z)^2 \end{pmatrix} \begin{pmatrix} 0 \\ -g \end{pmatrix} \\ &= \frac{k_{\varphi_{zz}}^0 (1-\mathcal{B}) \Delta \varrho_{\mathrm{f}} g}{2\eta \mathcal{B}(x^2 + z^2)} \left(x^2 - z^2, (x+z)^2 \right); \end{aligned} \tag{II.63}$$

note that $\vec{e}_z$ points downward and that $\Delta \varrho_{\mathrm{f}} < 0$. This confirms the result of Sleep (1984) and Phipps Morgan (1987) (see sect. II.4.2, p. 57) about dike orientation stating that melt moves away from the spreading center (assumed to be at $x = 0$) for $x < z$, *i.e.* $v_{\mathrm{seg}_x} > 0$; v_{seg_z} points always upward, if the melt is less dense than the matrix[7]. The streamlines for a corner flow model and the velocity vectors for melt ascending through dikes are plotted in figure II.10.

In the more general case of an asthenospheric triangle beneath a ridge, the coefficients of eq. II.56 are

$$A = B \sin^2 \phi_{\mathrm{r}}; \qquad B = \frac{2v_{\mathrm{r}}}{\pi - 2\phi_{\mathrm{r}} - \sin 2\phi_{\mathrm{r}}} \tag{II.64a}$$

(Spiegelman and McKenzie, 1987); ϕ_{r} is the angle between the surface and lithosphere base, and v_{r} is the horizontal velocity of the entire lithosphere. From this follow the matrix

---

[7]The derivation considers the $x, z > 0$ quadrant; for $x < 0$, the sign of v_{seg_z} changes for reasons of symmetry.

Figure II.10: Streamlines (solid) and melt segregation velocity vectors (arrows) for the right half of a corner flow ridge model, after eqs. II.56 and II.63; the ridge is at $x = 0$. The dashed line separates the "defocussing" $(x < z)$ and the "focussing" $(x > z)$ region.

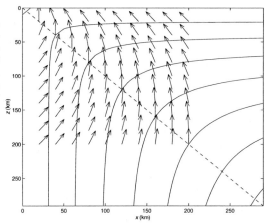

velocities

$$v_x(x, z) = -\frac{\partial \Psi}{\partial z} = 2B\left(\arctan\frac{x}{z} - \frac{xz}{x^2 + z^2}\right) \tag{II.64b}$$

$$v_z(x, z) = \frac{\partial \Psi}{\partial x} = A - 2B\frac{z^2}{x^2 + z^2}. \tag{II.64c}$$

Hence, the change in geometry only changes the magnitude of the stresses by the same constant, but not the form of the tensor.

II.5   THE GENERATION OF OCEANIC CRUST

II.5.1   *Mantle processes*

The melt generated in the supersolidus regions of the mantle is extracted and moved upwards to form the oceanic crust. Many observations suggest that ultramafites accumulate at the base of the crust and differentiate before a certain fraction of the material ascends further while the rest remains there and forms cumulate gabbros (Wilson, 1989) or a similar rock, depending on the depth of melt generation and crystallization (Farnetani *et al.*, 1996); the presence of a crystal mush with up to 30 % melt in the gabbroic lower crust at slow spreading centers deduced from the RAMESSES results is also supported by geochemical investigations at other ridges (Coogan *et al.*, 2000). The exact ratio of underplated vs. intruded and erupted magma is though uncertain, with estimates *e.g.* for oceanic islands ranging from 15 to 40 % (see White, 1993). Furthermore, petrological experiments suggest that fractionation is not limited to the crust–mantle transition, but might reach deeper and sometimes precede accumulation of MORB melts (Grove *et al.*, 1992). The ascending melt is piped by dikes into shallow-level magma chambers, where it may differentiate further and either erupts later or solidifies and forms new crust by intrusion, or it erupts directly.

The thickness h of the resulting crust depends on several factors; seismic studies suggest that normal oceanic crust has a globally remarkably uniform value of 6–7 km, unless the full spreading rate falls below 1.5–2 cm/a; in this case h decreases strongly to values as low as 3–4 km. Geochemical models result in values larger by 1–2 km, but this discrepancy is probably due to simplifications like the assumption of ideally fractional melting instead of retention of a melt fraction of *ca.* 1 % (White *et al.*, 1992, 2001; Parmentier, 1995). Reasons for the sharp decrease of h at very low spreading rates could be that conductive cooling from above suppresses melting (White *et al.*, 1992, 2001) or that light depleted material is not removed fast enough by the spreading plates to give way to upwelling fresh mantle. The amount of produced melt is controlled by the temperature and the height of the melting zone, which itself depends on $T(z)$ and the thickness of the lithosphere: a thick lithosphere shifts the top of the melting region downwards and will also modify the mean pressure and composition of the melts (Langmuir *et al.*, 1992). It must be emphasized that the thickness of the crust is very sensitive to the temperature, because T controls almost all parameters – in particular the degree of melting, the upwelling velocity and the composition – directly or indirectly.

Another important aspect is the upwelling velocity v_z of the mantle and its relation to the spreading velocity at the ridge, in particular if active upwelling comes into play (Scott and Stevenson, 1989; Li and Spohn, 1991). It controls the volume flux of mantle material processed in the melting region and also has some influence on the transport of the melt to the base of the lithosphere. While v_z depends on buoyancy and viscosity, the efficiency of the transport of the crust out of the region of accumulation depends on the spreading rate $2v_r$; however, both velocities are coupled to a certain degree, because v_r also controls the lateral density variations which contribute to the upwelling, apart from being the controlling quantity in passive upwelling.

It is not likely that all melt produced in the mantle is moved to the spreading center. Numerical models which attempt to model explicitly the migration of melt in the mantle, as those of Ghods and Arkani-Hamed (2000) or this study, show that a certain amount especially at the more remote parts of the melting zone will freeze again and be removed laterally by the mantle flow; Ghods and Arkani-Hamed (2000) report that suppressing crystallization increases the thickness of the crust in their models, thus it is worth noting that accounting for refreezing can force such models towards relatively high potential temperatures for the mantle ($T > 1400\,^\circ$C) as a remedy to achieve crustal thicknesses consistent with observations. For a relevant fraction of melt to actually be extracted from the mantle, sufficiently efficient focussing mechanisms in the melting regions as those described in section II.4.2 are necessary. The effect of the limited lateral extent of the crust generation region and of the focussing in the mantle therefore is a selective sampling of mantle material which is restricted to the inner and probably hotter parts of the upwelling.

II.5.2 *Formation and transport of the crust*

Once erupted or intruded in the already existing crust, the crustal material is transported with the drifting plate, unless the surface is stationary, which in most instances is not the case. The transport of crust of thickness h in general is described by

$$\frac{\partial h}{\partial t} = \frac{g\varrho_c \Delta\varrho_{mc} z_\eta^3}{\eta_0 \varrho_m}\nabla^2 h - \nabla \cdot (h\vec{v}_r) + \mathcal{G}_v \qquad (II.65)$$

(after Buck, 1991). The first term on the RHS describes flow in the crust due to horizontal pressure gradients which result from thickness contrasts, the second term combines advective transport and thinning or thickening due to the velocity field, and the last term $\mathcal{G}_v$ is the volcanic growth rate due to eruption or intrusion; the factor of the first term is a diffusivity and z_η the *e*-folding length for changes of the viscosity, which is assumed to increase exponentially with the height above the Moho. For a closer investigation, it is useful to consider the diffusion-free system

$$\frac{\partial h}{\partial t} + \nabla \cdot (h\vec{v}_r) = \mathcal{G}_v,$$ (II.66)

reduce it to the one-dimensional case and apply the chain rule to the term $\nabla \cdot (h\vec{v}_r)$:

$$\frac{\partial h}{\partial t} + h\frac{\partial v_r}{\partial x} + v_r\frac{\partial h}{\partial x} = \mathcal{G}_v.$$ (II.66a)

First, piecewise defined settings are analyzed. Two kinds of "eruption rates" are considered: a uniform $\mathcal{G}_v := \mathcal{G}_{v0}$ which is constant in space and time and characterizes a region $-x_v \le x \le x_v$, and a spatially varying $\mathcal{G}_v(x) := \mathcal{G}_{v0}(1 - {}^x\!/_{x_v})$, *i.e.* a rate which decays linearly over the interval and represents a somewhat focussed volcanic activity. Furthermore, $v_r(x)$ is assumed to be a simple function, ideally

$$v_r(x) = v_{r0}(\mathcal{H}(x) - \mathcal{H}(-x)) \quad \Rightarrow \quad \frac{\partial v_r}{\partial x} = 2v_{r0}\delta_0,$$ (II.67a)

but more realistically changing linearly over a finite interval $-x_r \le x \le x_r$:

$$v_r(x) = \begin{cases} \frac{x}{x_r}v_{r0}, & |x| \le x_r \\ v_{r0}\,\mathrm{sgn}(x) & \text{otherwise;} \end{cases} \quad \Rightarrow \quad \frac{\partial v_r}{\partial x} = \begin{cases} \frac{v_{r0}}{x_r}, & |x| \le x_r \\ 0 & \text{otherwise.} \end{cases}$$ (II.67b)

This form of $v_r(x)$ allows to simplify eq. II.66a and distinguish some different cases related to the different segments of the model. Later additional smooth continuous models will be considered.

Spreading center — In the middle of the spreading center ($x = 0$), the spreading velocity is zero: $v_r(0) = 0$. Thus, eq. II.66a simplifies to

$$\frac{\partial h}{\partial t} + \frac{v_{r0}}{x_r}h = \mathcal{G}_{v0}$$ (II.68a)

which has the solution

$$h(t) = \frac{\mathcal{G}_{v0}}{v_{r0}}x_r\left(1 - e^{-\frac{v_{r0}}{x_r}t}\right);$$ (II.68b)

obviously, the steady-state solution ($t \to \infty$) is

$$h_\infty = \frac{\mathcal{G}_{v0}}{v_{r0}}x_r.$$ (II.68c)

The solution is identical for both growth rate models.

Inner formation & spreading zone, $0 < |x| \le x_r, x_v, v_r = v_r(x)$ — In this case, eq. II.66a has to be solved in its full form, because the spreading velocity is not yet constant, and x lies in the volcanic zone. For the uniform rate, the differential equation is

$$\frac{\partial h}{\partial t} + \frac{v_{r0}}{x_r}h + \frac{v_{r0}}{x_r}\frac{\partial h}{\partial x}x = \mathcal{G}_{v0}.$$ (II.69a)

General solutions can be found using PDE handbooks (*e.g.* Kamke, 1956). The above equation is solved by

$$h(x,t) = h_0|x|^{\frac{x_r}{v_{r0}t_0}-1}e^{-\frac{t}{t_0}} + \frac{\mathcal{G}_{v0}}{v_{r0}}x_r\left(1 - e^{-\frac{v_{r0}}{x_r}t}\right),$$ (II.69b)

but the initial condition $h(x,0) = 0$ requires that $h_0 = 0$, so that the solution reduces to eq. II.68b or II.68c, respectively, *i.e.* the crustal thickness is uniform above the whole melting region. – The initial growth of the crust seems to have the same exponential form as in eq. II.68b in the other crustal regimes as well, but as it is not actually observed in nature, only the steady-state solutions ($\partial h/\partial t = 0$) will be given in the following.

For the spatially varying eruption rate, the differential equation is

$$\frac{\partial h}{\partial t} + \frac{v_{r0}}{x_r}h + \frac{v_{r0}}{x_r}\frac{\partial h}{\partial x}x = \mathcal{G}_{v0}\left(1 - \frac{x}{x_v}\right), \tag{II.69c}$$

and its steady state solution as determined by variation of constants is

$$h_\infty(x) = \frac{\mathcal{G}_{v0}x_r}{v_{r0}}\left(1 - \frac{x}{2x_v}\right). \tag{II.69d}$$

Outer formation zone, $x_r < |x| \leq x_v$, $v_r = v_{r0}$ — This case is a bit simpler than the previous one, because the spreading velocity has reached its constant value $\pm v_{r0}$; x still lies in the volcanic zone, though, so the different eruption rate models have to be examined separately. The uniform rate model is described by

$$\frac{\partial h}{\partial t} + v_{r0}\frac{\partial h}{\partial x} = \mathcal{G}_{v0}. \tag{II.70a}$$

Applying a continuity condition with eq. II.68c, the solution is

$$h_\infty(x) = \frac{\mathcal{G}_{v0}}{v_{r0}}x \tag{II.70b}$$

in the stationary case.

For the spatially varying eruption rate model,

$$\frac{\partial h}{\partial t} + v_{r0}\frac{\partial h}{\partial x} = \mathcal{G}_{v0}\left(1 - \frac{x}{x_v}\right) \tag{II.70c}$$

must be solved. The steady-state solution is

$$h_\infty(x) = \frac{\mathcal{G}_{v0}}{v_{r0}}\left(x - \frac{x^2}{2x_v}\right) = \frac{\mathcal{G}_{v0}}{v_{r0}}x\left(1 - \frac{x}{2x_v}\right). \tag{II.70d}$$

Outer spreading zone, $x_v < |x| \leq x_r$, $v_r = v_r(x)$, $\mathcal{G}_v(x) = 0$ — As an alternative, the formation zone may be narrower than the spreading zone. In this case, the solution of the homogeneous version of eq. II.69a,

$$\frac{\partial h}{\partial t} + \frac{v_r}{x_r}h + \frac{v_r}{x_r}\frac{\partial h}{\partial x}x = 0 \tag{II.71a}$$

is to be found. The limit for $t \to \infty$ is

$$h_\infty(x) = \frac{\mathcal{G}_{v0}}{v_{r0}}\frac{x_r x_v}{x} \tag{II.71b}$$

for the constant eruption case and, for continuity with eq. II.69d at $x = x_v$,

$$h_\infty(x) = \frac{\mathcal{G}_{v0}}{2v_{r0}}\frac{x_r x_v}{x} \tag{II.71c}$$

for the variable eruption case.

Remote parts, $v_r = v_{r0}$, $\mathcal{G}_v(x) = 0$ — In the remote parts of the plate, the plate velocity is constant, and no volcanic activity is present; this corresponds to the homogeneous version of eq. II.70a:

$$\frac{\partial h}{\partial t} + v_{r0}\frac{\partial h}{\partial x} = 0, \tag{II.72a}$$

i.e. the solution for the steady state is

$$h_\infty(x) = \frac{\mathcal{G}_{v0}}{v_{r0}}x_v \tag{II.72b}$$

for the uniform eruption model and

$$h_\infty(x) = \frac{\mathcal{G}_{v0}}{2v_{r0}}x_v \tag{II.72c}$$

for the variable eruption model, regardless if x_v or x_r is greater. Thus, only the width of the region of volcanic activity is important for the thickness of old oceanic crust, but not the width of the spreading zone. A more focussed volcanic activity leads to a smaller final thickness of the crust outside the production area if the other parameters are held fixed. – In the special case $x_r = x_v$, eqs. II.70 and II.71 are meaningless. For a uniform distribution of volcanic activity, this results in the constant steady-state thickness given by eq. II.68c or II.72b for all x.

By integrating the two different $\mathcal{G}_v(x)$ from 0 to x and setting this equal to the whole amount of melt per time available for extraction from the mantle source, one can show that $\mathcal{G}_{v0}$ is twice as high for the linear model than for the uniform one. Thus, for a given amount of melt and a given x_v, the far-ridge crustal thickness is the same for both $\mathcal{G}_v(x)$ models, while directly at the ridge, it is twice as high in the variable $\mathcal{G}_v$ case. This insensitivity of h for $x > x_r, x_v$ with respect to x_r and $\mathcal{G}_{v(x)}/\mathcal{G}_{v0}$ may be part of the explanation for the uniformity of the thickness of oceanic crust all over the world.

Smooth $\mathcal{G}_v$ and v_r — The general solution of the steady-state form of eq. II.66a,

$$h\frac{dv_r}{dx} + v_r\frac{dh}{dx} = \mathcal{G}_v, \tag{II.73a}$$

for continuous, piecewise differentiable functions $v_r(x)$ with $v_r(0) = 0$ and $\mathcal{G}_v(x) := {}^{d\mathfrak{G}}/{dx}$ is

$$h(x) = \frac{\mathfrak{G}(x) + h_0}{v_r(x)}, \tag{II.73b}$$

with a constant h_0. Hence, if $\mathcal{G}_v \sim {}^{dv_r}/{dx}$ and $h_0 = 0$ for simplicity, the solution of eq. II.73a is simply a constant, namely the proportionality factor between $\dot{\varepsilon} = {}^{dv_r}/{dx}$ and $\mathcal{G}_v$; thus, the crustal thickness is constant for all x. One example for this is the case $\mathcal{G}_v(x) = const.$, $x_r = x_v$ in the models above. – If a more complicated, non-linear relation between $\mathcal{G}_v$ and ${}^{dv_r}/{dx}$ is assumed, which is more likely when considering possible relations between strain rate, dike widths, and permeability, different crustal models *e.g.* with a central graben will result.

As a further example, where thickness and transport of the crust are described by a smooth model which fakes the spreading rate as a function of distance from the spreading center at $x = 0$, $v_r(x) = {}^{2v_{r0}}/{\pi}\arctan(a_r x)$ is chosen arbitrarily; this function is zero at $x = 0$, increases more or less rapidly until the far-field value v_{r0} is almost reached at a certain distance, and approximates this value asymptotically. The strain rate is then given by $\dot{\varepsilon} = 2v_{r0}\,{}^{a_r}/{\pi(1+a_r^2 x^2)}$, *i.e.* the maximum of the strain rate (and of the strain) appears at the spreading center. The spatial distribution of volcanic activity $\mathcal{G}_v(x)$ shall have the same form as $\dot{\varepsilon}$: $\mathcal{G}_v(x) = 2\mathcal{G}_{v0}\,{}^1\!/\pi(1+b_v^2 x^2)$; to provide some more generality, departing from the case considered in the previous paragraph, different sloping constants a_r and b_v are used for $v_r(x)$ and $\mathcal{G}_v(x)$. Inserting these functions in eq. II.73a yields

$$v_{r0}\arctan(a_r x)\frac{\mathrm{d}h}{\mathrm{d}x} + v_{r0}\frac{a_r}{1+a_r^2 x^2}\,h = \frac{\mathcal{G}_{v0}}{1+b_v^2 x^2}, \tag{II.74a}$$

which has the solution

$$h_\infty(x) = \frac{\mathcal{G}_{v0}}{v_{r0}b_v}\frac{\arctan(b_v x)}{\arctan(a_r x)}; \tag{II.74b}$$

at the spreading center and far from the ridge, respectively,

$$h_\infty(0) = \frac{\mathcal{G}_{v0}}{v_{r0}a_r}, \qquad \lim_{x\to\infty}h_\infty(x) = \frac{\mathcal{G}_{v0}}{v_{r0}b_v} \tag{II.74c}$$

i.e. $^1\!/a_r$ corresponds to x_r and $^1\!/b_v$ to x_v in the former examples (eq. II.68c). Obviously, in the case $a_r = b_v$ the crustal thickness has the constant value $\mathcal{G}_{v0}/a_r v_{r0}$ for all x, as predicted from the general approach.

A similar model, where strain rate and lava deposition have a similar functional form, had already been considered by Pálmason (1980); he proposed the functions $v_r(x) = v_{r0}\,\mathrm{erf}\,(x/\sqrt{2}x_r)$ and $\mathcal{G}_v(x) = \mathcal{G}_{v0}\exp\left(-x^2/2x_v^2\right)$, whereby the ridge width x_r and the volcanic width x_v are the standard deviations of the Gaussian distributions of strain rate and deposition, respectively. With these functions, eq. II.73a takes the form

$$v_{r0}\,\mathrm{erf}\left(\frac{x}{\sqrt{2}x_r}\right)\frac{\mathrm{d}h}{\mathrm{d}x} + \sqrt{\frac{2}{\pi}}\frac{v_{r0}}{x_r}\exp\left(-\frac{x^2}{2x_r^2}\right)h = \mathcal{G}_{v0}\exp\left(-\frac{x^2}{2x_v^2}\right) \tag{II.75a}$$

and has the solution

$$h_\infty(x) = \sqrt{\frac{\pi}{2}}\frac{x_v\mathcal{G}_{v0}}{v_{r0}}\frac{\mathrm{erf}\left(\frac{x}{\sqrt{2}x_v}\right)}{\mathrm{erf}\left(\frac{x}{\sqrt{2}x_r}\right)}; \tag{II.75b}$$

at the spreading center and far from the ridge, respectively,

$$h_\infty(0) = \sqrt{\frac{\pi}{2}}\frac{x_r\mathcal{G}_{v0}}{v_{r0}}, \qquad \lim_{x\to\infty}h_\infty(x) = \sqrt{\frac{\pi}{2}}\frac{x_v\mathcal{G}_{v0}}{v_{r0}}. \tag{II.75c}$$

For the smooth functions, h also depends on x_v, but not on x_r for old crust. The constant, linear, arctangent, and Pálmason (1980) $\mathcal{G}_{v0}$ models are plotted for $x_r > x_v$ and $x_r < x_v$ in figure II.11.

From the $h(x)$ formulae, one could get the impression that crustal thickness also depends on the reference growth rate $\mathcal{G}_{v0}$, which is a measure of the total flux of lava into the eruption zone and thus depends on the total melt production in the mantle melting region, $\dot{m}_f$, and on v_{r0}, which would be in contradiction to observations at mid-oceanic ridges throughout the world. However, in decompression melting, $\dot{m}_f$ is a function of v_z of the mantle, among others, and it must be kept in mind that at MOR, where upwelling is regarded as passive, v_z is proportional to v_r; this is most obvious from the corner flow equation II.58b. Assuming that over a large part of the melting interval $\dot{m}_f \sim v_z$, it is clear that $\mathcal{G}_{v0}$ is a linear function of v_{r0}, and both quantities will approximately cancel out in the expressions for $h(x)$.

Finally, it should be recalled that there are significant simplifications inherent in eq. II.65; *e.g.*, no assumptions are made about the mechanical properties of the crust, and an underlying assumption of its derivation is Newtonian rheology (Buck, 1991). Thus, it would be a

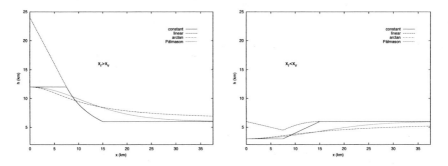

Figure II.11: Analytic solutions of crust formation at a mid-oceanic ridge of eq. II.73a for the constant, linear, arctangent, and Pálmason (1980) $\mathcal{G}_{v0}$ models. Left: $x_r = 15\,\text{km}$, $x_v = 7.5\,\text{km}$; right: $x_r = 7.5\,\text{km}$, $x_v = 15\,\text{km}$. $\mathcal{G}_{v0}$ was chosen for each model such that the crustal thickness for $x \to \infty$ is always 6 km.

mistake to expect this simple model to deliver a close representation of details of the crustal structure and in particular its topography, because the topography is controlled by a number of different factors acting on several length scales. As an example, on a large scale vertical mantle currents lead to deformation of the plates, which results in some kind of dynamical topography. On shorter lengthscales, factors such as the spreading rate have strong influence: there is a trend that slow-spreading ridges ($v_r < 2$–$2.5\,^{\text{cm}}/_{\text{a}}$) have median valleys, while fast-spreading ridges show an axial high (Small and Sandwell, 1989), although there are exceptions like the Reykjanes Ridge with its anomalously thick crust. Phipps Morgan and Chen (1993) explain the generation of the median valley by the stretching of strong axial lithosphere, whose thickness is controlled by the thermal structure resulting from magmatic heating and hydrothermal cooling. In terms of the model presented here, the median valley would indicate that stretching takes place in a narrower strip around the spreading axis than volcanic activity, which seems to be consistent with a strong lithosphere, but leaves open how melts can traverse a thick lithospheric lid. The axial high in turn demands a focussed volcanic activity in a broad stretching band; at least as far as the Reykjanes Ridge is concerned, it is plausible that along-ridge flow of plume material could concentrate volcanism near the axis, while at the same time the thermal effect could soften the adjacent lithosphere enough to widen the spreading region. Given the uncertainties and simplifications here, it seems though to be safest with respect to application in the numerical models presented in chapter V to impose the degenerate case $x_r = x_v$ and to judge crustal thickness from observations at $x > x_r, x_v$.

Further remarks — Although high-resolution 3D modelling of crust formation will not be attempted in this study, it is worth noting that there are details of crustal structure which can already be assessed from the simple two-dimensional model considered here. Pálmason (1980) gives the relation

$$-\left(\frac{\partial h}{\partial t}\right)_x \bigg/ \left(\frac{\partial h}{\partial x}\right)_t = \left(\frac{\partial x}{\partial t}\right)_h = v_r \qquad (\text{II}.76)$$

for the lava deposition rate, the slope of the basaltic layer and the horizontal velocity of the crust, which corresponds to the conventional spreading rate far away from the location

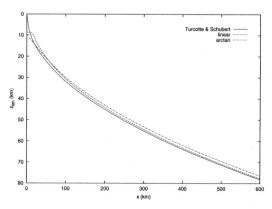

Figure II.12: Analytic solutions of oceanic lithosphere thickness after eqs. II.77 and II.79, for $\theta_{\text{lith}} := 0.9$.

of deposition. This would allow to predict dip angles from the $v_{\text{r}}(x)$ and $\mathcal{G}_{\text{v}}(x)$ used in the model.

As an aside it is worthwhile to recall the well-known $\sqrt{\text{age}}$ relation between the spreading velocity and the thickness of the thermal lithosphere, which for $x_{\text{r}} = 0$ is given by

$$z_{\text{lith}} = 2\,\text{erf}^{-1}(\theta_{\text{lith}})\sqrt{\kappa t} = 2\,\text{erf}^{-1}(\theta_{\text{lith}})\sqrt{\frac{\kappa x}{v_{\text{r}}}} \tag{II.77}$$

(*e.g.* Turcotte and Schubert, 1982), where θ_{lith} is an arbitrarily chosen temperature, scaled with the temperature difference between the mantle and the surface, which is defined to represent the bottom of the lithosphere; Turcotte and Schubert (1982) chose $\theta_{\text{lith}} = 0.9$, *i.e.* 90 % of the mantle temperature, so that $2\,\text{erf}^{-1}\theta_{\text{lith}} = 2.32$ (see figure II.12). Guðmundsson *et al.* (1994, app. B) extended this model to a stationary model of a ridge with finite width x_{r} and vertically constant horizontal velocity $v_x = v_{\text{r}}(x)$ overlying an incompressible mantle by solving the two-dimensional stationary temperature equation with $Di = 0$ (see eq. I.3a); their solution for the temperature is

$$\theta(x, z) = \text{erf}\left[\frac{z v_{\text{r}}(x)}{2\sqrt{\kappa \int_0^x v_{\text{r}}(x')\,\text{d}x'}}\right]. \tag{II.78}$$

For the linearly increasing $v_{\text{r}}(x)$ of eq. II.67b, inserting and solving for z with $\theta_{\text{lith}} = 0.9$ yields

$$z_{\text{lith}} = \text{erf}^{-1}(\theta_{\text{lith}}) \cdot \begin{cases} \sqrt{\frac{2\kappa x_{\text{r}}}{v_{\text{r}0}}}, & x \leq x_{\text{r}} \\ \sqrt{\frac{2\kappa(2x - x_{\text{r}})}{v_{\text{r}0}}} & \text{otherwise,} \end{cases} \tag{II.79a}$$

and the smooth arctangent version of $v_{\text{r}}(x)$ gives

$$z_{\text{lith}} = \frac{\text{erf}^{-1}(\theta_{\text{lith}})}{\arctan(a_{\text{r}}x)} \sqrt{\frac{\pi\kappa\left[2a_{\text{r}}x\arctan(a_{\text{r}}x) - \ln\left(1 + a_{\text{r}}^2 x^2\right)\right]}{a_{\text{r}}v_{\text{r}0}}}. \tag{II.79b}$$

Figure II.12 shows that eq. II.77 and eqs. II.79 are very similar for $x \gg x_{\text{r}}$, as expected, but that the latter give more reasonable results within the spreading zone.

CHAPTER III

NUMERICAL TREATMENT
OF THE CONVECTION AND MELT FLOW MODELS

Number rules the universe.

—PYTHAGORAS

The combined solution of the mantle convection and melt dynamical equations poses considerable problems to every algorithm, because phenomena on a very large range of spatial, and possibly also temporal, scales have to be accounted for in some way. On the one hand, mantle currents extend over hundreds or thousands of kilometers, and the flow and temperature field may undergo significant variations on a length scale of a few tens of kilometers; on the other hand, melt migration phenomena are partly controlled by the situation on the scale of mineral grains and boundaries, *i.e.* some millimeters to microns, and show variations over lengths of centimeters, meters or some kilometers. Thus, relevant processes cover a range of 11 or 12 orders of magnitude – obviously, it is impossible to model all these processes explicitly with one program with the necessary resolution, because this would require far more memory than even the largest computers have today and take unreasonably long time. This is especially true for three-dimensional models, which are the only reasonable approach for addressing the dynamical situation of a plume–MOR setting; Cartesian 2D models might be inadequate for reproducing the dynamics of a cylindrical plume due to differences *e.g.* in the temperature field, and a 2D cylindrical model cannot include a spreading ridge. – The range of temporal timescales depends on what processes are included into the model: large timescales of thousands to millions of years apply to mantle convection features and possibly to isotropic porous flow, and are thus far tractable in a normal convection model; if, however, rapid processes such as anisotropic flow through channels or dikes, which have timescales as small as years or hundreds of years, are included, the time increments in combined convection–melt migration models designed to investigate the temporal evolution would have to be so small to keep the run numerically stable at least if certain explicit time-stepping schemes are applied that, again, the calculation cannot be completed in any reasonable amount of time.

In order to achieve a useful approximation, certain simplifications have to be made. As this study focusses on the interaction of plume convection and melt dynamics rather than on the microscopic details of melt movement, the large scale of plume convection has to be given priority in the design of the numerical model. Thus, the numerical grid of the overall model will need to have dimensions of several hundreds of kilometers, with a grid point spacing in the range of not much less than 10 to 20 km. This coarse convection grid will provide the framework in which melt dynamics act.

Although melt generation and segregation has a significant influence on large-scale convection, as discussed in section I.2.1 and chapter II, it takes place only in a comparatively small part of the whole convection system; on the other hand, it is to be expected that the spatial resolution appropriate for large-scale convection is not sufficient for melt-related processes which have a smaller scale. For this reason, a second numerical grid for melt processes with smaller dimensions, but a higher spatial resolution is inserted into the coarse convection grid; this restriction of melt modelling to the region where melt actually exists helps reducing the computational expenses to a size which can be handled by computers available today, although resolutions as high as those used by other workers, who investigated two-dimensional models (*e.g.* Scott, 1988; Ghods and Arkani-Hamed, 2002) are still out of reach. Variables of the convection model relevant to melt dynamics – T, $\vec{v}$, p, and possibly chemical markers – are interpolated onto the melt model and used as boundary and starting conditions for the calculation of the instantaneous melt dynamics in each time step; variables of the melt model relevant to convection dynamics – T, f, φ – are transferred back to the convection model for use in the next time step.

All models presented in this study have been calculated on normal Linux PCs with one or two Intel Pentium III or Pentium IV processors and up to 1 GB of RAM; models of common size in this study needed between about 100 and 700 MB, depending on whether and how melting was included, and run durations ranged from about one day to four weeks, depending on model features such as the viscosity model, inclusion of melting, and on the time-stepping scheme used. Experience with different CPU architectures and compilers teaches that both memory requirements and computational speed also depend strongly on these two factors; in particular, it seems to be preferable with respect to performance to use processor vendors' compilers and exploit their architecture-specific optimizations than to stick to a uniform, cross-platform development tool such as the GNU compiler, convenient and deserving as it is in other respects.

III.1   THE NUMERICAL SOLUTION OF THE CONVECTION EQUATIONS

In this section, an outline of the numerical algorithm for the solution of eqs. I.1a, I.2d and I.10 used in this work is given. The solution of the fluid-dynamical equations basically comprises two stages:

The joint solution of the equation of mass conservation I.1a and the time-independent Stokes equation I.2d — The flow field resulting from the distribution of thermal and compositional instabilities can safely be assumed to be stationary in each timestep, thanks to the infinite Prandtl number approximation. The algorithm of the program used in this study (DECO3D, G. Marquart, pers. comm., 1998; Marquart *et al.*, 2000; Marquart, 2001) is a hybrid finite-difference/spectral method formulated in primitive variables, which makes it somewhat similar to the method of Gable *et al.* (1991) and distinguishes it from other hybrid convection codes such as the one by Christensen and Harder (1991). In the case of viscosity varying in all three space dimensions, which is assumed in most of this study, the Stokes equation has to be solved iteratively because of its nonlinear nature; this is because spectral methods cannot be applied directly to products of functions of location, which would be necessary for the horizontally spectral representation adopted here. To perform the iteration, the viscosity is first split into one depth-dependent term and one variational term which is also a function of x and y:

$$\eta(x,y,z) = \eta^*(z) + \Delta\eta(x,y,z). \qquad \text{(III.1)}$$

The terms containing $\eta^*(z)$ are kept on the LHS and those with the variation $\Delta\eta(x,y,z)$ on the RHS of the Navier–Stokes equation, whose three components essentially read as follows in the spectral form (G. Marquart, pers. comm.):

$$-ilk_x\overline{\tilde{p}} + \eta^*\left[\left(-l^2k_x^2 - m^2k_y^2\right)\overline{v_x} + \frac{\partial^2\overline{v_x}}{\partial z^2}\right] + \frac{\partial\eta^*}{\partial z}\left(\frac{\partial\overline{v_x}}{\partial z} + ilk_x\overline{v_z}\right) =$$

$$-\mathcal{F}\left[\Delta\eta\left(\frac{\partial^2 v_x}{\partial x^2} + \frac{\partial^2 v_x}{\partial y^2} + \frac{\partial^2 v_x}{\partial z^2}\right)\right.$$

$$\left. + 2\frac{\partial\Delta\eta}{\partial x}\frac{\partial v_x}{\partial x} + \frac{\partial\Delta\eta}{\partial y}\left(\frac{\partial v_x}{\partial y} + \frac{\partial v_y}{\partial x}\right) + \frac{\partial\Delta\eta}{\partial z}\left(\frac{\partial v_x}{\partial z} + \frac{\partial v_z}{\partial x}\right)\right]_{lm} \quad \text{(III.2a)}$$

$$-imk_y\overline{\tilde{p}} + \eta^*\left[\left(-l^2k_x^2 - m^2k_y^2\right)\overline{v_y} + \frac{\partial^2\overline{v_y}}{\partial z^2}\right] + \frac{\partial\eta^*}{\partial z}\left(\frac{\partial\overline{v_y}}{\partial z} + imk_y\overline{v_z}\right) =$$

$$-\mathcal{F}\left[\Delta\eta\left(\frac{\partial^2 v_y}{\partial x^2} + \frac{\partial^2 v_y}{\partial y^2} + \frac{\partial^2 v_y}{\partial z^2}\right)\right.$$

$$\left. + \frac{\partial\Delta\eta}{\partial x}\left(\frac{\partial v_x}{\partial y} + \frac{\partial v_y}{\partial x}\right) + 2\frac{\partial\Delta\eta}{\partial y}\frac{\partial v_y}{\partial y} + \frac{\partial\Delta\eta}{\partial z}\left(\frac{\partial v_y}{\partial z} + \frac{\partial v_z}{\partial y}\right)\right]_{lm} \quad \text{(III.2b)}$$

$$-\frac{\partial\overline{\tilde{p}}}{\partial z} + \eta^*\left[\left(-l^2k_x^2 - m^2k_y^2\right)\overline{v_z} + \frac{\partial^2\overline{v_z}}{\partial z^2}\right] + 2\frac{\partial\eta^*}{\partial z}\frac{\partial\overline{v_z}}{\partial z} =$$

$$Ra\overline{T} - \mathcal{F}\left[\Delta\eta\left(\frac{\partial^2 v_z}{\partial x^2} + \frac{\partial^2 v_z}{\partial y^2} + \frac{\partial^2 v_z}{\partial z^2}\right)\right.$$

$$\left. + \frac{\partial\Delta\eta}{\partial x}\left(\frac{\partial v_z}{\partial x} + \frac{\partial v_x}{\partial z}\right) + \frac{\partial\Delta\eta}{\partial y}\left(\frac{\partial v_z}{\partial y} + \frac{\partial v_y}{\partial z}\right) + 2\frac{\partial\Delta\eta}{\partial z}\frac{\partial v_z}{\partial z}\right]_{lm} . \quad \text{(III.2c)}$$

Together with these equations, in which incompressibility is implicit, the equation of mass conservation for an incompressible fluid is solved (G. Marquart, pers. comm., 1998):

$$ilk_x\overline{v_x} + imk_y\overline{v_y} + \frac{\partial\overline{v_z}}{\partial z} = 0. \quad \text{(III.2d)}$$

Here $\overline{\tilde{p}}$, $\overline{T}$ and $\overline{v_i}$ denote the spectral component (l, m) of the non-hydrostatic pressure, the temperature, and the ith flow velocity component, respectively, and $\mathcal{F}[\ldots]_{lm}$ is the corresponding component of the Fourier transform of the term in brackets on the RHS. $k_x = {}^{2\pi}/\lambda_x$ and $k_y = {}^{2\pi}/\lambda_y$ are the wavenumbers of the basic modes in x and y direction and λ_x and λ_y are the corresponding wavelengths; the latter are related to the corresponding model box dimensions x_m and y_m by $\lambda_x = x_m + \Delta x$ and $\lambda_y = y_m + \Delta y$, respectively, whereby Δx and Δy are the grid point spacings. In eq. III.2c, only the thermal Rayleigh number was included for simplicity, but the other buoyancy terms in eq. I.2d are appended in an analogous form.

If $\mathcal{F}_{lm}(\text{RHS})$ denotes the (l, m) spectral component of the RHS of eqs. III.2 and

$$f(\vec{v}) := \Delta\eta\nabla^2\vec{v} + 2\nabla\Delta\eta\,\nabla\vec{v} - \nabla\Delta\eta \times (\nabla \times \vec{v}) \quad \text{(III.3a)}$$

is introduced as an abbreviation for the derivatives on the RHS, the iteration scheme for the nth iteration is

$$\mathcal{F}(\text{RHS})_{lm}^n = Ra\overline{T} - \left(\frac{1 - c_\eta}{\sqrt{lk_x + mk_y}}\mathcal{F}\left(f(\vec{v})\right)_{lm}^{n-2} + \frac{c_\eta}{\sqrt{lk_x + mk_y}}\mathcal{F}\left(f(\vec{v})\right)_{lm}^{n-1}\right), \qquad c_\eta < 1$$

$$\text{(III.3b)}$$

with a damping factor c_η which is usually *ca.* 0.2 (G. Marquart, pers. comm.). The convergence criterion for the iteration is

$$(\vec{v}, \tilde{p})_n - (\vec{v}, \tilde{p})_{n-1} < \epsilon, \qquad\qquad\qquad \text{(III.3c)}$$

with ϵ commonly being $O(10^{-2})$.

An advantage of the hybrid method is that it decouples the horizontal layers of the model grid, making three-dimensional models tractable at reasonable cost; this also provides a way to approach parallelization of the convection algorithm. It must be noted, though, that the treatment of the Stokes equation in the above form cannot handle viscosity contrasts of more than about three orders of magnitude with acceptable accuracy. Hence, a study of plume channeling in MOR as the one by Albers and Christensen (2001) is beyond its possibilities; recently developped, more flexible methods which offer the possibility of adapting the spatial resolution to the problem even for very steep gradients resp. for largely different length scales without excessive memory requirements, *e.g.* the multigrid method of Albers and Christensen (2001) or the adaptive wavelet method of Vasilyev *et al.* (2001), seem to represent promising approaches for such a task.

The Fourier transforms are usually computed with highly efficient Fast Fourier Transform routines, which are available in several implementations; in this case, the FFTW package (Frigo and Johnson, 1998) was chosen, because it is open-source, portable code and allows for arbitrary array lengths (although efficiency will vary depending on prime factorization of the length), in addition to having a reputation of good performance.

The solution of the equation of energy conservation I.10 — This equation involves the calculation of temperature changes and the temporal evolution of T due to advection, diffusion and additional heat sources. In the frame of this study, two approaches have been followed: 1) an explicit 3D-FD time-stepping scheme with upwind being used for the advection terms if necessary to ensure stability (G. Marquart, pers. comm., 1998), and 2) a combined semi-Lagrange/ADI scheme with operator splitting. The time-stepping is performed with a variable increment, which is continually adjusted to the critical properties of the flow field and the stability and accuracy requirements resulting from it. The two methods are investigated and compared in more detail in sections III.1.2 and III.1.3.

In the context of advection problems, it should be recalled that the temperature is usually not the only quantity which is advected in the models presented in this study: in addition to the thermal advection–diffusion problem, there is pure advection of chemical fields (see eq. I.4 with the RHS set to zero, and eq. I.5), *e.g.* the depletion field related to melting processes or the advection of chemically different mantle regions, which both have, or at least can be designed to have, influence on buoyancy or on viscosity in addition to their control of the chemical pattern produced by the convection. The advection of these chemical fields is done with the same methods as the advection of T, so that the discussion in sections III.1.2 and III.1.3 also applies to this case.

III.1.1   Boundary conditions

The numerical solution of the convection equations demands the prescription of certain boundary conditions for the velocity and the temperature at the margins of the model box. The type of models investigated and the hybrid nature of the code used in this study leads to quite different boundary conditions for the upper and lower boundary on the one hand and the lateral boundaries on the other hand. While for the latter periodic boundaries seem to be the most natural choice due to the periodicity of the Fourier method, it is desirable

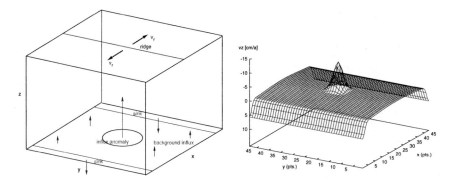

Figure III.1: Left: scheme of convection model box and boundary conditions. Right: example of the bottom boundary condition $v_z(x, y, z_{\mathrm{m}})$. The radius of the circular influx anomaly is $R_{\mathrm{P}} := 2r_{\mathrm{P}}$, whereby r_{P} is the distance from the center at which the amplitude of the anomaly has decreased to $1/e$ of the maximum value.

to prescribe boundary conditions derived from the physical and observational constraints at the top and bottom boundaries; this section presents a short outline of this topic.

Top boundary — The most frequent case of upper model boundaries in mantle convection models is the assumption of a free surface, which represents the surface of the earth or the ocean floor; nonetheless, one could of course assume some artificial or physical boundary in the mantle instead. In an oceanic environment, a stationary top boundary can only be assumed in exceptional cases; the usual situation will be that of a lithospheric plate moving with a spatially uniform, constant velocity. In special settings, as are mid-oceanic ridges, the kinematics are more complex, though: on each side of the ridge the plate moves in a different direction, so that there is a discontinuity, or at least a very steep velocity gradient, across the spreading center; things become even more difficult if the ridge itself is moving or if transform faults appear, so that the stationarity is dropped or the one-dimensional spatial variability is extended to both dimensions of the boundary. In the convection models of this study, where the focus lies on the processes around mid-oceanic ridges, but not on the influence of more remote structures like subduction zones, it is reasonable to prescribe the plate velocities explicitly as a kinematic boundary condition at the upper boundary of the model box (see figure III.1). The downside of such an imposed boundary condition is that it does not readily allow for modelling the interactions between the crust–lithosphere system and the plume in both directions, *i.e.* the overriding spreading crust–lithosphere system influences the plume, but the plume will not induce fundamental changes in the motion of the crust and lithosphere like a ridge jump as a direct physically consistent consequence of the model.

Bottom boundary — The bottom boundary in a large-scale model can be placed at virtually any depth down to the CMB, although for physical reasons it may be reasonable to let it coincide with a presumed thermal boundary layer or a petrological phase change. As this study concentrates on processes related to melt dynamics in the upper mantle and a compromise between a sufficient spatial resolution of the discrete model and the available computational resources must be found, the bottom boundary of the model box is placed either at the 410 km or at the 660 km discontinuity, *i.e.* it lies within or at the bottom of the upper mantle; this has the side effect that by preserving small to moderate aspect ratios, the spherical form of the real earth can be neglected.

At this boundary, the temperature will be prescribed by a fixed value; if a plume is included in the model, it will be superposed as an anomaly of a certain shape on the background bottom temperature. The horizontal velocity components can be described in terms of a no-slip or a free-slip boundary condition, while in the case of the vertical components mass conservation considerations have to made, because a non-zero condition at the model bottom signifies mass flux into or out of the model box. In the case of a plume it is particularly reasonable to prescribe such a velocity; the most straightforward choice is to use the same weight function as for the temperature anomaly in this case. Theoretical models for the internal structure of a plume (Loper and Stacey, 1983) suggest that a radially symmetric Gaussian curve is well-suited for use as a weighting function, and in fact it is used in several other studies (*e.g.* Ito *et al.*, 1996; Albers and Christensen, 1996; Allen *et al.*, 1999; Moore *et al.*, 1999), and it can also be shown that it is more convenient from a numerical point of view than other functions which might come to mind for this purpose (see appendix B.1); in this study, the radius of the anomaly is chosen to be $R_P := 2r_P$, *i.e.* twice the distance from the center at which the amplitude of the anomaly has decreased to $1/e$, which will usually be regarded as the plume radius. The volume flux resulting from this influx model is then

$$q_{\text{in}} = \left(1 - e^{-4}\right) \frac{\pi v_{0_P} R_P^2}{4} \qquad (\text{III.4})$$

(see appendix B.2 for details). These boundary conditions imply that the plume starts from the CMB, as no thermal boundary layer is present at the model bottom; however, this seems to be consistent with observations (*e.g.* White and McKenzie, 1995). Nonetheless, it is not meant here to argue in favour of layered or whole-mantle convection, but is primarily imposed by computational and technical circumstances. – Similarly, the new plate which is created at the top boundary introduces new mass into the model which has to be deleted at some boundary; it also induces shear flow in the whole model box, which can be approximately described by a Couette-like model (see appendix B.2.3). In this case, strips at the x boundaries of the model bottom are defined which act as sinks and suck as much material out of the model as necessary to compensate all influxes (see figure III.1). The flow circuit initiated in this way must be closed by providing a corresponding mass influx through the model bottom, whose (x, y) distribution can be shaped by using the corner flow solution of a spreading-ridge setting; a somewhat similar approach has already been taken by Scott (1992). The whole flux system is schematically shown in figure III.1.

Side boundaries — The use of the Fourier method for solving the Stokes equation implies periodic boundary conditions at the sides of the model box as the most natural ones, although different choices like outflow conditions are possible as well; it is therefore consistent to apply them also to the solution of the temperature equation, which is solved by finite differences. Although periodic boundary conditions are in principle often particularly easy to handle and straightforward to implement, a special feature of this type of convection models must be paid some further attention, because it can have certain influence on some variables: the model box contains two full convection cells with both upwellings (in the center) and downwellings (at the sides). In the case of a model based on a spreading ridge parallel to the y axis, which is the fundamental scheme of the models in this study, strong descending convection branches would form in the neighbourhood of $x = 0$ and $x = x_m$ and lead to rather cold slabs flowing back to the box center at the bottom, where they are likely to disturb the upwelling, if no outflow is provided; it is the very purpose of the lateral sinks at the model bottom discussed in the previous paragraph to swallow the background flow with the cold downgoing slabs and to compensate the influx of the plume and the background corner-flow-like influx, so that the mass budget of the model is balanced. However, at least in the case of temperature-dependent viscosity, this procedure is not without problems of several kinds: at the margins of the downwellings, the strong temperature and viscosity gradients, which are of the same order as beneath the lithosphere at the top, have an unfavourable effect on the numerics in that the computation of derivatives becomes more and more inaccurate; even worse, the iterative solution algorithm of the Stokes equation described by eqs. III.3b and c tends to lower the damping factor c_η so that the contribution of the updated field to the final solution of the iteration is strongly decreased in this case. For this reason, slices of a certain width are defined at the respective model sides where the temperature is forced to the values of the horizontally adjacent part of the model box after solving eq. I.10; a somewhat similar approach has also been taken by Ito *et al.* (1996), although they keep a certain part of the descending slab. Although this is a certain violation of energy conservation, tests indicate that this procedure affects the T field in the central parts of the model, on which interest is focussed, only to a minor degree; it even seems to improve the model in the remote off-ridge parts, because the T and η fields are smoother. Nonetheless, the side-strip boundary condition results in a downward forcing of warm material against its buoyancy, and there are visible differences in the velocity fields when comparing models with constant T side-strip and with cold downwellings; in particular, when a phase boundary such as the 410 km discontinuity is included, there is a tendency for the circulation through the entire upper mantle to be weakened, as far as passive flow under the normal ridge is concerned. On the other hand, the side-strip method shows a much more convincing behaviour in the iteration for the Stokes equation due to the smaller horizontal η variations and is therefore considered to yield a more consistent flow field; the faster overall convergence also improves performance. – As further alternatives, it has been tried to heat the cold slab linearly to bottom temperature over the lowermost fifth of the model height and dismiss or weaken the T dependence of η in the downwelling while keeping the T field as delivered by the temperature equation; however, both alternatives did not show substantial differences to, or improvements of, the cold-slab or the original side-strip boundary condition, respectively, and have therefore not been considered further.

Apart from these numerical considerations, care should also be taken to reduce the effects of the boundaries on the flow field in the center. Tests with challenging models with a depth of 410 km and widths of 1500 and 3000 km, respectively (models TB1 and TB2), and strong p–T-dependent viscosity indicate that at least for models with significant three-dimensional viscosity variations, it is advisable to place the x boundary with the downwellings far enough away from the ridge; on the other hand, the large memory requirements for these models limit this offset to not much more than twice the box depth, and computing time also increases greatly the larger the grid is. The profiles of T and v_z along the axis of a plume spreading beneath a ridge and across the plume in different depths in figure III.2 show the differences. Temperatures are quite similar in both cases and differ only notably in the shallowest parts of the spreading plume head due to differences in the dynamics of the plume, but centerline temperatures are virtually identical. These differences result from the higher ascent velocities of the plume in the narrow box, which apparently allows it to maintain a hot, focussed core region to shallower depths and lets it spread faster than the plume in the broad box; on the other hand, the depth levels at which a given velocity is reached do not differ too much between both models in several parts of the model, which is particularly important in the melting zone. Furthermore, in a deeper box, the influence of the boundaries should not be as large, because the mass is distributed in a larger volume, but nonetheless, there is the danger that a too narrow box results in too high plume ascent velocities and, accordingly,

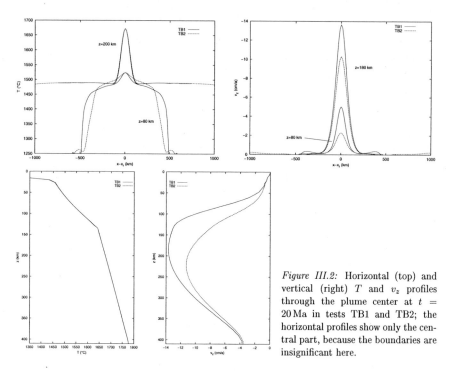

Figure III.2: Horizontal (top) and vertical (right) T and v_z profiles through the plume center at $t = 20\,\mathrm{Ma}$ in tests TB1 and TB2; the horizontal profiles show only the central part, because the boundaries are insignificant here.

artificially reinforces melt production; it then depends also on additional factors such as the position of the solidus depth and the assumed melting behaviour how sensitive to model geometry *e.g.* the crustal thickness in such models is. Besides, in broad models like TB2, the side-strip models remained more symmetrical than the cold-slab models, whereas both preserved symmetry at lower aspect ratios.

III.1.2   Advection–diffusion problem 1: Explicit method, and the computation of spatial derivatives

In the explicit method with upwind, the advection–diffusion problem for a quantity F,

$$\frac{\partial F}{\partial t} = \kappa_F \nabla^2 F - \vec{v} \cdot \nabla F + S_F \tag{III.5}$$

with some source term S_F, is discretized as follows at the grid point i (in one dimension) at timestep n:

$$\frac{F_i^{n+1} - F_i^n}{\Delta t} = \kappa_F \frac{F_{i+1}^n - F_{i-1}^n}{(\Delta x)^2} - v_{x_i} \frac{(1-c)F_{i+1}^n + 2cF_i^n - (1+c)F_{i-1}^n}{2\Delta x} + S_{F_i} \tag{III.6}$$

(central differences: $c = 0$; for upwind: $c = \operatorname{sgn} v_{x_i}$). The upwind method in the advection term provides numerical stability which is hardly or not at all achievable with central spatial finite differences (Peyret and Taylor, 1985), as long as the Courant–Friedrichs–Lewy (CFL) criterion

$$\Delta t_\mathrm{C} \leq \frac{\Delta x}{v_{i_{\max}}} \tag{III.7a}$$

is ensured to be fulfilled; in practice, where the full advection–diffusion problem eq. III.5 is solved on two- or three-dimensional grids with possibly different spacings in each direction, the stability criterion is considerably more restrictive than even this Courant timestep. This requirement, and the fact that at most grid points $v < v_{\mathrm{max}}$, has the consequence that the numerical diffusion of the time integration and the truncation error of the short operator do not cancel out, but that a certain amount of numerical diffusion remains and deteriorates the solution[1]. Furthermore, the mere existence of a stability criterion is a very strong handicap in models which are meant to cover a long timespan or where high velocities can be reached. Both problems have long been known and therefore motivated the quest for solutions or workarounds. For instance, Gable *et al.* (1991) did not apply upwind, but used central differences in space and included a second-order correction for the forward step in time to suppress the instability. Their scheme is approximately second-order in both space and time, but still subjected to a strict CFL-type stability criterion. – Apart from the stability criterion, an accuracy criterion for the diffusional term, stating that

$$\Delta t_{\mathrm{diff}} \lesssim \frac{3\ldots4\cdot\min[(\Delta x)^2, (\Delta y)^2, (\Delta z)^2]}{4\pi^2\kappa_F},\tag{III.7b}$$

has also to be observed; however, it is much less restrictive than the CFL criterion for mantle convection problems.

Proper spatial resolution of the model is crucial with respect to accuracy. Thus, it is important to make sure that the results are not affected seriously by the fact that all variables are represented on a grid whose discretization is rather coarse in the horizontal dimensions; it is known that spectral methods as used for solving the Stokes equation do not require such a dense spatial sampling as low-order finite-difference methods, *e.g.* the central second-order operators

$$\frac{dF_i}{dx} = \frac{F_{i+1} - F_{i-1}}{2\Delta x} + O(\Delta x^2)\tag{III.8a}$$

$$\frac{d^2F_i}{dx^2} = \frac{F_{i+1} - 2F_i + F_{i-1}}{(\Delta x)^2} + O(\Delta x^2)\tag{III.8b}$$

(*e.g.* Peyret and Taylor, 1985; Bronstein and Semendjajew, 1991) used here in many instances. If the energy conservation equation is solved by a pure FD method on a coarse grid as in the present case, however, there is a potential danger of substantial loss of accuracy, especially due to the upwind terms, whose truncation error is of $O(\Delta x)$ for the lowest order FD operator. Comparing the results for a critical model at different resolutions (see below) and using first-order and second-order FD operators of the form

$$\frac{dF_i^{(1)}}{dx} = \begin{cases} \frac{F_{i+1}-F_i}{\Delta x} + O(\Delta x), & v_i < 0 \\ \frac{F_i-F_{i-1}}{\Delta x} + O(\Delta x), & v_i > 0 \end{cases} \qquad \text{1st order} \tag{III.8c}$$

$$\frac{dF_i^{(2)}}{dx} = \begin{cases} \frac{4F_{i+1}-F_{i+2}-3F_i}{2\Delta x} + O(\Delta x^2), & v_i < 0 \\ \frac{3F_i-4F_{i-1}+F_{i-2}}{2\Delta x} + O(\Delta x^2), & v_i > 0 \end{cases} \qquad \text{2nd order} \tag{III.8d}$$

---

[1]For pure advection, the forward-in-time/central-in-space scheme is unconditionally unstable. This is important for the advection of chemical fields, where chemical diffusion is usually neglected. Farnetani and Richards (1995) remark that a certain amount of chemical diffusion helps reducing the numerical diffusion, albeit to the cost of artifically enhanced mixing. However, the use of more accurate advection techniques (see section III.1.3) should make it unnecessary to introduce such an unphysical diffusion term.

(*e.g.* Peyret and Taylor, 1985; Bronstein and Semendjajew, 1991), respectively, in the advection term suggests that a rather high resolution is necessary with the first-order operator, while a reasonable accuracy can already be achieved with the second-order operator at lower resolution; a comparison of the scheme in two dimensions with benchmark results for variable-viscosity models (Blankenbach *et al.*, 1989) confirms this finding (Marquart, pers. comm., 2000). However, on individual profiles it can be seen that using a long operator in the coarsely sampled horizontal dimensions as well as in the more densely discretized vertical direction results in small, but with respect to melt processes possibly non-negligible wiggles near steep gradients of the variable on which the operator acts. For this reason, a mixed operator – first order in the vertical direction, second order in the horizontal directions – seems to be a useful compromise which delivers reasonably precise solutions at not too dense spatial sampling. The small additional cost of the longer operator at a given resolution is more than compensated by the gain in computing speed and memory saving made possible by the use of a coarser grid.

Resolution tests — The accuracy of numerical models is endangered by several factors, such as too low spatial resolution and numerical diffusion. The logical conclusion would be to increase the spatial sampling as much as desired, but of course, this has several disadvantages, in particular in three-dimensional models: memory and data storage requirements grow tremendously, and computation time extends to an intolerable degree. Thus, an attempt was made to determine a sampling of the numerical grid which would ensure acceptable accuracy while delivering results within a reasonable time; in this context, the abovementioned considerations regarding different FD operators were given special attention. Apart from varying the resolution, different aspect ratios of the grid cells where also tested.

The test model was designed to be numerically rather challenging, more than most of the models expected to be realistic would be; the model parameters are listed in table III.1. Fifteen model runs were made: $60 \times 40 \times 66$ points with short, long, and mixed FD; $60 \times 40 \times 100$ points with short and long FD; $90 \times 60 \times 100$ points with short, long, and mixed FD; $108 \times 72 \times 100$ points with mixed FD; $120 \times 80 \times 100$ points with short, long, and mixed FD; $120 \times 80 \times 132$ points with short and long FD; and $150 \times 100 \times 100$ points with short FD. The quality of the runs was judged from the global root-mean-square convection velocity v_{rms} and the Nusselt number Nu as defined in the benchmark by Blankenbach *et al.* (1989) (figure III.3), and additionally for some selected cases by comparing some selected T, v_x and v_z profiles (figure III.4). Resolutions for which these observables converge are considered sufficient to yield meaningful models, although numerical diffusion as the general drawback of using upwind is undeniably present; as mentioned earlier, upwind was only used at points with elevated velocities, but nonetheless, this applied for fractions between one fifth and half of the grid points in the model at some time.

Inspection of figure III.3, which shows v_{rms} and Nu for the short, long, and mixed upwind operator separately, reveals notable differences among some models, in particular in v_{rms}. For all models, Nu does not vary strongly with resolution or FD operator length, with exception of the coarsest grid ($60 \times 40 \times 66$ points; the densest grid ($150 \times 100 \times 100$ points) displays some irregularities of unclear origin, though, and is therefore deemed suspicious. For the other models, differences in Nu are $< 5\%$. – In contrast, v_{rms} variability reaches *ca.* 15 % for the whole range of models and therefore seems to be a more sensitive criterion for accuracy. From figure III.3a one sees at once that v_{rms} has larger values at a given resolution if a short upwind operator is used and that lower resolution also increases v_{rms}. Models with the same operator converge against a certain value, but this value is apparently not the same for different operators, although the differences are less than 5 %. Considering the curves for

(x_m, y_m, z_m)	convection model dimensions	$(1500\,\mathrm{km}, 1000\,\mathrm{km}, 660\,\mathrm{km})$
η_0	reference viscosity	$5 \cdot 10^{22}\,\mathrm{Pa\,s}$
g	gravity acceleration	$10\,\mathrm{m/s^2}$
ϱ_0	mantle reference density	$3660\,\mathrm{kg/m^3}$
c_p	isobaric specific heat	$1350\,\mathrm{J/kg\,K}$
$\varkappa$	heat conductivity	$3\,\mathrm{W/m\,K}$
α	mantle thermal expansivity	$3.331 \cdot 10^{-5}\,\mathrm{1/K}$
$\mathfrak{T}$	mantle potential temperature	$1420\,^\circ\mathrm{C}$
v_r	half spreading rate	$1\,\mathrm{cm/a}$
ΔT_P	maximum plume excess temperature	$300\,\mathrm{K}$
r_P	plume radius	$125\,\mathrm{km}$
v_P	influx velocity at plume center	$1\,\mathrm{cm/a}$
n_{itmax}	maximum iteration count for Stokes equation	20

Table III.1: Model parameters for convection model resolution tests.

all operators, it seems that a grid with 100 points in the vertical direction and something between 90 (60) and 120 (80) points in x (y) direction should provide a reasonable accuracy; the curve for the $108 \times 72 \times 100$ points mixed FD model supports this assumption.

In view of the theoretical truncation errors given in eqs. III.8c and III.8d and the benchmark comparisons one would expect the long operator to be optimal. However, long operators tend to introduce "overshoot" and thereby introduce noise into the model. This can be seen in the temperature profiles in figure III.4: the T reduction above the plume top in the $T(z)$ profile and at the sides of the plume is clearly a numerical artefact and is feared to affect severely models in which an accurate representation of the T field is essential, which is the case in melt models. As the higher order of the truncation error is not as important in the vertical direction, where the resolution is higher anyway, the mixed FD operator was tried as a compromise to allow for a reasonable accuracy at resolutions as low as possible. The case of $108 \times 72 \times 100$ points mixed FD model apparently is the best in the series to combine these relative virtues; nonetheless, real-problem runs (without melting) with such a grid usually take several days on an ordinary PC. – Note that all of these resolution tests, as well as those for the segregation part, have been done with the upwind version of the program. As for resolution issues the one-sided, comparatively short upwind operators are probably the most critical items, it seems justified to assume that the representation in the semi-Lagrangian formulation, which uses more sampling points, will relax the problem, so that no further resolution tests seem necessary.

III.1.3   Advection–diffusion problem 2: (Semi-)implicit method and operator splitting

As discussed in the previous section, explicit time-stepping with upwind has severe short-comings with respect to accuracy and performance. For this reason, an alternative class of time integration schemes was considered, especially one for handling the advection problem in a more advantageous way in terms of both performance and accuracy. In this context, methods considering the characteristic of a flowing particle seem to be of particular interest: for instance, in numerical weather prediction, where accuracy and performance are vital, the class of so-called semi-Lagrangian methods, which combine the geometrical advantages of fixed-point (Eulerian) grids with the stability of schemes in which the reference frame follows the path of a particle (Lagrangian schemes), has been developed in the 1980s and 1990s, and are receiving considerable attention (see review by Staniforth and Côté, 1991).

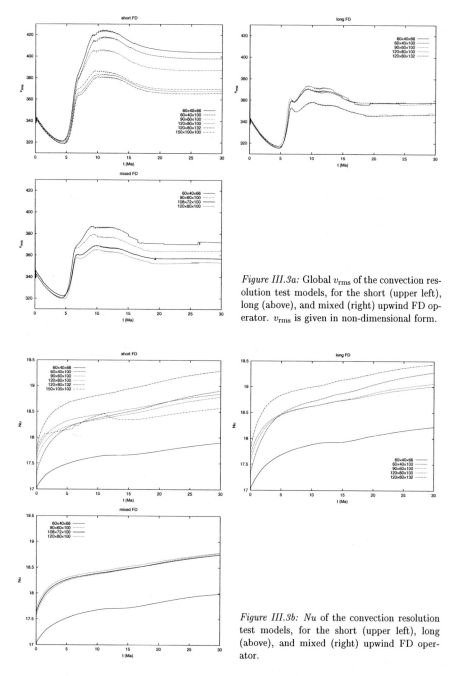

Figure III.3a: Global v_{rms} of the convection resolution test models, for the short (upper left), long (above), and mixed (right) upwind FD operator. v_{rms} is given in non-dimensional form.

Figure III.3b: Nu of the convection resolution test models, for the short (upper left), long (above), and mixed (right) upwind FD operator.

Figure III.3: Convection resolution test results: global benchmarking quantities

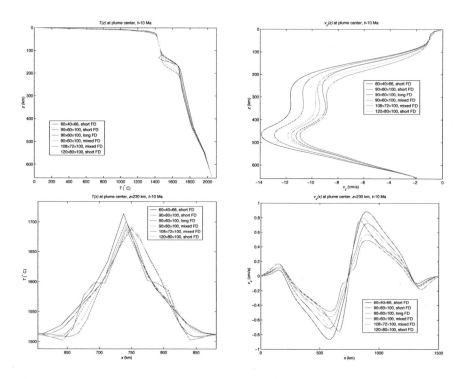

Figure III.4: Convection resolution test results: selected profiles. The $T(z)$ and $v_z(z)$ profiles (upper row) are taken at the center of the model and the plume at $t \approx 10\,\mathrm{Ma}$; slight differences in the time increment preclude measurements at exactly identical times, but the differences are less than about 18 ka at worst. The $T(x)$ and $v_x(x)$ profiles (lower row) are taken at the same time and are located at $y_m/2$ at 230 km depth, *i.e.* slightly below the plume top; note that the $T(x)$ profile is zoomed into the close surroundings of the plume.

Using a semi-Lagrangian method, the complete advection–diffusion problem of eq. III.5 can then be solved by splitting its RHS into a disjoint advection and diffusion problem and solving each subsequently; tests indicate that the order in which the subproblems are solved is unimportant.

Semi-Lagrangian advection — In this study, a two-level semi-Lagrangian advection scheme is used for the calculation of the advection of T, f and X fields; in two-level schemes, only two time levels, t_n and t_{n+1}, are considered, in contrast to the less efficient three-level schemes, which additionally include t_{n-1}. The principle of the scheme is depicted in figure III.5. For the grid point x_m corresponding to particle position C, an initial guess for the displacement $\hat{x}/2$ between points B and C and the time-averaged velocity $v^{n+1/2}$ is made using v_m^{n+1}, and is then refined iteratively, whereby linear interpolation of the velocities is found to be sufficient; two iterations lead to an acceptable estimate (Staniforth and Côté, 1991). The advected field F is then evaluated at time t_n at the upstream location $x_m - \hat{x}$ (point A) determined as $v^{n+1/2}\Delta t$, whereby cubic interpolation is commonly regarded as a reasonable compromise between accuracy and efficiency, and is then shifted to (x_m, t_{n+1}) (point C).

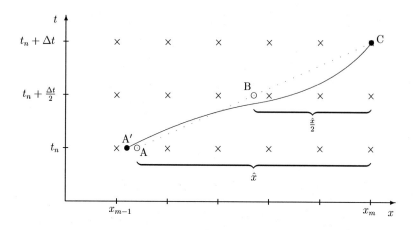

Figure III.5: Two-level semi-Lagrangian advection in one dimension (redrawn from Staniforth and Côté, 1991). A particle located at A' (between grid points x_{m-1} and x_m) at time t_n traverses the distance $\hat{x}' = \int_t^{t+\Delta t} v(x) \, dt$ in the time interval Δt following the solid line in x-t space and eventually arrives at C (grid point x_m); its trajectory is approximated by the straight dotted line, the approximated traversed distance from A to C is $\hat{x} = v^{n+1/2} \Delta t$.

For the implementation, a set of ready-to-use subroutines from Spiegelman (2000) for the source-free, one-dimensional problem with periodic boundary conditions has been used as a starting point. In the framework of the present model setting, periodic boundaries are only present at the x and y faces of the model cube, so that the routines had to be extended for use with non-periodic boundaries. It is not necessary to formulate the algorithm explicitly for higher-dimensional grids; instead, one can apply the one-dimensional advection successively for each direction of the model (Spiegelman, 2000).

Application of the semi-Lagrangian method did not yield any indication for the rise of numerical instability, even at timesteps more than one order of magnitude above the CFL criterion. However, Staniforth and Côté (1991) report the criterion

$$\Delta t < \frac{1}{\max\limits_{i,j=1...3}\left(\left|\frac{\partial v_i}{\partial x_j}\right|\right)} := \Delta t_{\mathrm{SL}} \tag{III.9}$$

for ensuring accuracy of the iterative approximation of the displacement at a given grid point. However, this is in general not likely to be a critical issue in the mantle, because the derivatives of $\vec{v}$ are small almost everywhere. For instance, one can consider two-dimensional corner flow (see section II.4.3, p. 61): inserting the velocities eq. II.58a and II.58b in eq. III.9 shows that $|\partial v_i/\partial x_j|$ would yield a $O(\Delta t) = 10^6 \ldots 10^7$ Ma for most of the mantle, assuming $v_\mathrm{f} = 1\,\mathrm{cm/a}$. Only in a small region within a few kilometers of the origin of the coordinate system would Δt drop to small values; at (0,0), $\partial v_x/\partial z$ has a pole. These restrictions are not a problem, though, because the velocities are very low there, and the critical region is smaller than the usual resolution of the model grid. A two-dimensional implementation of the method has been tested against isoviscous benchmark models from Blankenbach *et al.* (1989) for different Ra values and was found to yield satisfactory results, although energy conservation was found to be violated notably at high Ra, so that care has to be taken in certain parameter constellations (Marquart, pers. comm., 2002). As the violation of conservation laws is not only relevant to energy, but possibly also to compositional variables, this topic is discussed in some more detail below.

Another source of errors is the velocity at time t_{n+1}, which is needed for approximating the displacement at the half-timestep, but is not known at the time when the advection problem is solved. Therefore, $\vec{v}^{n+1/2}$ has to be extrapolated using $\vec{v}^n$ and $\vec{v}^{n-1}$:

$$\vec{v}_m\left(t + \frac{\Delta t}{2}\right) = \vec{v}_m(t) + \frac{\vec{v}_m(t) - \vec{v}_m(t - \Delta t_{\text{old}})}{2\Delta t_{\text{old}}}\Delta t + O(\Delta t^2). \tag{III.10}$$

This is a generalization of eq. 13 of Staniforth and Côté (1991) allowing for varying time increments; in practice, reasons like the temporal evolution of the flow field or simply the requirement of data output at times which are not integer multiples of Δt can require a change of the timestep. Temperton and Staniforth (1987) compared this method with others and found a three-term extrapolator which also used $\vec{v}^{n-2}$ to give better results, whereas time extrapolation of $\vec{v}$ along a trajectory is less accurate; nonetheless, a three-term extrapolator is considered too costly in terms of memory requirement, because for every grid point, three velocity components have to be stored for every additional time level. It is important to note that the error of the time extrapolation enforces a reduction of the time increment to a value significantly below Δt_{SL} for the computation to deliver accurate results (see tests on page 86).

Finally, the presence of steep gradients in F can cause the cubic interpolation to produce "overshooting", leading to artefacts or even unphysical values like negative concentrations near advection fronts. For this reason, the original semi-Lagrangian routines have also been extended by the QMSL (quasi-monotone semi-Lagrangian) algorithm (Bermejo and Staniforth, 1992) which was designed to suppress artificial wiggles. The algorithm consists in the computation of the departure point (A in figure III.5) and the corresponding grid cell, the cubic interpolation of F, determination of the local maximum and minimum v^+ and v^- of the adjacent grid points, and assignment of v^+ (v^-) to the result of the cubic interpolation if it is larger (smaller) than these bounds. Bermejo and Staniforth (1992) defined the QMSL algorithm for two dimensions, whereas here it was reduced to the one-dimensional case. It should though be noted that in principle, the elimination of the numerical oscillations violates mass conservation to somewhat larger extent than the basic method and also introduces some numerical diffusion; therefore, it is important to ensure a sufficiently dense sampling to prevent oscillations from becoming very large. Several methods have been proposed to suppress oscillations and enforce mass conservation while limiting the numerical diffusion (*e.g.* Gravel and Staniforth, 1994), but there seems to be no optimal method for all problems (Oliveira and Fortunato, 2002). It is therefore left for future extensions of the algorithm to evaluate the most appropriate correction scheme for mantle convection (and melt transport) problems.

Implicit time integration of the diffusion problem — The diffusive part of the temperature equation is solved with a Crank–Nicolson scheme applied successively to each spatial direction before solving the advection problem for the same direction; source terms like adiabatic heating and dissipation are added during the solution for the first dimension. Hence, the solution of the diffusion problem is one-dimensional and can be cast into the form of a linear equation system with a tridiagonal, symmetric, positive definite, real matrix,

$$
\begin{pmatrix}
\ddots & \ddots & \ddots & 0 & 0 & 0 \\
0 & -\beta & 2(1+\beta) & -\beta & 0 & 0 \\
0 & 0 & -\beta & 2(1+\beta) & -\beta & 0 \\
0 & 0 & 0 & \ddots & \ddots & \ddots
\end{pmatrix}
\begin{pmatrix}
\vdots \\
T_{i-1}^{n+1} \\
T_i^{n+1} \\
\vdots
\end{pmatrix}
=
$$

$$
\begin{pmatrix}
\ddots & \ddots & \ddots & 0 & 0 & 0 \\
0 & \beta & 2(1-\beta) & \beta & 0 & 0 \\
0 & 0 & \beta & 2(1-\beta) & \beta & 0 \\
0 & 0 & 0 & \ddots & \ddots & \ddots
\end{pmatrix}
\begin{pmatrix}
\vdots \\
T_{i-1}^{n} \\
T_i^{n} \\
\vdots
\end{pmatrix}
\tag{III.11}
$$

with $\beta := {}^{\kappa \Delta t}/_{\Delta x_i^2}$ (*e.g.* Spiegelman, 2000), which can be solved very efficiently for the T^{n+1} with existing routines (*e.g.* dptsv from LAPACK, Anderson *et al.*, 1999). There is no limitation to stability, but eq. III.7b is a restriction for the accuracy of the method. However, to take advantage of this form, special attention has to be paid at the boundaries.

At the top and bottom boundaries, T is kept at a constant value (Dirichlet condition); for other problems, it is possibly desired to prescribe a heat flux at the bottom (von Neumann condition), but this is not of interest here. With a constant T at the top, the first two lines of eq. III.11 have the following form:

$$
\begin{pmatrix}
1 & 0 & 0 & 0 \\
0 & 2(1+\beta) & -\beta & \ddots
\end{pmatrix}
\begin{pmatrix}
T_1^{n+1} \\
T_2^{n+1}
\end{pmatrix}
=
\begin{pmatrix}
T_1^{n} \\
T_2^{n} + \beta T_1^{n}
\end{pmatrix};
\tag{III.12}
$$

the manipulation of the second row is necessary to preserve the tridiagonal structure of the matrix, and it is possible, because $T_1^{n+1} = T_1^{n}$.

At periodic boundaries, the tridiagonal form is not preserved in the first and the last row, because $A_{1N} = A_{N1} = -\beta$ in the LHS matrix of eq. III.11. Fortunately, the Sherman–Morrison formula (see Press *et al.*, 1992, for a description) provides a method for handling this case, although the solution of an additional $N \times N$ system makes it roughly twice as expensive.

The introduction of the adiabatic and dissipative source terms poses a problem, because they both depend on T and $\vec{v}$, and ideally, one would have to use the T and $\vec{v}$ at $t_{n+1/2}$, which are not known; the problem is even more acute if the viscosity is assumed to be T-dependent. For now, it has been chosen to use future values where available, *i.e.* $\vec{v}^n$ and $\vec{v}^{n+1}$, but stick to the values of the nth timestep otherwise; the relatively small magnitude of the source terms and the choice of a rather moderate time increment are expected to prevent the error from becoming severe.

Additional accuracy considerations — As already mentioned, semi-Lagrangian advection is not conservative by itself, and tests indeed demonstrate violations *e.g.* of mass conservation. Although Staniforth and Côté (1991) quote the accuracy criterion eq. III.9, the need to extrapolate $\vec{v}$ into the future if it is not known beforehand complicates the issue and imposes an additional restriction on Δt which can be more severe than the criterion from the computation of the displacement, as shown in the abovementioned tests by Marquart (2002, pers. comm.). To illustrate the problem in some more detail, four simple models have been set up within a typical framework of this study: a dynamically passive marker sphere with a radius of 50 km was placed in a 3000 km×1000 km×660 km box with a spreading ridge and a plume rising in its center; the other values were essentially chosen as in table III.1, but allowing for a maximum of 30 iterations, and with $\mathcal{T} = 1410\,°C$ and $\Delta T_P = 250\,K$. In the test cases TC1a–TC1c, the sphere was initially positioned with its center at (1475 km, 525 km, 550 km), *i.e.* 110 km above the model bottom and *ca.* 35 km off the plume axis, so that it would be strongly dragged and sheared by the rising plume as soon as it reached it; in test case TC1d, such a sphere was placed at (1475 km, 800 km, 300 km), *i.e.* slightly above mid-depth and slightly off-ridge beneath a MOR segment little affected by plume ascent. In models TC1a, TC1b, and TC1d, the "concentration" of the marker material in the sphere was initially 1 within the sphere and 0 outside, whereas in model TC1c, it was Gaussian. To test the quality of mass conservation, the concentration was integrated in each timestep; if the scheme were conservative, the integral should not change during the run, because no sources nor sinks were present. The use of different distribution functions for the material should reveal if the presence of a sharp step at the sphere boundary results in deterioration of conservation properties. A timestep of $0.5\Delta t_{SL}$ was used in models TC1a, TC1c, and TC1d; in model TC1b, the factor was 0.1.

Figure III.6 shows the result of this test. All models, but in particular cases TC1a and TC1c, show a strong artificial mass increase, which begins at the time where the sphere is picked up by the rising plume; when a calmer stage of plume evolution is reached some time after impinging on the lithosphere and the marker sphere is in a region with less shear, conservation laws are fulfilled quite well again. Hence, a significant error can be introduced locally by large timesteps, even if the criterion III.9 is still well fulfilled, which indicates that time extrapolation is the most problematic part of the formalism; the reason lies in strong temporal changes of the local velocity field related to steep velocity gradients at the front of the ascending plume, which cannot be predicted appropriately by linear extrapolation if the timestep is too long. As proven by case TC1b, a shorter timestep is able to limit the error, so that it seems indicated to use timesteps no larger than one tenth of the accuracy criterion in models as challenging as the present test case; if the velocity field is less variable, and therefore better predictable, larger timesteps are safe, and mass conservation is well fulfilled, as suggested by case TC1d. To prevent timesteps from becoming too large, one could try to control them automatically by including a threshold value for the allowable violation of conservation of a monitor tracer field, but experiments with such improvements are left for future work, as is the already mentioned alternative to use extrapolation involving more timesteps than the previous one (Staniforth and Côté, 1991). – The shape of the anomaly appears to be only secondary, as the strong artificial mass growth in the Gauss-weighted sphere of model TC1c shows, which underpins the suspicion that this is indeed an extrapolation problem and not, or not directly, related to overshooting in the interpolation of the field and its suppression by QMSL. The fact that model TC1c is even worse than model TC1a, which might be unexpected at first sight, could be due to an even less favourable distribution of its masses relative to the zone of maximum shear at the margin of the plume.

A second test series has been conducted to clarify the importance of the time increment for the consistency of a typical model setup. The series TC2 has a similar setup as the TC1 models, but the box is only 410 km deep and the plume excess temperature is 200 K. A time step of $0.4\Delta t_{SL}$ was used in model TC2a and of $0.1\Delta t_{SL}$ in TC2b. The key result is displayed in figure III.7, which shows some T and v_z profiles through the plume, where variability is expected in the first place: the $T(z)$ profiles along the center axis of the plume of both models are virtually identical, and so are the $v_z(z)$ profiles. This suggests that the temperature field is not as sensitive to the timestep as the chemical marker field in the previous test, but from the results of series TC1 it seems nonetheless recommendable to be very moderate in the choice of the timestep.

As an additional issue, an assessment of the importance of the basal influx velocity for

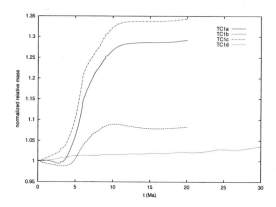

Figure III.6: Whole-model integral of the marker field in the TC1 models as a function of time, normalized to the respective initial value.

whole-plume dynamics, and consequently, for melt production was attempted by extending the series by model TC2c, in which the maximum influx velocity at the bottom was increased from $3\,\mathrm{cm/a}$, as in TC2a, to $6\,\mathrm{cm/a}$. Compared with model TC2a, model TC2c maintains the higher velocity from its influx to about mid-depth (see figure III.7); however, in the upper third of the box, where melting takes place, the upwelling velocity is not dependent on the basal influx velocity anymore. Nonetheless, the larger mass flux in the faster plume has to be accommodated, and the $T(x)$ and $v_z(x)$ profiles through the stem at mid-depths and through the plume head suggest that this happens by broadening of the whole structure in model TC2c. The ramifications of such flux variations will be revisited in more detail in section V.2.2, with special attention to Iceland.

III.2   THE NUMERICAL SOLUTION OF THE MELT DYNAMICAL EQUATIONS

This section describes shortly the algorithm for the solution of eqs. II.35a, II.38a and the term of II.42 related to melting/freezing after Schmeling (2000), which has been rewritten and extended to three dimensions; a more detailed accounting for some physical properties of the melt and the conditions of melting and crystallization have also been added. The computation of anisotropic permeability from section II.4.3 is also shortly described. The computational domain where the melt calculations are performed is the higher-resolution grid inserted into the convection grid which was mentioned in the introduction to this chapter (see figure III.8).

III.2.1 Solution of the transport equations

After the computation of the temperature field in the convection grid, T, $\vec{v}$, and if applicable, certain compositional fields, are interpolated linearly on the dense melt grid, and additional variables, in particular η, are determined as applicable. The melt dynamics itself is computed in one or two steps, depending on the desired level of complexity:

Depletion and melting In the first, mandatory step, the depletion and porosity fields are advected with the velocity of the matrix, resulting in new distributions of f and φ; the advection problem is solved with the semi-Lagrangian technique in most models, but was dealt with using an upwind method in previous versions of the algorithm. Independently, the T field would result in a petrologically derived melting degree which

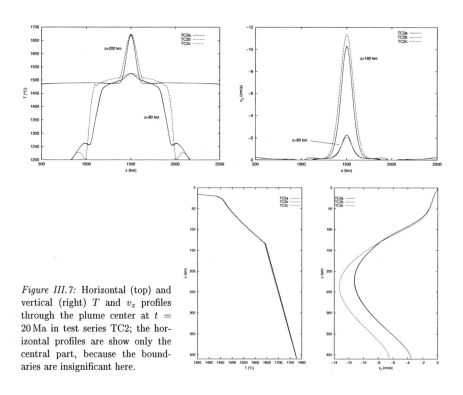

Figure III.7: Horizontal (top) and vertical (right) T and v_z profiles through the plume center at $t = 20\,\mathrm{Ma}$ in test series TC2; the horizontal profiles are show only the central part, because the boundaries are insignificant here.

Figure III.8: Scheme of the two nested grids used in combined mantle convection and melt dynamical models. Note that the melt grid encloses not only the plume head, but also the whole melting region of the ridge.

can be computed by parameterizations such as those by McKenzie and Bickle (1988) (eq. II.28a) or Katz *et al.* (2003) (eq. II.32). The difference between the latter, thermochemically appropriate, and the advected f determines how much melting or crystallization has to take place at the respective grid point; this approach represents an attempt to account for the upward shift of the solidus with increasing depletion, but also allows for recrystallization of melt in a unified form. Along with the concomitant change in f and φ, *i.e.* the melting/freezing rate, the temperature change due to the latent heat of melting resp. freezing is calculated. If the consideration of melting is desired to be simple, the newly produced melt is kept and the fraction of it exceeding the threshold of extraction φ_{max} is taken out of the model; otherwise, the porosity field is dismissed, because it will be determined again in the segregation step in the framework of the more complex mechanisms involving relative movement.

Segregation The segregation algorithm (Schmeling, 2000, and pers. comm.) consists of an iteration loop comprising the calculation of the buoyancy, viscous stress, and compaction terms in melt momentum conservation eq. II.38a and the subsequent solution of the equation of melt mass conservation, eq. II.35, in the compaction Boussinesq approximation. The latter part involves the melting/freezing rate determined previously in the first step and repeats the advection of porosity, this time in the context of melt segregation. After having found the solution, possible further freezing is considered, and then the excess melt is extracted. In the older version of the program, which used an upwind scheme, the melt distribution resulting from an iteration step was directly fed back into the next one, whereas in the newer, semi-Lagrangian formulation, it seemed to be more plausible to compute the segregation velocities for porosities not exceeding the extraction threshold, implying that any excess melt is immediately removed from the isotropic porous flow domain by some kind of fast transport.

In the solution algorithm, it is of interest to consider the density and the viscosity of the melt and their dependence on p–T conditions in some detail instead of always assuming that they are constant. In the current version of the program, density changes are taken into account by determining the isothermal compression of the melt at the temperature of the reference density during the initialization of the melting calculations according to lithostatic pressure and to compute the actual melt density at each grid point from this vertical density profile from the thermal variations, assuming a constant thermal expansivity (see section II.1.1). Hence, compression is only calculated once, so that it is possible to use one of the more widely applicable isothermal equations of state discussed in section II.1.2 instead of the simple linearized eq. II.4 with negligible effort. – In the solution of the equation of melt momentum in the segregation part, the viscosity of the melt also comes into play. In most models in this study, it is assumed to be constant, but as an alternative, variability is allowed for via the definition of Rt in eq. II.38a using the formalism by Bagdassarov (1988) (section II.1.4); the strong decrease of melt viscosity resulting from the latter, while potentially significant, can make the numerical solution more difficult to achieve and is considered in more detail in section V.1.8.

After computing the generation and motion of the melt within the mantle and determining the amount to be extracted, this material is added to the top of the whole model to form the crust (see section III.3). The new f and φ distributions as well as the T field, which was modified by melting or freezing enthalpy, are eventually transmitted back to the convection grid.

From the numerical point of view, it should be noted that the solution of eq. II.35 is split by applying the chain rule; in the newer program version, the advective subproblem is then solved by the semi-Lagrange method (in earlier versions with a simple upwind method), whereas the other parts are treated with finite differences applied to a melt distribution corresponding to the average of the old timestep and the iterated guess of the new timestep. Although this was not investigated, it is reasonable to suspect that numerical diffusion in the use of upwind schemes can adulterate the porosity field in such a way that its boundaries are too broad and have too low melt contents; this might have the particular disadvantage of artificially shrinking the volume where the threshold for melt extraction is reached. Furthermore, it seems advisable to apply the chain rule to the deviatoric stresses in eq. II.38a as well, yielding

$$\frac{\partial \sigma_{ij}}{\partial x_j} = \frac{\partial}{\partial x_j}\left[\eta\left(\frac{\partial v_i}{\partial x_j} + \frac{\partial v_j}{\partial x_i}\right)\right] = \left[\frac{\partial \eta}{\partial x_j}\left(\frac{\partial v_i}{\partial x_j} + \frac{\partial v_j}{\partial x_i}\right) + \eta\left(\frac{\partial^2 v_i}{\partial x_j^2} + \frac{\partial^2 v_j}{\partial x_i \partial x_j}\right)\right]; \qquad \text{(III.13)}$$

the first derivatives and the plain second derivatives can then be computed by central finite differences as in eqs. III.8, whereas the mixed second derivatives can be calculated by a "cross-over" central operator of the form

$$\frac{\partial^2 F_i}{\partial x \partial y} = \frac{F_{i-1,j-1} + F_{i+1,j+1} - F_{i+1,j-1} - F_{i-1,j+1}}{4\Delta x \Delta y} + O[(\Delta x + \Delta y)^2], \qquad \text{(III.14)}$$

so that all derivatives are determined with truncation errors $O(\Delta x^2)$ and $O[(\Delta x + \Delta y)^2]$, respectively. This should result in a higher accuracy than first determining the stress from the first derivatives of the velocity field and then differentiating the already truncated result again with an FD operator. As an aside, it is noted that for isoviscous mantle models, the first term in the brackets on the RHS of eq. III.13 can be dropped, which offers a possibility to reduce computing time; however, one will probably seldomly want to run models with an isoviscous mantle.

Resolution tests — As in section III.1.2 for the convection models, resolution tests were made for the melt grid; for these tests, the upwind version had been used. However, no other FD operators apart from the short (eq. III.8c) were used; only resolution and grid cell aspect ratio were varied. The model parameters of the melt grid were chosen to represent an average situation and are listed in table III.2, those of the associated convection model

(x_f, y_f, z_f)	melt model dimensions	(567 km, 983 km, 167 km)
η_b	bulk viscosity (constant)	$5 \cdot 10^{22}\,\mathrm{Pa\,s}$
η_m	melt viscosity	$2\,\mathrm{Pa\,s}$
ϱ_{m0}	melt reference density	$2663\,\mathrm{kg/m^3}$
$\Delta\varrho_{dp}$	density reduction of depleted material	$150\,\mathrm{kg/m^3}$
L_m	melting enthalpy	$6 \cdot 10^5\,\mathrm{J/kg}$
α_m	melt thermal expansivity	$6.3 \cdot 10^{-5}\,\mathrm{1/K}$
K_T	melt isothermal bulk modulus	$22.3\,\mathrm{GPa}$
K_T'	pressure derivative of melt isothermal bulk modulus	7
a	grain size	$2\,\mathrm{mm}$
b	geometry factor in eq. II.8	648
n	exponent in eq. II.8	3
φ_c	extraction threshold porosity	0.01
n_{itmax}	maximum number of iterations for solving eq. II.38	200

Table III.2: Model parameters for melt model resolution tests.

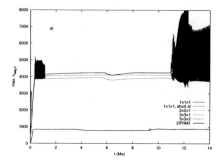

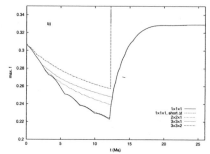

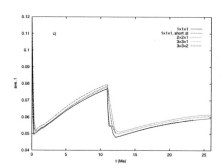

Figure III.9: Selected results of melt dynamics resolution tests; the numbers in the legend show the degree of grid refinement with respect to the convection grid. Top left: maximum segregation velocity; top right: maximum melting degree; left: average melting degree. The v_{seg} plot is only shown for $t < 14\,\text{Ma}$, because the oscillatory pattern does not change fundamentally at later times; it includes an oscillation-free example from section V.2.1 (where a is 1 mm). The "average melting degree" is calculated as the melting degree at each point where $f < 0$, divided by the number of such points.

are the same as in the convection resolution tests of section III.1.2; most models were run with a time increment of $0.3\Delta t_C$, except one, where $0.2\Delta t_C$ was used. The melting degree function of McKenzie and Bickle (1988) eq. II.28a was used.

Figure III.9 shows some essential results of the resolution tests; as characteristic variables of greatest interest, the segregation velocity $\vec{v}_{seg}$ and the melting degree have been chosen. The velocity plot shows two phases, one calm, where melting takes place under normal MOR only, and a strongly oscillatory with period $2\Delta t$, whose beginning coincides with the onset of plume melting. In the calm phase, the velocity is quite uniform for all models, suggesting a reliable result; in the oscillatory phase, it is obvious that the algorithm is at the edge of instability or does indeed become unstable, as in the case with the highest resolution, $3 \times 3 \times 2$. The fact that the model with the more restrictive timestep has oscillations with a much smaller amplitude indicates that for the forthcoming models, a shorter time increment should be used; in the models presented in chapter V, especially those of section V.2.1, the timestep has therefore been chosen such that no such oscillations appear (see curve DPRM1 in figure III.9a as an example). As far as resolution is concerned, one can deduce from the melting degree plots that a threefold refinement of the grid in the horizontal directions leads to useful results; further refinement in the vertical direction might be desirable, but, as seen, is apparently more prone to instability in the test setup. With respect to the excessive memory requirements and runtimes of models with very dense spatial sampling, a grid with a refinement of $3 \times 3 \times 1$ seems to be an acceptable compromise; it would result in roughly similar grid cell dimensions in the fine grid with the convection grid design used here.

Time-stepping considerations — From the numerical point of view, the restrictions on, and rules of, the advection problem discussed in the previous section also apply to the advection of depletion and porosity resp. melt, irrespective of the time integration method used. As far as the semi-Lagrange method is concerned, it is possible that strong permeability variations, which would lead to large velocity gradients, could potentially result in very restrictive Δt_{SL} values, although this has not actually been observed. The appearance of a source/sink term in the equation of melt conservation II.35a gives rise to a particular problem in the combined iterative solution of the equations of melt mass conservation and melt momentum, eqs. II.35a and II.38a, if excess melt is extracted instantaneously, as often assumed here. The source term provides a certain amount of new melt generated every timestep, the more, the longer the timestep is; this amount is then fed back into the determination of the segregation velocity in the iteration loop and will result in larger velocities when a large amount of melt is produced, because the segregation velocity depends on permeability, which in turn is a function of porosity, *i.e.* melt content. Hence, the segregation velocity is essentially controlled by the time increment of the model, which is imposed in part by model properties not related to melt generation and in part by the modeller's choice of the fraction of Δt_{SL} to be used. If excess melt is not extracted from the whole model after each timestep, this will probably be physically consistent anyway, because the melt would accumulate and reach significant volume fractions, and the iterative solution should ensure a meaningful spatial distribution of the melt; however, if a fast extraction mode is postulated on the basis of connectivity thresholds (section II.1.5) and the presence of melt channels (section II.4.2) and is implemented as instantaneous extraction after every timestep, it is possible that the porosities reached by every melting increment in the iteration before the extraction step strongly exceed the threshold chosen on the basis of observations and result in segregation velocities which are incompatible with the maximum porosity implied by the threshold, and are, in fact, arbitrary. If the threshold is low and the timestep is large enough, this was observed to happen quite easily, especially if the semi-Lagrangian algorithm is applied, because it allows for relatively large timesteps; the upwind scheme is not immune against this problem, but in the numerical experiments conducted in this study, the self-control of the model through CFL criteria for convection and melt movement has apparently prevented the porosity from exceeding the threshold so far that large errors have been introduced. To circumvent the problem, it was decided to use $\min(\varphi(x, y, z), \varphi_{\mathrm{c}})$ instead of $\varphi(x, y, z)$ in the computation of $\vec{v}_{\mathrm{seg}}$, in particular in the permeability–porosity relation.

III.2.2 Determination of anisotropic permeability

In section II.4.3, a method for the computation of the anisotropic permeability applicable to melt flow in oriented channels or dikes was outlined and applied to a two-dimensional corner flow model. Here, the implementation of the computation is shortly described. Note that the computation can be done *a posteriori* using the stored $\vec{v}_{\mathrm{s}}$ dataset of a model run.

The permeability tensor can be calculated straightforwardly from the definition of the stress according to eq. II.52 and the rotation eq. II.55. As the viscosity is assumed to be isotropic, one can actually use the strain rate tensor instead of the stress tensor for the computation of the rotation matrix, because they differ only by a constant scalar factor; otherwise, the different components of the viscosity tensor would have to be stored as well and used in the computation. Different model permeabilities for $\mathbf{k}_{\varphi}^{0}$ in the rotated local coordinate system can be implemented, *e.g.*

$$\mathbf{k}_{\varphi}^{0} = \begin{pmatrix} 0 & 0 & 0 \\ 0 & k_{\varphi}^{0} & 0 \\ 0 & 0 & k_{\varphi}^{0} \end{pmatrix}, \tag{III.15}$$

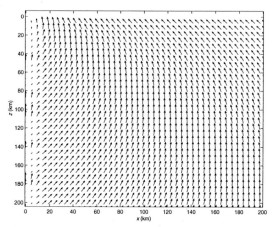

Figure III.10: Numerically determined melt segregation velocity vectors for the right half of a corner flow ridge model; the ridge is at $x = 0$. This image is the numerical counterpart to the analytic solution in figure II.10.

for an array of vertical tabular channels extending infinitely far in y direction; "infinitely far" of course means "as long as the grid point spacing in y direction", which would nonetheless be much larger than the width of the channel. The strain rates $\dot{\varepsilon}_{ij}$ at each relevant grid point are computed with central finite differences (eq. III.8a), and the eigenvalues of $\dot{\varepsilon}$ are then computed using the subroutine dspev from the LAPACK library (Anderson *et al.*, 1999). To save time, the velocity field from the part of the coarse convection grid coinciding with the melt grid is used, and it is only the components of the permeability tensor determined by rotating $\mathbf{k}_\varphi^0$ with eq. II.55 that are interpolated onto the dense melt grid; efficiency can be improved by restricting the computation to those grid points where melt is actually present, unless the formation of dikes in unmolten rock, *e.g.* in the lithosphere, is to be included. Furthermore, if the melt velocity in the channels is calculated independently on the basis of purely buoyancy-driven flow as in the two-dimensional example of eq. II.63, then only the z component of the driving pressure gradient is nonzero, and only the $k_{\varphi_{iz}}$ components of the rotated tensor are needed; if the permeabilities are used in a fully-fledged iterative algorithm for solving the equations of mass and momentum conservation like the one described in section III.2.1 for the isotropic-flow case, they are needed many times and have to be stored in a grid-sized array, so that it is desirable to take the opportunity to reduce memory requirements for the tensor by a factor 2. It should be noted the computation is done independently at each grid point; if this part of the algorithm turns out to be a performance bottleneck, it would be a good target for parallelization.

As a test example, the algorithm has been applied to the flow field of a corner flow as in figure II.10, using the model permeability eq. III.15 and assuming that the dike plane is normal to the minimum compressive stress. An x–z slice of the right half of the resulting direction field of the melt velocity is shown in figure III.10; the numerical solution does only deliver a negligibly small spurious y component of the velocity. The numerical solution agrees well with the pattern from the analytic solution in figure II.10, although the accuracy seems to deteriorate a bit near the spreading center ($x = 0$) in some instances: the vectors should approach an orientation of 45° for $x \to 0$, but on the other hand *e.g.* eq. II.59 indicates that the stresses disappear there, so that no channels would form.

III.3 FORMATION AND TRANSPORT OF THE CRUST

The most obvious surface expression of the interaction and combination of large-scale mantle convection and melt generation and segregation as described in the former sections is the formation of a crust from the extracted melt and its feedback on the processes in the mantle. The transport of the crust at the surface of the model, which was discussed in section II.5.2, is also included in the model.

The variable to be modelled is the crustal thickness h, whose temporal and spatial evolution is described by eq. II.66, *i.e.* the diffusional transport mode of eq. II.65 is not included due to lack of tight constraints; this will result in an overestimate of the thickness of older parts of crust anomalies. It is represented on a two-dimensional grid of the same discretization as the melt model, but of the same spatial extent as the convection model; this enables a better resolution of the resulting crustal topography with respect to the distribution of the melt sources and the ridge region than a coarse grid, where the values for several melt model points would be averaged and lead to smaller observed values of h. For solving eq. II.66, the chain rule is applied to the term $\nabla \cdot (h\vec{v}_r)$. The term $h\nabla \cdot \vec{v}_r$ is solved with central differences and the term $\vec{v}_r \cdot \nabla h$ with the semi-Lagrangian scheme described in section III.1.3; it is worth recalling that no velocity extrapolation in time is necessary here, because the velocity of the plate is prescribed as a kinematic boundary condition and thus known beforehand, so that the advection problem is solved with a higher accuracy here than in the other instances described before. Time stepping is done with the general time step for the whole model.

III.4 RHEOLOGICAL, COMPOSITIONAL AND THERMODYNAMICAL ROCK PROPERTIES

III.4.1 Rheology

Rock rheology — To be able to reproduce the dynamics of mantle convection in a numerical model, it is desirable to include viscosity variations of several orders of magnitude as observed in, or inferred for, the mantle, which are due to a variety of mechanisms, as discussed to some extent in the review sections I.2.4 and II.2.1. Ideally, the most general viscosity law as represented by eq. II.10 would be implemented and applied to numerical convection models, allowing for different rheologies and yielding a large range of viscosities, but unfortunately this is precluded by the difficulty of an accurate numerical treatment and, as mentioned in section III.1, the method used in this study in particular will not handle viscosity contrasts of much more than three orders of magnitude properly. Furthermore, the numerical treatment of the Stokes equation with a stress-dependent rheology as applicable to dislocation creep is much more complicated than the Newtonian case, because the Stokes equation then becomes non-linear. The numerical method applied in this study is therefore restricted to linear rheologies, but allows for inclusion of pressure (resp. depth) and temperature dependence and the effects of melt and water; it should not be concealed that even this feature, which leads to a three-dimensional viscosity pattern, increases the complexity of the algorithm considerably and is also more expensive in terms of computation time, because an iterative solution scheme is applied. The limitation to linear rheologies is regrettable, because on the basis of laboratory experiments dislocation creep is considered the dominating creep mode in the melting region and the shallow, hot mantle, which is in the focus of this study, but it is hoped that the existing flexibility already gives a useful approximation to the true behaviour of the material; on the other hand, the applicability of laboratory results on rock rheology to the real earth is also an open issue, because they have to be extrapolated over several orders of magnitude. Furthermore, possible rheological heterogeneity due to changes in the modal composition of the rock is not considered either, because the uncertainty in the mixing rules seems to be still too large at present (see I.2.4), apart from the fact that the compositional changes themselves are also difficult to constrain. If they were known, it would though be straightforward to include them into the rheology model.

In principle, eq. II.10 could be implemented with $n = 1$, but it would be necessary to impose maximum and minimum limits on the viscosity to ensure accuracy. For this reason, and to have a better, more direct control on the viscosity law, an alternative, simplified rheological law of the form

$$\eta(z, T, \varphi, C_{H_2O,s}) = \eta_0 \left(\frac{C_{H_2O,s}(0)}{C_{H_2O,s}(f)} \right)^{b_C} e^{b_z z - b_T T - s\varphi} \tag{III.16}$$

(Schubert *et al.*, 2001; Hirth and Kohlstedt, 1996; Hall and Parmentier, 2000; Mei *et al.*, 2002) is used; note that z and T are non-dimensional here. As the algorithm is restricted to Newtonian rheology resp. diffusion creep ($n = 1$), it is appropriate to set $s = 26 \ldots 28$ (see section II.2.1). Clearly, the viscosity variations are much smaller with this law, but with a reasonable choice for the reference viscosity η_0 and the parameters $b_{z,T}$ it should still be possible to achieve a viscosity resembling that of the mantle under most of the p–T conditions of the model; usually, the cold lithosphere tends to be too soft with this law, but it is sufficient to meet the most important requirement, namely to make the lid mechanically strong enough to respond to the kinematic top boundary condition approximately uniformly. Figure III.11 shows an $\eta(p, T, \varphi, 0)$ distribution for typical parameters. With regard to the study of the influence of the individual factors as in section V.1.4, it is convenient that they are formally independent from each other in expressions like eq. III.16 or II.10; nonetheless, there is of course *e.g.* a dependence of φ and $C_{H_2O,s}$ (via f) on the p–T conditions at supersolidus temperatures.

The bulk viscosity of the partially molten mantle can be set to a constant value or it is calculated from the shear viscosity of the solid phase and the porosity by the semi-analytical law

$$\eta_b = b_{b1} \eta_s \frac{b_{b2} - \varphi}{\varphi} \tag{III.17}$$

with the empirical parameters b_{b1} and b_{b2}, which are 1.16 and $0.75\pi a_{inc}$, respectively, for a material with tubules and ellipsoids with an aspect ratio a_{inc} of 0.1 (Schmeling, 2000). Obviously, η_b will be larger than η by at least one order of magnitude at the low porosities normally inferred for most of the melting region in the mantle.

Melt viscosity — As discussed in section II.1.4, melt viscosity also depends on p, T, and composition, and should therefore also be treated as a variable. On the other hand, data on natural mantle melts are still rather scarce, and within the framework of the present study, it is not yet possible to include the complicated details of melt chemistry. Hence, a constant viscosity is assumed in many models, and where an attempt is made to include at least p–T dependence, eq. II.7 (Bagdassarov, 1988) is used with a Saucier–Carron parameter $\mathcal{P}_{SC}$ fixed at 150. This results in a spatially variable Rt in eq. II.38a.

III.4.2 The effect of water

As mentioned in previous sections, water has a particularly strong effect on the properties of the mantle in that it softens the rock and lowers the solidus temperature. Therefore, an attempt has been made to account for these effects at least in a simplified manner. The fundamental assumption made here is that the mantle is nowhere oversaturated with water,

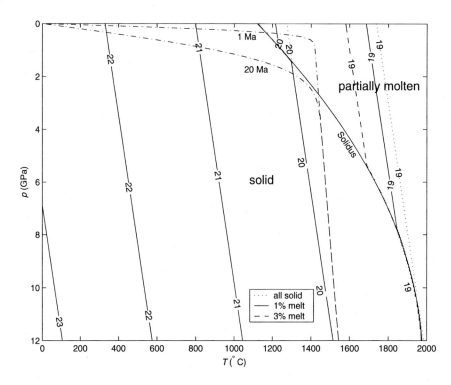

Figure III.11: $\log \eta(p, T)$ with $\varphi = 0$, 0.01, and 0.03, respectively, for the stability field of α-olivine in an idealized, anhydrous mantle, according to eq. III.16. The solidus is again the formula by Hirschmann (2000, see eq. II.25c). The partially molten mantle is assumed to contain a constant retained melt fraction; the dotted contours are for the case of no melting. The parameters used in eq. III.16 are $\eta_0 = 5 \cdot 10^{22}$ Pa s, $b_z = \ln 10 = 2.30258$, $b_T = \ln 1000 = 6.90776$ and $s = 26$. Two idealized geotherms with a potential temperature of 1410 °C are also plotted. For the pressure-to-depth conversion for eq. III.16, $g = 9.9\,\text{m/s}^2$ and $\varrho_0 = 3500\,\text{kg/m}^3$ have been used, and z has been normalized with an upper-mantle thickness of 660 km.

but that water is structurally bound in the crystals; hence, water is not present as a free phase which would move by two-phase flow similar to melt. This should be acceptable for sub-MOR and plume mantle, where water contents are likely to be low (see section II.3.2).

In the simplest case, a homogeneous initial distribution of water is assumed, *i.e.* a certain concentration $C_{\text{H}_2\text{O,s0}}$ is prescribed and can only change by melting processes; the hydrous melting parameterization by Katz *et al.* (2003) in its modified form (see section II.3.3) is used if water is included in the models. Given the reference concentration, the parameterization can be used directly to determine the solidus of undepleted mantle, which is the precondition to calculating the melting degree; if the mantle element is already at a certain $f > 0$, its solidus has already raised and must also be determined to decide whether further melting or recrystallization occurs. To do this, the current local water concentration in the melt is determined with the formulae of Shaw (1970) from the current local melting degree; as non-modal, near-fractional melting is assumed, the corresponding version of eq. II.24b is applied with a set of modal and melting proportions and taking into account the retained melt fraction as described in section II.3.2 on p. 41. Altogether, the solution involves eqs. II.24b, II.31b, II.32, II.29, and II.29a. In order to reduce the complexity a bit, f is recalculated with the anhydrous version of the parameterization if a melting degree beyond cpx-out was computed; it is safe to assume that at this stage, no appreciable amount of water is present in the solid anymore. Eq. II.24b can always be used to determine the concentration of water in the melt or the matrix from f; as f is given on both numerical grids, it can straightforwardly be used to determine η, so that it is not necessary to allocate additional memory to store the spatial distribution of $C_{H_2O,s}$.

If different sources with individual water contents are introduced as a chemical initial condition, the above procedure has to be extended by keeping track of the different sources. This can be achieved by chemical marker fields, whose temporal evolution is computed by solving eq. I.4 without the diffusion term, but possibly with a source; note that this field does not change by melting. This marker field, which is also interpolated onto the melt grid, allows for the application of different hydrous solidi according to the presence or absence of the marker. In case of mixing of mantle of different source types, a weighted average of the corresponding water concentrations is used for determining the solidus.

It should be noted here that the water is not explicitly tracked together with the melt either; as soon as it has been removed from the solid phase, it is assumed to be coupled with the melt and to have no effect on, nor interaction with, the solid. In particular, the "chromatographic effect" of melt percolation is not included here, because the water is defined to travel with the velocity of the melt instead of the effective velocity given by eq. II.43. In the following, a short assessment of the consequences of this simplification is made. Assuming purely buoyancy-driven Darcy flow, *i.e.* eq. II.36 with $\nabla p = 0$, and applying the permeability law eq. II.8, one can rewrite eq. II.43 as a function of solid velocity and porosity only:

$$v_{f_z} = v_{s_z} - \frac{\Delta \varrho_f a^2 g}{\eta_f b} (1 - \varphi) \varphi^{n-1} \qquad \text{(III.18a)}$$

$$\Rightarrow v_{C,\text{eff}_z} = v_{s_z} - \frac{\Delta \varrho_f a^2 g}{\eta_f b} \frac{\varrho_f (1 - \varphi) \varphi^n}{\varrho_f \varphi + \varrho_s (1 - \varphi) D}; \qquad \text{(III.18b)}$$

v_s is assumed to be independent of φ, because the contribution of retained melt to buoyancy of the bulk mantle is of only minor importance if φ is not larger than a few percent. While it has been deduced from eq. II.43 that a component with $D \gg \varphi$ travels hardly faster than the solid and one with $D \ll \varphi$ about as fast as the melt (*e.g.* Spiegelman and Elliott, 1993), it must be kept in mind that, at least for equilibrium porous flow, $\vec{v}_f$ is a function of φ. Hence, even moderately incompatible components such as water ($D^{H_2O} = 8.97 \cdot 10^{-3}$) would not be transported substantially faster than the solid if φ is very low, as can be seen from eq. III.18b and figure III.12, because the melt itself is not much faster. The figure shows that therefore, in a stationary reference frame, there is a finite porosity at which the ratio of the effective velocity and the melt velocity reaches a minimum. The lower the matrix velocity is, the smaller are the porosity corresponding to the minimum and the deviation of $v_{C,\text{eff}}$; the plot for $1\,\text{cm}/\text{a}$ in figure III.12 is representative for the mantle in *ca.* 80 km depth below a slow-spreading ridge, the plot for $5\,\text{cm}/\text{a}$ maybe for a plume ascending beneath it at the same depth. In terms of segregation, *i.e.* of separation from the matrix, the minimum is at $\varphi = 0$, of course, and water will move considerably slower relative to the matrix than the melt at the low porosities of around 1 % used in most models of this study (see chapter V). However, as the main purpose of considering water in some of the models of this study is the effect on the viscosity, and as an additional, significant fast disequilibrium transport mode is also

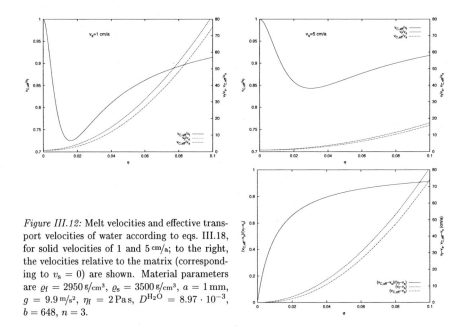

Figure III.12: Melt velocities and effective transport velocities of water according to eqs. III.18, for solid velocities of 1 and 5 cm/a; to the right, the velocities relative to the matrix (corresponding to $v_s = 0$) are shown. Material parameters are $\varrho_f = 2950\,\mathrm{g/cm^3}$, $\varrho_s = 3500\,\mathrm{g/cm^3}$, $a = 1\,\mathrm{mm}$, $g = 9.9\,\mathrm{m/s^2}$, $\eta_f = 2\,\mathrm{Pa\,s}$, $D^{\mathrm{H_2O}} = 8.97 \cdot 10^{-3}$, $b = 648$, $n = 3$.

included in these models, the current treatment should be sufficient to capture the relevant effects and has in fact been used by others in a similar way (*e.g.* Ito *et al.*, 1999; Braun *et al.*, 2000); it is though probably not appropriate for modelling chemical characteristics of the melts.

III.4.3   Solid-state phase transitions

The proper implementation of phase transitions poses some difficulties because they depend on several thermochemical variables and would in principle require to account for reaction kinematics; the probably best-known dependence is that on p–T conditions, which is the one taken into account here. Furthermore, the width of the transitional two-phase zone is in some cases of the order of grid point spacing, which can lead to numerical inaccuracies due to the sharp step in variables; for instance, the 410 km discontinuity is maybe as thin as about 8 km (Wood, 1995), although most authors seem to favour a thickness between 11 and 20 km (Katsura and Ito, 1989; Stixrude, 1997; Gaherty *et al.*, 1999), and the transition of β- to γ-olivine at about 520 km is estimated to be about 12 km wide (Weidner and Wang, 2000). Therefore, it is attempted to represent the transition between two phases as an interval at least two or three grid point spacings wide, following the implementation by Marquart *et al.* (2000).

The complex pattern of olivine phase changes in the transition zone is represented in a simplified way as a linear shift in p–T space directly from the α to the γ phase extending over a pressure interval of more than 3 GPa resp. a depth interval of about 100 km, but the two transitions of the aluminous phases are represented individually. For the olivine transitions and for the plagioclase–spinel transition, constant Clapeyron slopes have been assumed, *i.e.* the stability fields are separated by straight boundaries in p–T space. By contrast, the garnet–spinel transition, which changes its Clapeyron slope significantly near the solidus (O'Hara *et al.*, 1971; Hall, 1996) is parameterized by a more complicated function (see appendix D.1.6). Altogether, all phase transitions have to be prescribed by two parallel functions bracketing a certain p interval.

Phase transitions change the density of the rock and hence have a certain impact on buoyancy, which is accounted for by the terms $Rc_{X_k}X_k$ in the Stokes equation, eq. I.9; they also cause consumption or release of latent heat (depending on the sign of the Clapeyron slope), which is expressed in the energy conservation equation, eq. I.10, by the terms $\Delta S_{X_k}\left(\partial/\partial t + \vec{v}_{\mathrm{s}}\cdot\nabla\right)X_k$. The phase information is represented as a special compositional field whose integer part serves as a flag which indicates the mineral assemblage stable under the local p–T conditions[2] and whose fractional part corresponds to the X_k, *i.e.* to the fraction of a phase transformed in the transition zones. This approach has the advantage that it saves memory and requires only the advection of one variable. Difficulties are though expected if several phase transitions occur at the same or various neighbouring points, which would probably be the case if the transitions of pyroxenes, garnet, and majorite (see section I.2.2) are also included; under the assumptions made in this study, however, these critical circumstances never arise. A point of concern with the transitions of aluminous phases in the uppermost mantle is their consumption and possible exhaustion by melting, as remarked in section II.3.2, which is expected especially in plumes. While this effect is neglected in the Stokes equation assuming that all compositional changes are already covered by the depletional buoyancy term, an attempt has been made to account for it in the temperature equation at least for the gt-sp transition, assuming that the consumption of garnet would reduce or suppress the temperature decrease usually related to the transition. The amount of garnet still present at a melting degree f is calculated with eq. II.33b; models assessing the importance of this effect are presented in section V.1.6. The plagioclase–spinel transition is not treated in a similar way due to lack of information on appropriate melting proportions.

III.4.4   Rock composition and thermoelastic rock properties

In general, many simplifying assumptions are made in convection models concerning the geochemical and petrological characterization of the rock and its thermodynamic properties; a few such features and attempts to achieve a closer approach to reality have been outlined in section I.2, but the extraordinary complexity of the dependence of all properties on p, T, and composition and the variability of composition itself make a numerical treatment of the convection problem extremely difficult and preclude a truly comprehensive modelling of all these aspects in the near future. However, the composition and the corresponding spatial distribution of parameters such as α, $K_{T,S}$, c_p, κ etc. according to their p–T dependence are fundamental for any attempt to combine convection models with geophysical and geochemical observations and have considerable impact on the style of convection, as shown by several studies (Hansen and Yuen, 1994; Ita and King, 1994; Dubuffet *et al.*, 1999, see sect. I.2.3). Therefore it may be worthwhile to make an attempt to derive these properties from modelled p and T distributions, be it for a simplified composition or even for a quite complex and detailed one. In a further development of the method for inclusion of a p–T-dependent α in the convection models presented by Schmeling *et al.* (2003), such an attempt was made in this study, and the corresponding algorithm is detailed here.

The basic approach for the determination of α, ϱ, K_T and c_p as functions of p, T, and composition at a given grid point is to start with the p–T conditions given by the model,

---

[2]−2: plagioclase lherzolite, −1: spinel lherzolite, 0: garnet lherzolite, 1: peridotite in the ringwoodite stability field, 2: lower mantle (pv+mw, not used in this study)

determine the phase stability field the point lies in, and then use parameterizations of experimental data to determine some properties of pure mineral endmembers. The parameterizations used are isobaric for normal pressure; in a subsequent step, isothermal compression is calculated to reach the p conditions in the mantle. Eventually, simple mixture rules are applied to the endmembers to determine the properties of the composite. In general, the database from Saxena *et al.* (1993) is used here, because it provides a very comprehensive set of experimental mantle mineral data for most relevant minerals.

Thus, the first step is isobaric heating of the pure mineral endmembers at normal pressure (resp. $p = 0$). For $\alpha(T)$ and $K_{T0}(T)$ the parameterizations from Saxena *et al.* (1993) yield the density

$$\varrho(T) = \varrho_0 \exp\left(-\int_{T_0}^{T} \alpha(T')\,\mathrm{d}T'\right) \tag{III.19}$$

and the Anderson–Grüneisen parameter

$$\delta_{T0} = -\frac{1}{\alpha K_{T0}}\left(\frac{\partial K_{T0}}{\partial T}\right)_p \tag{III.20}$$

at zero pressure. The second step is isothermal compression of the endmembers at T. K'_{T0} at T is determined from K'_{T0} at the reference temperature using a constant T derivative. The Vinet equation of state (see sect. II.1.2, eq. II.3c, Vinet *et al.*, 1987) is used to determine $\varrho(p,T)$ from $\varrho(0,T)$; this is a time-consuming step, because a root-finder has to be used to solve for ϱ. As a by-product, this would yield the Anderson–Grüneisen parameter at p:

$$\delta_T = \delta_{T0}\left(\frac{\varrho_0(0,T)}{\varrho(p,T)}\right)^{c_\delta}; \tag{III.21}$$

$c_\delta = 1.4$ is considered appropriate for all materials of interest (Anderson, 1995). It is now also possible to calculate the thermal expansivity at p:

$$\alpha(p,T) = \alpha(0,T)\exp\left\{-\frac{\delta_{T0}}{c_\delta}\left[1 - \left(\frac{\varrho_0(0,T)}{\varrho(p,T)}\right)^{c_\delta}\right]\right\}. \tag{III.22}$$

For the purpose of variable-α modelling (see sections I.2.3 and V.1.3), this result would already suffice. If, however, compressible modelling were done, K_T would have to be determined; if a p–T-dependent c_p had to be included, it could also be computed here; and if an assessment of seismic velocities is desired, K_T and c_p are needed for the calculation of K_S. Therefore, and for the sake of completeness, the necessary extensions are summarized below.

For K_T, a pressure derivative K'_{T0} constant with p and a constant temperature derivative of K'_{T0} are assumed:

$$K_T(p,T) = K_{T0}(T) + \left[K'_{T0} + \frac{\partial K'_{T0}}{\partial T}(T - T_{\text{ref}})\right]p. \tag{III.23}$$

$c_p(T)$ of the endmembers at normal pressure is again determined from the parameterizations of Saxena *et al.* (1993) and extrapolated to high pressures using

$$\left(\frac{\partial c_p}{\partial p}\right)_T = -\alpha^2 VT\left[1 + \frac{1}{\alpha^2}\left(\frac{\partial \alpha}{\partial T}\right)_p\right] \approx -\frac{\alpha^2 T m_{\text{mol}}}{\varrho}(1 + 2\delta_T - K'_T); \tag{III.24}$$

here the approximation is actually used, which is based on the assumptions $c_V = const.$ and $T > \Theta_{\mathrm{D}}$ (Pankov *et al.*, 1998).

So far, only endmember properties have been derived. In a first step toward composite properties, the individual minerals, which are thought of as solid solutions of two or three endmembers have to be considered. With given molar proportions of the endmembers, the density is given by the mole-weighted average of the molar volumes of the endmembers (see app. D.1.2, eq. D.1a; Ita and Stixrude, 1992) if ideal mixing is assumed; for simplicity, c_p is mixed in the same manner (Putnis, 1992). Using the definitions of α and K_T, resp. β_T, one can derive the mixing rules

$$\varrho_{\mathrm{ss}} = \frac{m_{\mathrm{ss}}}{\sum\limits_{i=1}^{n} \frac{X_i m_i}{\varrho_i}} \Rightarrow \begin{cases} \alpha_{\mathrm{ss}} = \frac{\sum X_i V_i \alpha_i}{V_{\mathrm{ss}}}, \\ K_{T\mathrm{ss}} = \frac{V_{\mathrm{ss}}}{\sum \frac{X_i V_i}{K_{T_i}}} \end{cases} \tag{III.25}$$

for a solid solution with n endmembers (see appendix D.1.3 and D.1.4). For weighting, indicators such as Mg# or Cr# (in the case of spinel) are useful choices for the X_i. The easiest method would be to use constant values, *e.g.* Mg#=0.89 for olivine and Cr#=0.091 for spinel (cf. II.34) for unmolten mantle. If melting is included, it would though be more consistent to account for compositional changes of the solid solutions, such as preferred loss of Fe and Al.

After having determined these variables for all minerals present under the prevailing p–T conditions, the rock can be assembled as a composite of mineral grains. With given weight proportions for the n mineral modes, the density is given by another mixing rule, from which α and K_T again can be derived using the respective definitions:

$$\varrho_{\mathrm{mc}} = \frac{1}{\sum\limits_{i=1}^{n} \frac{X_i}{\varrho_i}} \Rightarrow \begin{cases} \alpha_{\mathrm{mc}} = \varrho_{\mathrm{mc}} \sum \frac{X_i}{\varrho_i} \alpha_i, \\ K_{T\mathrm{mc}} = \frac{1}{\varrho_{\mathrm{mc}} \sum \frac{X_i}{\varrho_i K_{T_i}}}, \end{cases} \tag{III.26}$$

which actually correspond to the Reuss limit valid for complete intergranular relaxation; this should be an acceptable assumption for mantle convection (also see Stacey, 1998). c_p is calculated as the weight average of the mineral heat capacities. The modal weight proportions can be taken from petrological assessments of the modal composition of the mantle (see appendix D.1.1, table D.1); for melting mantle, these modal proportions can be changed with parameterizations of the modal composition as a function of f such as the one by Kostopoulos and James (1992) (see section II.3.3) or using other experimental data on melting proportions.

For the determination of seismic velocities, the Reuss limits of α and K_T are not sufficient to achieve a useful estimate. The Voigt limits also have to be determined (see eqs. D.5c and D.9c). The relations

$$c_V = c_p - \frac{K_T T \alpha^2}{\varrho}, \qquad K_S = K_T \frac{c_p}{c_V} \tag{III.27}$$

then yield the corresponding adiabatic bulk moduli, and their arithmetic mean is the Voigt–Reuss–Hill average, which is a common estimate for the bulk modulus of the composite. Using this information, the seismic parameter and, with an assumption for the Poisson ratio, the seismic velocities can be estimated:

$$\Phi_{\mathrm{seis}} = \frac{K_S}{\varrho}, \qquad v_{\mathrm{P}} = \sqrt{\frac{K_S + \frac{4}{3}\mu}{\varrho}}, \qquad v_{\mathrm{S}} = \sqrt{\frac{\mu}{\varrho}}, \qquad \mu = \frac{3K_S(1 - 2\nu)}{2(1 + \nu)}; \tag{III.28}$$

obviously, this last step is error-prone due to the uncertainty on ν. A further potential by-product of these results is the Grüneisen parameter

$$\gamma = \frac{\alpha K_T}{c_V \varrho}. \tag{III.29}$$

For partially molten mantle, these results would apply to the solid matrix, but an analogous procedure can be followed to determine the properties of the melt. The bulk properties of the partially molten rock would then be derived from a combination of the solid and the liquid properties according to porous-media models (Gassmann, 1951; Schmeling, 1985a).

Although the prospect of a more comprehensive accounting for compositional aspects is attractive, computational performance remains an issue to be considered. In this context, it must be emphasized that the calculation even of the smaller, essential set of variables for $\alpha(p, T)$ models needs a significant amount of time, the more the greater the compositional complexity is; *e.g.*, for the 18-endmember system in the examples of this method in appendix E, figures E.1 and E.2, the root-finder for the compression step in the calculation takes about 40 % of the CPU time of the whole subroutine call. The most straightforward measure to reduce the problem of course is to compute the bulk expansivity values only once per timestep and to store them in a dedicated array; however, if more variables are treated as p-T-dependent and have to be stored, this can lead to considerably larger memory requirements, which will enforce tighter limits for the size of models in the three-dimensional case. Another possibility is the replacement of the Vinet equation of state by one which can be solved in the form $\varrho(p)$ without a root-finder; here linear forms like eq. II.4 or approximations *e.g.* to the Birch–Murnaghan equation of state like in Schmeling *et al.* (2003) can be used, but their validity in the pressure range of the model has to be ensured carefully. – It is worth noting that the calculation of the thermoelastic properties can in principle be carried out for each grid point independently, which implies a potential for efficient, scalable parallelization.

Finally, the implementation of the spatially varying α in the solution of the Stokes equation remains to be solved; as detailed in section III.1, the convection problem is treated with a method which uses a spectral representation of the variables in the horizontal direction, and this also affects the thermal buoyancy term in eq. I.9 resp. III.2c, where Ra now becomes a function of p and T, or of position. Instead of the usual way with constant Ra of transforming T into the spectral domain and then multiplying the spectrum with Ra, now the whole term $Ra\,T$ is Fourier-transformed.

CHAPTER IV

ICELAND: THE CURRENT STATE OF RESEARCH

"It was like a huge wall!," said a
blind man.
"Oh, no! It was like a big tree," said
another blind man.
"You are both wrong! It was like a
large fan!," said another.
After listening to the blind people,
the Lord said, "Alas! None of you
have seen the elephant!"

—EAST-INDIAN FOLKLORE

The ridge-centered plume *par excellence* is presently the Iceland plume, and it has therefore received much attention over the past decades. As much of this study focusses on ridge-centered plumes and several numerical models presented in chapter V attempt to reproduce observations of the Icelandic mantle (section V.2), this chapter reviews and comments on the results of previous work and tries to sketch a comprehensive, though preliminary, picture of the plume. As this study is mostly concerned with the current structure of, and processes in, the mantle and treats the crust as a second-order structure with only minor direct effect on the mantle, this chapter concentrates on results dealing with the mantle and with crust–mantle interactions and interdependence.

IV.1 THE CRUST

In Iceland and the North Atlantic, there are essentially three different types of oceanic crust: normal oceanic crust of average thickness (6–7 km) with transform fractures, generated at basically undisturbed MOR; anomalous oceanic crust of increased thickness without fractures and a possibly non-orthogonal spreading direction relative to the MOR; and crust generated by the plume (White, 1997). The second and, even more, the third type has several unusual features which have been investigated during the last decades, but are partly not yet understood. One of the central controversial topics is the question how thick the Icelandic crust produced by the plume is and which temperature it has, or, putting it in another way, whether a layer between *ca.* 10–15 km and *ca.* 28–30 km with $v_P = 7 \ldots 7.4 \, \mathrm{km/s}$ is part of the crust or the top of a mantle with anomalously slow seismic velocities.

It is far beyond the scope of the following review to summarize and comment on all aspects of recent research on the Icelandic crust; most results on phenomena or structures restricted to the crust alone without broader relation to the mantle are omitted.

IV.1.1 Geotectonic evolution and present morphology

In the north Atlantic, the plume and the oceanic spreading center have been interacting more or less since continental break-up in the late Paleocene and early Eocene 58 Ma ago, although it is commonly agreed that the plume has not always been located directly beneath the ridge, but hit the lithosphere more to the west, probably below south central Greenland.

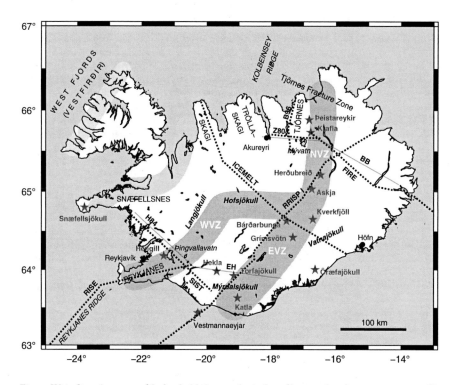

Figure IV.1: Overview map of Iceland. Major geophysical profiles are also shown: RRISP77 profile I (Gebrande *et al.*, 1980), BB (Beblo and Björnsson, 1980), Z80 (Zverev *et al.*, 1980b), H84 (Hersir *et al.*, 1984), EH (Eysteinsson and Hermance, 1985), SIST (Bjarnason *et al.*, 1993), FIRE (Staples *et al.*, 1997), ICEMELT refraction (Darbyshire *et al.*, 1998), B96 (Menke *et al.*, 1998), RISE profiles A and B (Weir *et al.*, 2001). Gray-shaded strips: neovolcanic zones (extinct zones in lighter gray); ⋆: major volcano; •: city; slanted bold text: glaciated area; dotted lines: seismic profiles; gray lines: MT profiles.

It is not yet clear whether the plume arrived under eastern or central Greenland (*ca.* 62–64 Ma ago) before rifting in the east started, but there seems to be consensus that rifting preceded plume-related volcanism in western Greenland (Skogseid *et al.*, 1992; Larsen *et al.*, 1992; Chalmers *et al.*, 1995; White, 1997; Scarrow *et al.*, 2000) and that the subsequent massive volcanism and final break-up in the east were due to the melting of the plume, which preferredly took place at "thinspots" of the lithosphere. The exact position of the plume during late Cretaceous/early Tertiary time is still controversial, though: while Lawver and Müller (1994) and Torsvik *et al.* (2001) located it beneath the southeastern or central part of the Greenland craton during initial break-up and beneath the Greenland–Iceland Ridge at 37 Ma, White and McKenzie (1989) positioned it beneath the Scoresby Sound area and at the spreading center, respectively; as Clift *et al.* (1995) point out, the Lawver and Müller (1994) model, which they favour, can explain better the anomalously slow subsidence shortly after rifting observed in several ODP borehole cores, while the White and McKenzie (1989) model accounts more readily for the thickened crust of the Iceland–Faeroe ridge.

Torsvik *et al.* (2001) suggest that the plume coincided with the spreading center about 20 Ma ago and induced a change in plate motions, which made Eurasia drift to the northeast while North America continued moving to the northwest. An initial position under central Greenland is also in better agreement with the appearance of picrites and flood basalts in western Greenland at about the same time, which was probably caused by splitting resp. redirecting of the plume upwelling by the lithospheric keel of the Greenlandic craton (*e.g.* Larsen *et al.*, 1992; Riisager and Abrahamsen, 1999; Nielsen *et al.*, 2002). The eruption rate during the generation of the flood basalts in the North Atlantic Igneous Province has been estimated as more than $1^{km^3}/a$, compared to 0.023–$0.038^{km^3}/a$ for Iceland today (Richards *et al.*, 1989); the total volume of the volcanics amounts to $6.6 \cdot 10^6$ km^3 distributed over an area of $1.3 \cdot 10^6$ km^2 (Condie, 2001). The relative motion of the plume towards the ridge is reflected by a shift of CFB generation from western Greenland to the Scoresby Sound region in the east after *ca.* 6 Ma (*e.g.* Larsen *et al.*, 1992; Gill *et al.*, 1992) and an asymmetry of subsidence with regard to the spreading center (Clift *et al.*, 1995).

The style of interaction, which is controlled by this relative movement, caused the tectonic evolution of Iceland to be extremely variable: on the island itself, whose formation began more than 20 Ma ago, there is evidence for still ongoing eastward jumps of the ridge from now extinct rifts in northwestern Iceland, where activity ceased 15 Ma ago, and on Snæfellsnes to the still active Western Volcanic Zone (WVZ) 7–9 Ma ago (Hardarson *et al.*, 1997), and from the WVZ to the present Northern and Eastern Volcanic Zones (NVZ, EVZ), accommodated by the Tjörnes Fracture Zone (TFZ) in the north and the "bookshelf"-type faults of the South Iceland Transfer Zone in the south (figure IV.1); the EVZ began to form *ca.* 3 Ma ago. Similar shifts have happened 3–7 Ma years ago from a rift on the Skagi peninsula to the present NVZ and from the ~20 Ma-old western fjords to Skagi earlier (Sæmundsson, 1979). Hardarson *et al.* (1997) estimate one rift zone cycle to last about 12 Ma, including activity overlap during relocations.

On a topographic/bathymetric map of Iceland and the North Atlantic (figure IV.2), the different types of oceanic crust are easily visible: normal oceanic crust is present south of the Reykjanes Ridge, beyond *ca.* 58° N, and in principle also to the north of Iceland at the Kolbeinsey Ridge; the Reykjanes Ridge itself features anomalous oceanic crust characterized by V-shaped structures, increased thickness, a weak relief, the almost complete absence of transform faults, and the substitution of the usual median valley by an axial high (Applegate and Shor, 1994; Weir *et al.*, 2001), explained by anomalously high crustal temperatures, which precludes the common brittle rheology normally resulting in the formation of the central graben (White, 1997); Iceland itself forms the top of a roughly circular rise with a maximum topography of *ca.* 2.8 km above the surrounding seafloor lying in the center of a positive geoid anomaly with an amplitude of 60 m. The topographic high extends into the NW–SE-striking Greenland–Iceland–Faeroe Ridge created by the excessive melt produced by feeding of hot material from the plume into the spreading center when the plume itself was still off the MOR (Vink, 1984). Analysis of seismic, topographic, and bathymetric data suggests that the plume-generated crust of Iceland and the oceanic aseismic ridge has a much greater thickness than normal oceanic crust due to the elevated mantle temperature (White, 1997, also see sect. IV.1.2); for instance, seismic profiles shot between the Faeroe Islands and Iceland yield a thickness of 25–33 km for the aseismic ridge (Richardson *et al.*, 1998; Smallwood *et al.*, 1999). Literature values of the geoid–topography ratio of the Iceland area are around 1.5 m/km, which is a rather low value compared with other thermal swells, probably due to a strong contribution of the Airy-compensated shelf (Sandwell and MacKenzie, 1989; Marquart, 1991), but a more detailed investigation of the regional geoid taking into account the effects of several plateaus and of the thermal structure of the lithosphere re-

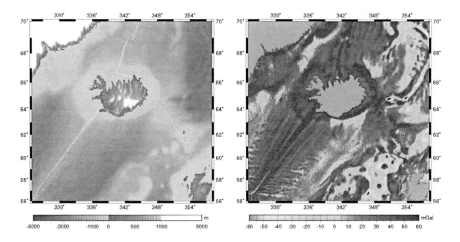

Figure IV.2: Topography/bathymetry from ETOPO5 (left) and free-air gravity from Sandwell and Smith (1997) (right) of Iceland and its surroundings.

vealed a geoid high centered beneath northwestern Iceland with $\vartheta/\iota \approx 7\,\mathrm{m/km}$, which demands a compensation depth below the base of the lithosphere and is hence most probably due to dynamic support by the Iceland plume (Marquart, 1991). Analysis of the admittance, *i.e.* the spectral ratio of geoid and topography, also showed that Iceland and the western Faeroe Plateau are supported by a deep, sublithospheric anomaly (Heller and Marquart, 2002).

It is a striking feature of Iceland that the formation of anomalous crust is much more pronounced to its south than to its north: the Kolbeinsey ridge has an unusually shallow bathymetry as well, but the crust basically resembles normal oceanic crust more both structurally as geochemically than the more southerly regions (see section IV.2.2); the boundary coincides with the TFZ in northern Iceland, an analogue of the common oceanic transform fault, but broader than most of them. Due to the influence of the plume, the crust of the Reykjanes Ridge has a significantly greater thickness than normal oceanic crust (8.4–11 km), because mantle temperatures are some tens of degrees above the normal value, and decays to the normal value over a distance of *ca.* 1000 km from the plume center (Smallwood *et al.*, 1995; Weir *et al.*, 2001). This has led to the very particular feature of southward-pointing V-shaped ridges opening towards Iceland, which are visible in bathymetric and gravity, but not in magnetic, maps, and are often explained by the action of low-viscosity pulses from the plume with a period of 5–10 Ma, flowing along the ridge axis (*e.g.* Vogt, 1971, 1976, see section IV.2.3) and producing excess crust; White *et al.* (1995) and White (1997) estimate that the temperature variations corresponding to the observed excess thickness of about 2 km are 30–50 K. Jones *et al.* (2002) determined an along-ridge propagation velocity of the pulses of 20–25 cm/a for distances $\lesssim 700$ km and of 10–15 cm/a beyond from bathymetric, short-wavelength gravity, and magnetic data; additionally, they could also extract a weaker pulsing with a period of only 2–3 Ma from their data and also reported northward-pointing V-shaped structures from the area between the Kolbeinsey Ridge and Jan Mayen, although these are smaller and much more blurred. – This view of course implies that the pulses are controlled by the plume and caused by some variability in its flux or temperature; as an alternative, though, it is possible that the relative movement of a constant plume and the overriding lithospheric plate, which has a particular thermal structure and underwent several rift relocations, is responsible for the variability in crust formation expressed by the V-shaped ridges (Hardarson *et al.*, 1997). These authors note that the age of the troughs between the three most pronounced V-ridges (15 and 5 Ma, respectively) coincides approximately with the time of rift relocations. It is still a matter of debate whether rift relocations and pulses are initiated by the plume or whether the varying productivity is just a response of the melting mantle to the changing lithospheric conditions.

IV.1.2   Seismics and gravimetry

Several seismic profiles have been shot in different parts of Iceland since the early 1960s in order to achieve a detailed picture of the crustal structure and the thickness of the hotspot crust. The data have been interpreted in different ways, though, thus partly giving rise to the controversy whether the Icelandic crust is thin or thick. This concerns especially the lower crust – in the tradition of an early investigation by Pálmason (1971) often termed "layer 3" –, whereas there seems to be a consensus that the thickness of the upper crust (layers 0–2), which is characterized by strong $v_{P,S}$ gradients due to the successive closure of fractures, is in general 3–6 km, with larger thicknesses usually corresponding to higher age (Flóvenz and Gunnarsson, 1991).

Considering older studies from the 1970s, the controversy can be highlighted to some extent by comparing the results of the RRISP77 experiment (Angenheister *et al.*, 1980; Gebrande *et al.*, 1980) and those of Zverev *et al.* (1980a,b) [see figure IV.1, lines RRISP77 I and Z80; the profile of Zverev *et al.* (1980a) coincides roughly with the central third of the SIST line]. While the latter find evidence for crustal structures down to at least *ca.* 20 km, the former interpret a relatively shallow reflector around 10 km depth as the Moho and the layer beneath it reaching down to 30 km as mantle with anomalously low velocity (see table IV.1), which they ascribe to the presence of partial melt; although Gebrande *et al.* (1980) do not ignore the possibility of other explanations such as gabbroic material, they prefer the thin-crust model in view of the magnetotelluric measurements, which image a good conductor at that depth (Beblo and Björnsson, 1980, see sect. IV.1.4). It should be noted, though, that Gebrande *et al.* (1980) were mostly concerned with the profile I of the RRISP77 experiment, a large part of which was located along the neovolcanic zone and thus does not sample very much of the older Icelandic crust. Based on these early seismic and electromagnetic results as well as gravity data, Schmeling (1985b) attempted an assessment of melt content and geometry for the depth interval from 10 to 50 km beneath central Iceland, which he considered being part of the mantle, and concluded from application of theoretical models of the effect of melt on seismic velocities and electrical conductivities that in the uppermost part ($z < 20$ km) porosities are as high as 10–20 %, while φ is rather less than 2 % at greater depths. However, his whole model set is biased by the assumption that the mantle does indeed reach up to 10 or 15 km, the alternative of a thick, gabbroic crust being left unconsidered in spite of its possible ability to account for the density difference; thus, it remains open whether this alternative could even remove the need of accounting for melt. The fact that the porosity is found to be low in the assumed mantle could as well be regarded as evidence that this region does not consist of mantle, but of crustal material, and that the good conductor has another cause than a high melt content. – A reinterpretation of the RRISP77 dataset by Menke *et al.* (1996) even resulted in a crustal thickness of 20–30 km. Moreover, the previously observed high attenuation was attributed to local factors, whereas the lower crust as a whole was found to be much less attenuating.

The southwest of Iceland has later been investigated in the SIST crustal tomography experiment (Bjarnason *et al.*, 1993) along a 170 km NW–SE traverse across the South Iceland Seismic Zone (see figure IV.1). The SIST data confirm the finding of earlier experiments that the Icelandic crust is structurally similar to oceanic crust, although thicker; table IV.1 shows the P-wave velocities and thicknesses for the crustal levels and the uppermost mantle. At the base of layer 3, there is a strong, slightly westward-dipping reflector interpreted as the Moho, at a depth of 20–24 km. From the velocities measured directly above and below this reflector, Bjarnason *et al.* (1993) infer that it lies in unmolten rock, probably gabbro at temperatures of only 600–900 °C. However, no reliable information on Moho topography can be derived from this profile, which is unreversed. More recently, the RISE profile along the axes of the Reykjanes Peninsula and Ridge and oblique to the ridge (Weir *et al.*, 2001, see fig. IV.1) has extended the image of the southwestern Icelandic crust away from the plume. Combined seismic refraction and gravity models from the RISE profiles confirm the presence of a thickened, but solid crust gradually approaching values typical of normal crust southward.

Another large profile is the FIRE profile (Staples *et al.*, 1997), a partly marine reflection/ refraction profile from the Faeroe Islands through Reydarfjörður to Krafla in northeastern Iceland crossing the NVZ (see figure IV.1). Basically, the FIRE profile shows the northeastern Icelandic crust to be similar to the southwestern: a low-velocity upper crust with a steep velocity gradient is separated from a very thick lower crust with a weak-to-moderate velocity increase by a transition layer of 2–4 km thickness (figure IV.3, upper part). Similar to the SIST profile, the high lower crust velocities are ascribed to either gabbro or an olivine-rich residue from which a tholeiitic melt has separated; this notion is also in qualitative agreement with petrological models of magma evolution at hotspots and v_P estimates derived from them, which suggest that picritic melts from depths corresponding to $p = 2 \ldots 3$ GPa form cumulates of melagabbros or gabbros by fractional crystallization (Farnetani *et al.*, 1996). The strong reflector interpreted as the Moho dips away from the neovolcanic zone with an astonishingly steep angle of *ca.* 30°: Moho depths below the NVZ lie around 19 km, but go down to 35 km under eastern Iceland (Staples *et al.*, 1997) and 25–31 km below the west flank of the NVZ (Menke *et al.*, 1998) (profile B96 in figure IV.1). In particular, a pronounced shallowing of the Moho and a strong updoming of the lower crust, which extends from about 3 to 8–10 km depth, are observed beneath Krafla, and at its top traveltime delays and S-wave attenuation indicate the existence of a magma chamber, while the crust below it does not yield evidence for the presence of melt, thus suggesting episodical melt transport in the crust (Brandsdóttir *et al.*, 1997; Menke *et al.*, 1998). Staples *et al.* (1997) conclude that the temperature at the base of the crust is less than 800 °C, because otherwise such a strong Moho topography could not persist long enough to fit the observations; they presume that it is not a stationary feature anyway and relate it to thickening due to extrusive flows and variations in plume activity and melt productivity, maybe modulated by the jumps of the spreading center. The strong reflectivity of the Moho is attributed to a substantial velocity contrast of about 0.9 km/s. Poisson's ratios of 0.26–0.27 have in general been derived for the lower crust and indicate that it is cold, solid, and essentially free of melt (Menke *et al.*, 1998; Staples *et al.*, 1997).

Between these profiles, a third, long refraction profile has been shot as part of the ICEMELT experiment from the Skagi peninsula across the central highlands, the plume center and Vatnajökull (Darbyshire *et al.*, 1998, see fig. IV.1). It yields a crustal thickness of 25–30 km for the northwestern half of the profile and of 38–40 km beneath central Iceland and Vatnajökull, where the upper crust is though only 3 km thick, but the lower crust forms a root, where a maximum v_P of 7.2 km/s is reached (figure IV.3, lower part). As Darbyshire *et al.* (1998) point out, this shows that melt production by the plume and underplating ex-

layer	v_P (km/s)	z_{bot} (km)
"Layer model" after Pálmason (1971)		
upper crust (layer 0–2)	2.75–5.08	0–2.5
lower crust (layer 3)	6.5	≤ 6.2
mantle (layer 4)	7.2	–
RRISP77		
upper crust	≤ 6.5	~ 5
lower crust	6.5–6.9	10–15
anomalous mantle	7–7.4	30
SIST		
upper crust (layer 2A)	3.5	0.7–3
middle crust (layers 2B,C)	5–6.5	3–7
lower crust (layer 3, "4")	6.5–7.2	20–24
mantle	7.6–7.7	–
FIRE		
upper crust	3–6.2	4
transition		6
lower crust	6.5–7.35	19–35
mantle	$> 7.9 \ldots 8$	–
ICEMELT		
upper crust	3.2–$6.3 \ldots 6.7$	2–11 (av. 5)
middle crust	6.6–6.9	9–12
lower crust	$> 6.9 \ldots 7.4$	19–41
RISE		
upper crust	≤ 6.8	4.5–7.3
lower crust	6.5–7.2	11–21
HOTSPOT receiver functions		
upper crust	$< 7 \ldots 7.2$	6–14
	< 6.5	6.5–11
lower crust	7.2–7.4	20–42
	6.5–7.2	20–40

Table IV.1: Representative P-wave velocities and depths to bottom (z_{bot}) of the crustal levels for the crust models of Pálmason (1971) and Flóvenz (1980), the RRISP77 profile I (Gebrande *et al.*, 1980), SIST (Bjarnason *et al.*, 1993), FIRE (Staples *et al.*, 1997), the ICEMELT refraction profile (Darbyshire *et al.*, 1998, 2000b), RISE profiles A and B (Weir *et al.*, 2001) and for receiver functions of individual stations throughout Iceland from the HOTSPOT campaign (Schlindwein, 2001; Du and Foulger, 1999, 2001; Du *et al.*, 2002); in the receiver function values, the first rows for each crust segment are from Schlindwein (2001), the second rows are calculated from v_S using $v_P/v_S = 1.7$. The mantle v_P of 7.7 km/s from SIST is an apparent velocity. The RISE values are for zero-age crust, with smaller values applying to the ridge profile.

ceeds crustal thinning by the rifting process. Calculated Poisson's ratios did not indicate the presence of melt in the crust and suggested that the crustal temperatures lie clearly below the basalt and gabbro solidi.

Menke (1999) pointed out that the inferred thick crustal root of Iceland would normally be expected to cause a high topography, which is though not observed. Under the assumption that the topography is locally isostatically compensated and that thickness variations are restricted to the lower crust, one can attempt to determine the densities of the lower crust and uppermost mantle under Iceland in order to explain this apparent contradiction. Menke (1999) estimated that for the lower crust $\varrho_{lc} = 3060 \pm 50$ kg/m³, and that the density jump across the Moho is 89 ± 12 kg/m³, leading to a mantle density as low as $\varrho_m = 3150 \pm 60$ kg/m³, much lower than the commonly used values around 3300 kg/m³ (see table D.3); it is noteworthy that the values of Menke (1999) bracket the density of 3100 kg/m³ proposed for the "layer 4"

Figure IV.3: Top: crustal model for the NVZ from FIRE (Staples *et al.*, 1997); bottom: crustal model for the ICEMELT refraction line across central Iceland (Darbyshire *et al.*, 1998). Note that only disjoint segments of the Moho are imaged (from Darbyshire *et al.*, 1998, © 1998 Blackwell, reproduced with permission).

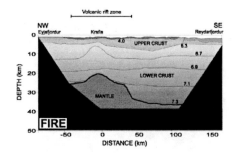

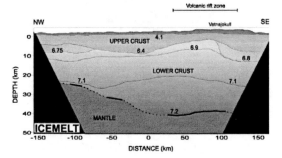

(Flóvenz, 1980; Pálmason, 1971), whose nature is in the focus of the controversy. He stated that even the combined effects of thermal expansion, depletion, and entrainment of basaltic material into the uppermost mantle can hardly account for such a large discrepancy. This gap is narrowed a bit by estimates for $\Delta\varrho_{\mathrm{m-lc}}$ between 130 ± 30 and $154\,\mathrm{kg/m^3}$ (Allen *et al.*, 2002b; Kaban *et al.*, 2002). On the other hand, Staples *et al.* (1997) get low ϱ_{m} values only for the NVZ, while outside they estimate $\varrho_{\mathrm{m}} = 3240\,\mathrm{kg/m^3}$, which can be explained by a temperature contrast between cool lithospheric mantle and a low-density sub-Moho wedge narrowing downward; the lowermost crust is required to have a very high density. Thus, it seems natural to consider the alternative of a mantle with only slightly reduced density in the plume in combination with a thick, unusually dense crust. This has been done by Guðmundsson (2003) in a series of models based on crustal thickness, topography, and gravity information and led to the conclusion that the density of the plume mantle is reduced by only up to $20\,\mathrm{kg/m^3}$ relative to normal mantle, but that the density contrast between the lower crust of Iceland and the adjacent ridges (with $\varrho_{\mathrm{lc}} = 2980\,\mathrm{kg/m^3}$) could be as high as $140 \ldots 170\,\mathrm{kg/m^3}$, assuming smooth lateral gradients from normal regions to Iceland; this would also result in the low density contrast needed to reconcile a thick crust and the rather low topography. He argues that the enhancement of melting and its deeper onset due to high temperatures and the effect of water would enrich the crust in olivine and raise the Fe content of crustal minerals, both of which increase the density; furthermore, the plume source might contain more iron than normal mantle by itself (see section IV.2.2), and a phase transition from plagioclase to garnet in the thickest parts of the crust could also contribute to the inferred density. He estimates v_{P} to be still lower than $7.5\,\mathrm{km/s}$, which would match observations (cf. table IV.1).

In an attempt to summarize the available information from seismics and gravimetry, Darbyshire *et al.* (2000b) collected data from several profiles, among them SIST, ICEMELT,

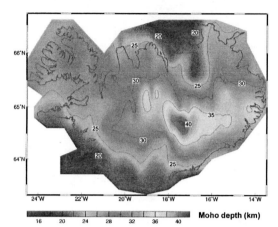

Figure IV.4: Map of the Mohorovičić discontinuity of Iceland (from Darbyshire *et al.*, 2000b, © 2000 Elsevier, reproduced with permission). *See plate section for colour version.*

FIRE, and modelled it together with Bouguer data for the whole of Iceland (Eysteinsson and Gunnarsson, 1995) and some information from receiver functions in order to construct a Moho map (figure IV.4). According to this map, a maximum crustal thickness of 40–41 km is reached beneath Vatnajökull, but Moho depths of more than 25...30 km are ubiquitous between the northwestern peninsula and the eastern coast; the thinnest crust is located beneath southern Iceland, reaching a minimum of about 16 km under Reykjanes, and the NVZ. As a whole, this Moho map provides a useful picture of the crustal thickness of Iceland, although it must be kept in mind that for some areas in the northwest, northeast and southeast, the database is sparse and often restricted to gravity at the best and that the seismic profiles are much longer than the lengths of the Moho segments imaged by them. Their gravity modelling makes some strong simplifications: although they demonstrated the need of a cylindrical low-density body in the uppermost mantle, they did not investigate the effect of a plume head, although its existence is likely according to mantle dynamical models and surface wave studies (Allen *et al.*, 2002a). It is unclear, how far this would affect the calculated Moho depths and crustal densities. – Subsequently, Allen *et al.* (2002b) and Kaban *et al.* (2002) also published Moho maps of Iceland, and as one would expect, there exist some differences between the three maps, which reach a magnitude of some 5 km in some cases. While the Darbyshire *et al.* (2000b) map shows a thickness of less than 25 km under the EVZ in southern Iceland, more than 30 km are drawn in the other two maps; in the Tertiary areas of eastern Iceland, Darbyshire *et al.* (2000b) also find a slightly thinner crust than the other two studies. In contrast, Allen *et al.* (2002b) have a much thinner crust beneath Snæfellsnes than the other two studies, but a thicker one above the plume center. The general picture emerging from the maps, however, is quite consistent.

IV.1.3 Seismology and petrophysics

According to the v_S model of the HOTSPOT experiment (Allen *et al.*, 2002b), significant velocity anomalies with Δv_S as large as -7% in some places such as the Bárðarbunga–Grímsvötn area are observed in the upper crust along the neovolcanic zones, suggesting the presence of local shallow magma bodies; on the contrary, the only low-velocity anomaly in the lower crust is a circular region beneath Vatnajökull reflecting the position of the plume in the mantle. In general, however, the calculated v_P/v_S ratios, which are in good agreement with independent v_P models, do not show evidence for extensive presence of partial melt beneath 15 km depth. The absence of melt is also supported by $Q_P = 200\ldots300$ and $Q_S = 150\ldots600$ found for mid- to lower-crustal depths in central and southwestern Iceland, indicating temperatures between 700 and 775 °C if gabbroic material is assumed (Menke and Levin, 1994; Menke *et al.*, 1995).

An important indicator for the thermal structure is the depth distribution of earthquake hypocenters, because rupture is associated with brittle behaviour and correspondingly low temperatures; therefore, earthquakes are not expected to take place in hot, ductile crust. Several evaluations of hypocenter depths of Icelandic earthquakes show that there are hardly any quakes at depths greater than *ca.* 12 km within a distance of *ca.* 100 km from the volcanic zones and that the maximum depth of hypocenters increases in older crust (*e.g.* Stefánsson *et al.*, 1993; Rögnvaldsson *et al.*, 1998; Tryggvason *et al.*, 2002), indicating that at that level the crust is still considerably below its solidus; Tryggvason *et al.* (2002) estimate the temperature at this depth to lie between 580 and 750 °C on the basis of borehole heat flow measurements. This is also supported by models of the viscosity structure beneath the Krafla region, which give best results for an elastic plate with a thickness of 12 km and a lower crust viscosity of about $3 \cdot 10^{19}$ Pa s (Pollitz and Sacks, 1996). On the other hand, Kaban *et al.* (2002) find an elastic thickness for the lithosphere derived from short-wavelength (< 200 km) gravity which is at most half as large; nonetheless, this is not interpreted as an argument supporting a hot layer 4, because they estimate the temperatures within this layer to lie between 650 and 950 °C. To resolve the dilemma, they resort to the notion of a thick transition zone between crust and mantle consisting of a mixture of mafic and ultramafic rocks, which had already been proposed by other workers.

Recently, the stations of the HOTSPOT network have served for Iceland-wide studies of crustal thickness and velocity structure (Du and Foulger, 1999, 2001; Du *et al.*, 2002; Schlindwein, 2001). The $v_P(z)$ and $v_S(z)$ profiles derived from the inversion of receiver functions and surface waves show as common features a strong positive velocity gradient in the upper few kilometers corresponding to the upper crust, nearly constant velocities in the lower crust, and sometimes another positive gradient at the crust–mantle transition; however, these features are not always clearly visible: in several cases, the boundary between upper and lower crust is not sharp, and the Moho, which is found at depths between 20–33 km below older crust and up to 40 km in the junction of the neovolcanic zones, is also frequently smeared over a certain depth interval. Du and Foulger (2001) note that in general the upper crust thins and the lower crust thickens towards the rifting zones. At various sites in western and northern Iceland, a slight LVZ is present in the lower crust (Du and Foulger, 2001; Du *et al.*, 2002; Schlindwein, 2001).

Darbyshire *et al.* (2000a) used eight SIL broadband stations distributed over the WVZ and NVZ for investigating the velocity structure of the crust. They confirmed some essential results from seismic refraction surveys, *e.g.* the crustal thicknesses and the fact that the upper crust is thickest on the flanks of the volcanic zones in the western, northern, and central parts, where lava flows from different rifts have been piled up onto each other at different times due to ridge jumps, and that it is thinnest in the direct neighbourhood of volcanic centers. A particular feature observed in the $v_P(z)$ from the receiver functions is the existence of local low-velocity zones at some sites, which are not visible in refraction profiles, because no head wave is generated in the upper layer.

An important result of the comprehensive receiver function study of Schlindwein (2001) is the apparent absence of a contiguous, sharp Moho below Iceland; this must be borne in mind when using Moho maps of the kind presented by Darbyshire *et al.* (2000b). The study of Schlindwein (2001), which includes 28 stations of the HOTSPOT network from all parts of Iceland, reveals that the Moho is only well visible in some areas, whereas a gradational change from crust to mantle velocities seems to occur in others; she remarks that indeed the refraction studies do not yield proof for a continuous Moho either, because the corresponding deep reflectors are only imaged on relatively short segments of the profiles. A broad transition from crust to mantle would also be in agreement with the model of Kaban *et al.* (2002).

IV.1.4 Electromagnetics

Electromagnetic deep sounding has also been performed in Iceland throughout the 1980s and 1990s. One of the major early experiments was a short-period (15 s–1 h) magnetotelluric campaign in northeastern Iceland by Beblo and Björnsson (1980), with one profile across the NVZ and the adjacent Tertiary areas (line BB in figure IV.1) and another one roughly along its strike, following the RRISP77 profile I. The essential result of their measurements is the presence of a 5–10 km thick well-conducting layer[1] with a resistivity of only 15 Ω m between two more resistive layers with about 100 Ω m. In the rift zone, the top of the good conductor is as shallow as 10 km, while it dives to about 20 km at greater distances from the NVZ and also becomes thinner; these results were confirmed by Thayer *et al.* (1981) in a magnetotelluric study in the Krafla–Mývatn area, which localized the good conductor ($\Delta h \leq 4$ km, $\rho \leq 10\,\Omega$ m) at a depth of 8 to 15 km, which they state to be the base of the crust without considering alternatives. The high-ρ layer with 70–100 Ω m beneath it reaches down to at least 100 km, indicating a small geothermal gradient; its lower boundary is uncertain, though. Beblo and Björnsson (1980) conclude from comparison with laboratory data from partially molten basalt and ultramafites that the low resistivity is due to temperatures of 1000–1100 °C, at which a basaltic partial melt with $\varphi \leq 0.1$ would be present. Similarly, Thayer *et al.* (1981) estimated $\varphi \leq 0.075 \ldots 0.15$ for $\Delta h \leq 4$ km, assuming interconnected channels; for the seismic layer 3, which is made from dikes, they estimate $\varphi \leq 0.02 \ldots 0.05$, *i.e.* only a small part of the material is molten at any time. They also provided estimates for the melt fraction at $z > 15$ km, which according to their model is part of the upper mantle. For interconnected channels, they establish an upper limit of 1.7–3.3 % for the depth range between 15 and 30 km, and of 1.2–2.5 % below; as these estimates assume an ideally insulating matrix, the true porosities are lower and thus support the notion of low porosities throughout the melting region. Assuming a thick crust, it seems though that the values for $z < 30$ km could also be explained by a dike model similar to that for the seismic layer 3 if the high-conductivity layer is due to something else than magma accumulation.

Similar observations of a well-conducting layer were made by Hersir *et al.* (1984) in southwestern Iceland on a profile close to the SIST line (H84 in figure IV.1) and across the WVZ, where resistivities between 2.5 Ω m and, in older regions, 50 Ω m have been measured, and on a continuation profile of Eysteinsson and Hermance (1985) across the EVZ (EH in figure IV.1), although with the notable difference that the good conductor is not found in older crust below eastern Iceland.

As a whole, these results are in blatant contrast to several, although not all, findings from seismic and seismological investigations, because according to them the depth range of 10–15 km corresponds to lower crust levels rather than to the top of the mantle. Therefore, alternatives to the shallow-melt model have been sought by several investigators. One frequently proposed is the presence of low-grade metamorphic minerals with high conductivities such as zeolites in the lower crust. If the crust is indeed cool, as it would be if its upper part were formed by subsiding lava layers, it might be possible that zeolites exist at

---

[1] In many studies, another good conductor is detected in the uppermost crust, at depths above *ca.* 5 km; this layer is though readily explained by hydrothermal fluids, and is not further considered here.

depths of 10–15 km (*e.g.* Smallwood *et al.*, 1998); further long-term cooling would explain why the high-conductivity layer descends to greater depths with increasing distance from the spreading center.

IV.1.5   Geochemistry and petrology

In principle, the crust is not only a permeable lid through which melt ascends, but can itself modify the composition of erupted melts by processes like differentiation, crystallization and contamination in magma chambers. Therefore it is of interest to determine whether erupted melts have undergone modification in the crust. Indeed, the fact that the relative movement of the plume and the ridge continually transports old crust above the hotspot and has led to a number of ridge jumps let Óskarsson *et al.* (1985) conclude that the chemistry of several volcanic centers in central Iceland is influenced by reworking of crustal material; they even claim that the variability of Icelandic lavas is entirely due to different degrees of crustal contamination, although this radical point of view has been questioned by several authors, who emphasized the role of mantle source heterogeneity – even on a scale of 10–20 km – as expressed by the isotopic signature and the effect of rift zone relocation on the degree and type of melting (*e.g.* Meyer *et al.*, 1985; Nicholson and Latin, 1992; Furman *et al.*, 1995; Stecher *et al.*, 1999; Breddam, 2002; Stracke *et al.*, 2003). A synopsis of Nd, Sr and O isotopy (Hémond *et al.*, 1993) led to the conclusion that a happy medium of significant mantle source heterogeneity in combination with pervasive, although usually minor contamination with altered Icelandic crust is an adequate explanation for the chemical variability.

Geochemical investigations have also been used to constrain the thickness and thermal structure of the crust by geothermobarometry. For central Iceland, Breddam (2002) derived eruption temperatures of up to 1240 °C and initial crystallization pressures and temperatures of 1–1.5 GPa and 1270 °C, respectively, which gives a rough estimate of 30–45 km for the lithospheric thickness above the plume center. For the NVZ, Nicholson and Latin (1992) determined a crustal thickness of *ca.* 22 km for the Krafla region, which is in fair agreement with the findings from the FIRE experiment (Staples *et al.*, 1997; Brandsdóttir *et al.*, 1997) and an independent geochemical study by Maclennan *et al.* (2001b). In this latter investigation, however, the observed range of crystallization pressures and temperatures (0.3–1 GPa and 1160–1350 °C) indicated that crust formation does not only take place at shallow magma chambers as those seen in seismic profiles, but happens over a large depth range from 10 km down into the uppermost mantle at 30 km.

IV.1.6   Geodynamical modelling

Of the attempts to model the evolution and structure of the Icelandic crust numerically, those by Pálmason (1973, 1980) have been pioneering. Pálmason (1980) developed a two-dimensional kinematic model of the crust to investigate the relation between the widths of the lava deposition and strain zones, lava production rate and spreading velocity in the axial area and the dip angle and deposition rate of the Tertiary basalt pile. His models suggest that the width of the volcanic zone ($\sim$ 40–125 km) and the spreading rate have not changed significantly in the last 10–15 Ma; this is not a contradiction to the notion of plume pulsing and related variations in erupted volumes, because the local deposition rate does not allow conclusions about total lava production rate. From his reproduction of the dipping basaltic layers he assesses that the visible part of the Tertiary extrusives was emplaced outside the innermost 50 km of the volcanic zone. The fraction of dikes in the upper crust and the sharpness of the transition to the intrusive lower crust is controlled by the ratio of the widths of the crustal strain zone and of the lava deposition zone and by the amount of strain taken up by normal faults.

Menke and Sparks (1995) followed a more comprehensive approach including the region of melt generation in the uppermost mantle and the crust formed from it in a 160 km deep two-dimensional FD model to address the problem of crustal thickness and temperature. The results from their preferred model with a shallow, rather narrow accretion zone show that the combined action of upwelling and spreading is not strong enough to produce a hot crust, especially if hydrothermal cooling in the upper crust is accounted for; models with a broader accretion zone or vertically uniform dike injection lead to reduced downwelling or a hotter crust, respectively, but cannot explain observations equally well. In particular, the preferred model does not yield a large, shallow layer of partial melt. Of course, the inclusion of hydrothermal cooling by assuming a higher thermal conductivity in the active zone is not very well constrained, and it is unclear if values such as a five- or eightfold increase as used for oceanic crust are equally appropriate for the subaerial crust.

Total crustal thickness is also a result of some mantle convection models further described in section IV.2.3 as well as those of this study (see section V.2). Details of crustal structure cannot be resolved by such models, but the comparison of resulting total thickness with measured Moho depths provides an important constraint for the evaluation of mantle dynamic models of the mantle below Iceland and melt production therein. Recent models by Ito *et al.* (1999) agree with most of the observations described in this chapter in that they yield a thickness of 33 km.

IV.1.7 Synopsis

The series of large field experiments conducted since the early 1990s has contributed much to the clarification of the picture of the Icelandic crust and led to a preference for a model of a rather thick, cool crust; crustal thickness estimates from modelling of mantle melting also tend to larger values. This notion is most strongly supported by the results of seismological investigations of several kinds; in particular, the results of the RRISP77 campaign, which had stimulated the discussion in the first place, turned out to allow for a thick-crust interpretation as well, and much of the controversy was probably due to a too narrowed view on those results. The results summarized in table IV.1 suggest that the thickness of the crust ranges from about 20 km beneath some parts of the neovolcanic zones to about 40 km at the plume center and in the older Tertiary regions of eastern Iceland.

Nonetheless, seismological and gravity studies, which even resulted in Moho maps, also put some restrictions on this seemingly neat and simple solution. It remains a fact that the Moho is not clearly seen in many places and that the lowermost layer of the thick crust seems to have some unusual properties (*e.g.* Schlindwein, 2001; Kaban *et al.*, 2002), which suggest that it does not consist of usual gabbroic lower-crust material, but that it partly has a transitional character. For this reason, several workers have pointed out that there is no classical sharp Moho throughout Iceland, but that there is a crust–mantle transition zone with a thickness of some kilometers. Thus, Moho maps of Iceland, while useful, have to be taken *cum grano salis*, and the clarification of the fine structure and the lithology of the lowermost crust remains an issue.

The strongest contradiction to the thick-crust model still comes from magnetotelluric measurements, which had already strongly influenced the interpretation of the RRISP77 data. Given the ambiguities of the method, *e.g.* the problem to distinguish between a moderately good conductor of considerable thickness and a very thin high-conductivity layer, and in view of the variety of results from other methods, it seems, however, necessary to consider alternatives to the usually invoked layer of partial melt like the presence of certain metamorphic phases with greater effort. The special character of the Icelandic crust might make it difficult to develop such explanations, because analogies from continental crust, where a similar good conductor is also observed, have to be made with care.

The conclusion that the crust is thick and clearly below its solidus in general also has implications for magma transport and the generation of new crust. The low attenuation even at deeper crustal levels and the minor degree of chemical contamination suggest that magma transport in the crust to the site of final deposition is relatively fast in many cases, and that larger magma bodies are rather local structures and not ubiquitous.

IV.2   The mantle beneath Iceland

In this section, several results concerning the mantle beneath Iceland are reviewed. In particular since the early 1980s, a great deal of studies has been conducted to shed light on the questions whether the Iceland plume is hot and narrow or broad and cool – a conflict between observations and numerical models which has only recently approached a resolution (see section IV.2.3) –, how heterogeneous the plume is and what sources are involved, and whether the plume emerges from the CMB or from the bottom of the transition zone. In particular the last point is one of general interest to geodynamics, because Iceland is regarded as one of the classical examples for a whole-mantle plume. Some key variables of the plume from several studies are listed in table IV.2 to highlight the large range of results. However, when comparing T anomaly or radius estimates from field observations with those from numerical models, one should bear in mind that the authors frequently do not make clear for which depth their temperature estimate was made, or that the estimates are not all for the same depth level; the temperature contrast may well vary at different depths due to conductive cooling upon ascent and enthalpic cooling during melting.

IV.2.1   Seismology

Regional tomography — One of the most common tools for imaging the deep structure of the Iceland mantle plume is seismic tomography, which has been used by several groups to determine the width, depth, and temperature contrast of the plume.

The first mantle tomographic image of the Iceland plume came from the early investigation of Tryggvason *et al.* (1983), which used teleseismic data from a permanent seismic network, which had though sparse to no coverage in western and central Iceland, but nonetheless achieved the mapping of a clear P-wave anomaly of 2 to more than 3 % in some parts down to 375 km depth.

In recent years, the ICEMELT experiment (Bjarnason *et al.*, 1996; Wolfe *et al.*, 1997) has provided a picture of the plume in the upper mantle which was widely paid attention to. From the ICEMELT data Wolfe *et al.* (1997) derived a P- and S-wave velocity anomaly model for the upper mantle between 100 and 400 km depth (figure IV.5) displaying an approximately circular and vertically continuous structure centered beneath the junction of the three neovolcanic zones and Vatnajökull; the maximum of the P-wave anomaly is about −2 %, that of the S-wave anomaly about −4 %. Both anomalies have basically similar shapes, although the S-wave anomaly extends a bit more northward and has a second maximum at 100–125 km depth under central Iceland, which might be related to incipient melting of the plume, although other effects related to seismic anisotropy could also play a role. As an upper bound to the radius of the plume, Wolfe *et al.* (1997) suggest 150 km; they estimate the excess temperature to be *ca.* 200 K.

T anomaly, ΔT_P, K	75	numerical	Ito *et al.* (1996), broad
	93	numerical	Ribe *et al.* (1995)
	100	numerical	Keen and Boutilier (2000)[a]
	100–200	seismics/numerical	Nielsen *et al.* (2002)[a]
	140	lab/numerical	Feighner *et al.* (1995)
	$\gtrsim 150$	P–S differential times	Shen *et al.* (1998, 2002)
	170	numerical	Ito *et al.* (1996), narrow
	180	numerical	Ito *et al.* (1999)
	< 200	regional tomography	Foulger *et al.* (2001)
	~ 200	regional tomography	Wolfe *et al.* (1997)
	~ 200	seismics	Smallwood *et al.* (1999)
	200–300	global tomography	Bijwaard and Spakman (1999)
	~ 250	chemistry	Nicholson and Latin (1992)
	263	analytical/chemistry	Schilling (1991)
	$\lesssim 300$	seismology	Allen *et al.* (1999)
radius, r_P, km	40	numerical	Albers and Christensen (2001)
	50	He isotopy	Breddam *et al.* (2000)
	60	numerical	Ito *et al.* (1996), narrow
	60–100	seismology	Allen *et al.* (2002a)
	100	numerical	Ito *et al.* (1999), Ito (2001)
	100	seismology	Bjarnason *et al.* (1996); Allen *et al.* (1999)
	100	regional tomography	Tryggvason *et al.* (1983)
	100–125	regional tomography	Foulger *et al.* (2001)
	≤ 150	regional tomography	Wolfe *et al.* (1997)
	~ 150	chemistry/analytical	Maclennan *et al.* (2001a)
	< 200	P–S differential times	Shen *et al.* (1998)
	≤ 250	global tomography	Bijwaard and Spakman (1999)
	300	numerical	Ito *et al.* (1996), broad
	> 300	numerical	Ribe *et al.* (1995)
volume flux, q, km³/a	0.7884	analytical	Ribe and Delattre (1998), case 2
	1.26	analytical/numerical	Steinberger (2000)
	1.43	analytical/chemistry	Schilling (1991), best estimate
	2.1	numerical	Ito *et al.* (1996), narrow
	2.2	gravity/bathymetry	Ito and Lin (1995)
	2.23	analytical	after Sleep (1990)[b]
	2.5	seismology	Allen *et al.* (2002a)
	4.94	laboratory/numerical	Feighner *et al.* (1995)
	6.08	numerical	Ribe *et al.* (1995)
	6.1	numerical	Ito (2001), average
	6.4	numerical	Ito *et al.* (1999)
	12	numerical	Ito *et al.* (1996), broad
waist width, y_{P0}, km	850	laboratory	Feighner and Richards (1995)
	870	numerical	Ito *et al.* (1996), narrow
	900	numerical	Feighner *et al.* (1995)
	923	analytical/chemistry	Schilling (1991)
	1620	numerical	Ito *et al.* (1999)
	2300	numerical	Ito *et al.* (1996), broad

[a]initial plume head
[b]including Jan Mayen; using $\alpha = 3 \cdot 10^{-5}$ 1/K, $\varrho_0 = 3300$ kg/m³, $\Delta T = 200$ K

Figure IV.5: Horizontal and vertical cross sections of the P-wave (left)
and S-wave (right) seismic velocity anomaly models of the ICEMELT
experiment (slightly modified from
Wolfe *et al.*, 1997, © 1997 Nature
Publishing Group, reproduced with
permission).

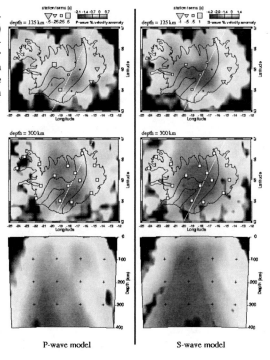

P-wave model   S-wave model

Few years later, the HOTSPOT experiment (*e.g.* Allen *et al.*, 1999, 2002a; Foulger *et al.*,
2001) has provided an even larger dataset and new tomographic images which also show a
strong low-velocity anomaly beneath Iceland, reaching down to at least 450 km. It essentially confirmed the ICEMELT estimates for r_P and ΔT_P of 100–125 km and less than 200 K,
respectively, as well as the location of the plume center, but finds slightly larger velocity reductions of $\Delta v_P = -2.7\%$ and $\Delta v_S = -4.9\%$ compared to IASP91 (Foulger *et al.*, 2001). An
interesting feature is the N-S asymmetry of the plume head, about which some information
could be retrieved especially in the study of Allen *et al.* (2002a), which included Love waves
and used the crust model of Allen *et al.* (2002b) to remove the crustal signal: Foulger *et al.*
(2001) remarked that while there seems to be a branch of the plume beneath the Reykjanes
Ridge, which is also visible in the global tomography by Zhao (2001), the head is apparently
sharply bounded in the north at the position of the TFZ, although one may doubt that its
influence reaches as deep as imaged. Another remarkable feature is the presence of a zone in
the uppermost mantle ($z < 100$ km) above the plume core, *i.e.* in the main melting region,
where the v_S anomaly is 2% *weaker* than in the outer parts of the plume head. Allen *et al.*
(2002a) suggested that this is due to particularly high degrees of depletion in the recent
past of the plume, although in this case it cannot actually affect this whole depth range,

---

Table IV.2 (preceding page): Preferred values of some characteristic variables of the Iceland plume
in the upper mantle from different published field observations, numerical models, and laboratory
experiments.

because the onset of major melting seems to lie not much deeper than about 100 km (Shen and Forsyth, 1995); besides it supports the notion of fast and efficient melt extraction, as the effect of melt on velocities does not seem to play a major role here.

As far as the temperature estimate of Wolfe *et al.* (1997) is concerned, they did not take into account quantitatively the influence of melt, water content, anelasticity and anisotropy on wave velocities, which would lead to a lower excess temperature. On the other hand, they remark that wavefront healing by rays diffracted around the plume could have faked shorter traveltimes and thus have masked a larger ΔT_P value; this is generally confirmed by modelling the contributions of attenuation and diffraction to the observed traveltime delay of the HOTSPOT data with independent methods, where a maximum S-velocity anomaly of -12% was derived for the best-fit model plume with $r_P = 100$ km (Allen *et al.*, 1999). In this context it is worth noting that recent studies, which accounted for the finite wavelength of seismic waves better than traditional ray-based method can do, also found the anomaly to be stronger than the older estimates by as much as a factor of two (*e.g.* Hung *et al.*, 2003). Allen *et al.* (1999) observe a distinct pattern of positive *and* negative traveltime shifts in their data, which can only be explained by adding a significant diffractory component to the overall anelastic delay. One shortcoming of their model is that they estimated the effects of attenuation and diffraction separately, which makes their combination in their best-fit model, in their own opinion, problematic; furthermore, it seems that their two-dimensional diffraction modelling did not account for ray bending, thereby introducing a small error, nor did it include a plume head. Nonetheless, it can be expected that the importance of wavefront healing is a robust result, although a three-dimensional anelastic wave propagation model would have been more appropriate. Unfortunately, their method does not constrain the magnitude of the temperature anomaly very well, so their estimate of $\Delta T_P \lesssim 300$ K must be regarded as a very soft upper bound. – In the interpretation of the HOTSPOT data, the effects of depletion and anisotropy were not taken into account quantitatively either due to the poor constraints on their effect on the velocities and the limited knowledge of the orientation of the dominant flow field. However, Foulger *et al.* (2001) remark that their observed v_P/v_S ratios cannot be explained by the thermal effect alone, but require additionally the presence of a few tenths of a percent of partial melt at depths between 100 and 300 km; while the deeper bound might still be in agreement with hydrous damp melting, the shallower is probably an overestimate due to the limited resolution in the uppermost mantle, because partial melts are to be expected at shallower depths as well. On global-scale Q_S models based on Rayleigh waves (Selby and Woodhouse, 2002), there is also a high-attenuation region visible around southern Greenland in the depth range from about 70–300 km, supporting the notion of a thermal anomaly maybe containing partial melt; the model has though only resolution up to spherical harmonic degree 8 (~ 5000 km) and is not reliable at greater depth, which precludes more detailed conclusions.

Although the ICEMELT and HOTSPOT images of the plume are quite impressive, there are still some problems with the results from tomography. As Wolfe *et al.* (1997) themselves concluded from tests of their inversion method, the inversion produces a broadened image of the plume, which is why they regard their r_P as an upper bound, but it does not propagate the real structure to greater depth artificially. In contrast, Keller *et al.* (2000) found in an independent test that the plume is not notably broadened, but that the dataset would produce an image with strong downward smearing effects, so that the inversion results for a broad shallow structure could hardly be distinguished from those for a narrow deep one; strong downward smearing can also be seen in inversions of numerical convection models, where they shift velocity reductions related to melt to unreasonable depths (Ito *et al.*, 1999), and it was also mentioned as a deteriorating factor in the HOTSPOT tomography. A particular

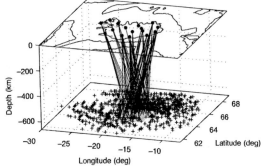

Figure IV.6: Stations and ray piercing points of P660s ray paths at 660 km for the ICEMELT experiment (from Shen *et al.*, 1998, © 1998 Nature Publishing Group, reproduced with permission).

problem of tomographic experiments of the ICEMELT or HOTSPOT type is the distribution of the seismic stations: the ICEMELT stations, which had a spacing of about 75 km, were all on the island, while the rays used for tomography were from teleseismic events and had a steep angle of incidence similar to the raypaths in figure IV.6. Therefore, the rays did not cross-cover sufficiently the shallowest part of the upper mantle, where most of the melting would be expected, and could not reveal the possible hat-like shape of the plume top ($z \lesssim 100\,$km); the plume structure at depths below 400–450 km could not be imaged either. The rays also did not sample a large part of the vicinity of the plume, so that there remains some uncertainty whether the outermost imaged parts are indeed unaffected mantle valid as a reference or if the tomographic image rather shows the innermost part of a much broader anomaly. As pointed out by Keller *et al.* (2000), it is even possible that the whole conic shape of the plume image is due to the small aperture of the ICEMELT geometry; however, Wolfe *et al.* (2002) defended their results by publishing additional resolution tests and noting that Keller *et al.* (2000) have made several simplifying assumptions *e.g.* concerning the geometry of raypaths, and reaffirmed their conclusions. The solution for this problem would be to establish a much larger network including marine stations, although apart from the logistical difficulties, data deterioration by wave-related noise is to expected; this is a problem always present in data recorded on islands.

Li and Detrick (2001) analyzed ICEMELT and HOTSPOT surface wave traveltimes and amplitudes to gain information about the depths too great for controlled-source seismics and too shallow for the regional tomography and found evidence for the presence of melt in the crust and the upper mantle, especially beneath the rift zones. On a larger scale, and with a much lower resolution than body waves, Rayleigh and Love waves show group velocity reductions between *ca.* 5 and 9 % for periods between 100 and 150 s beneath Iceland, which also indicates the presence of a strong anomaly in the upper mantle (Levshin *et al.*, 2001).

It was only recently that detailed anisotropy studies have been performed (Bjarnason *et al.*, 2002). The most remarkable results are that the S-wave anisotropy distribution reflects roughly the plate geometry and that neither a spreading-parallel nor a radial pattern, which could be assigned to the plume, are obvious from the data. The traveltime difference between the fast and the slow S phase is larger in the east than in the west and would correspond to a 100–200 km thick anisotropic layer in the east. The divide between the anisotropy domains is located up to 100 km to the west of the NVZ and EVZ, which is interpreted as a manifestation of former locations of the spreading center still conserved due to the sluggishness of crystal reorientation. The observed anisotropy can be explained by the shear of the diverging plates relative to a roughly NNW-directed flow with a velocity of *ca.* 3 cm/a. – By contrast, surface wave anisotropy studies at the Reykjanes Ridge (Gaherty, 2001) indicate that small-scale convection in the melting zone induces a mostly vertical orientation of olivine a axes at $z < 100$ km, which is frozen in when the material moves outward and results in increasing anisotropy, whereas at greater depth the usual horizontal fabric of passive spreading of older lithosphere is observed. Gaherty (2001) interprets this as reinforced thermal convection due to the influence of the nearby hotspot.

Global tomography — Global seismic tomography plays a key role in the determination of the deep structure of plumes, but compared with regional seismic tomography, it has a much smaller resolution, which is critical when it comes to mapping of rather thin structures like plume conduits. Nonetheless, several authors have made attempts to find evidence for whole-mantle plumes in global tomographic models. Recently, Bijwaard and Spakman (1999) and Zhao (2001) published P-wave models spanning the whole depth range of the mantle beneath the northern Atlantic, which display a twisted and tilted anomaly of varying amplitude. At its base at the CMB beneath Iceland, the Bijwaard and Spakman (1999) model is very broad and has only very small values, but it rises to 0.5 % for most of the lower mantle and develops a complex, widely spread structure with connections to low-velocity anomalies as distant as beneath central Europe (figure IV.7); the Zhao (2001) model also starts at the CMB beneath the southern tip of Greenland and shows the same eastward tilt, but seems to be a bit weaker in the depth range between 2000 and 1400 km. Above 1000 km both models show a broad -0.5 % region, and in the upper mantle, the anomaly is significantly stronger. In the model of Bijwaard and Spakman (1999), it has two branches, one linking up to the lower mantle and the other extending beneath Greenland, the latter being in agreement with an upper-mantle tomographic study by Zhang and Tanimoto (1993); in the model of Zhao (2001), the cold, fast Greenland craton reaches deeper, and there is only a small slow patch in the uppermost lower mantle beneath it. In the tomographic S-wave model of Grand (2002), the anomaly shows lateral extensions parallel to the ridge direction down to *ca.* 350 km and becomes more circular below, with the center remaining under Iceland or being displaced only slightly to the southwest. Its amplitude decays to some 0.7–0.8 % to the bottom of the transition zone and stays on that value to *ca.* 1150 km, whereby the anomaly becomes wider. At depths between 1150 and *ca.* 1800 km, there appears to be a shift of a wider anomaly beneath northwestern Europe, which is seen in a similar style in the horizontal slices of Bijwaard and Spakman (1999), while only a very weak signal remains under Iceland itself; however, the vertical slices in the Bijwaard and Spakman (1999) and Zhao (2001) models suggest that the weaker signal is the actual plume. The anomaly remains at very low amplitudes almost down to the CMB directly under Iceland.

The notion of a relatively narrow, significant low-velocity structure in the upper mantle beneath Iceland underlain by a weaker, at best diffuse anomaly in the lower mantle is confirmed by other global tomographic models (*e.g.* Grand, 1994; Ritsema *et al.*, 1999; Mégnin and Romanowicz, 2000), although these images do not always show a contiguous structure for the whole depth range of the mantle; on the other hand it must be kept in mind that *e.g.* the model of Mégnin and Romanowicz (2000) has a resolution of only 450–850 km in the horizontal and 100–300 km in the vertical direction, so one cannot expect a sharp signal from a supposedly narrow structure like a plume from it. An attempt by Pritchard *et al.* (2000) to pinpoint the anomaly in the mid-mantle gives some hints at a structure with a radius of no more than 125 km and $\Delta v_P \leq -1.5$ % some 1500 km beneath the Iceland–Faeroe Ridge, but the authors remark that the result is ambiguous, because only raypaths from a very limited azimuth range could be used. Depth sections between 1400 and 2200 km in the global S-wave models of Grand (1994, 2002), Ritsema *et al.* (1999) and Mégnin and

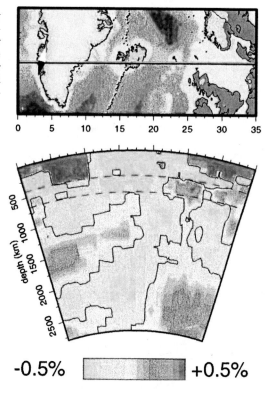

Figure IV.7: Seismic tomography cross-section through the mantle beneath Iceland. Note that the colour scale is clipped at ±0.5 % to enhance the low-velocity anomaly in the lower mantle (modified from Bijwaard and Spakman, 1999, © 1999 Elsevier, reproduced with permission).

Romanowicz (2000) seem to confirm the presence of a weak low-velocity anomaly in the mid-lower mantle in that region; however, a common feature of all tomographic models is that the structure is partly weaker at mid-mantle depths. The position of the source region of the plume beneath Iceland is corroborated by the detection of an ultra-low-velocity zone with a diameter of *ca.* 250 km by Helmberger *et al.* (1998); such a structure is also visible in several tomographic models at the CMB beneath Iceland, although the observed amplitude of the anomaly varies. The radius estimate of Bijwaard and Spakman (1999) is ≤ 250 km for the upper and 200–350 km for the lower mantle. A more precise assessment is not possible due to the limited resolution, but it is basically in agreement with the results from regional tomographic studies, as is the temperature estimate of 200–300 K, for which the anomaly was entirely ascribed to thermal effects; assuming that the plume is broader than the seismologically detectable ultra-low-velocity zone at the D'', the radius value also agrees well with the one by Helmberger *et al.* (1998). Proper separation of the large plume head and the stem in the upper mantle is precluded by the poor vertical resolution.

Altogether, it must be admitted, that an anomaly amplitude of at most 0.5 %, as observed in the lower mantle, does yield some indication for the existence of a whole-mantle plume, but it is certainly not an impressive proof. The lack or weakness of a clear plume signature in the lower mantle in global tomography models has hence provoked doubts about a lower-mantle origin of the Iceland plume (*e.g.* Foulger *et al.*, 2001; Foulger and Pearson, 2001).

Receiver function analysis — Although it does not give direct information about structural details of the lower mantle, the analysis of travel-time differences of P–S conversions from the discontinuities at 410 km and 660 km can also contribute to answering the question for the depth of origin of the plume, because the presence of a thermal anomaly crossing the transition zone of the mantle shifts the stability fields of the olivine polymorphs, such that the transition zone is narrowed a bit in a plume (see section I.2.2); as a result, the difference of the arrival times of P–S conversions from these boundaries is reduced.

The ICEMELT and HOTSPOT data have been used to investigate the existence and magnitude of differential times for arrivals of P–S conversions from the 410 and 660 discontinuities, and a narrowing of the transition zone relative to IASP91 was indeed observed beneath Iceland (Shen *et al.*, 1998, 2002, see fig. IV.6). The latter study showed that the effect was strongest in a region *ca.* 1° south of the junction of the volcanic zones, reaching a thinning by about 20 km; a possible extension of the anomaly further to the south is suggested by the fact that its observed maximum is at the southernmost border of the covered area, but it is probably not well constrained, because relatively few receiver functions per stack were available there. The fact that no significant deviation of the position of the discontinuities relative to IASP91 was found outside the region ascribed to the plume suggests that the plume indeed is rather narrow and that an adequate reference was used in the ICEMELT tomography as well, in spite of the steep incidence of the rays used there; it also indicates that the plume originates from the lower mantle and not from a hypothetical thermal boundary layer at the rw–(pv+mw) transition. Shen *et al.* (1998, 2002) estimate that the plume excess temperature at the depth level of the transition zone is at least 140–150 K and its radius at most 200 km; with respect to the possible influence of the pv → mj transition at *ca.* 660 km, which would reduce the deflection caused by the pv → rw transition, this excess temperature should be regarded as a lower bound and would yield evidence that the plume is hotter than 1800 °C at the bottom of the upper mantle (Hirose, 2002). The horizontal displacement between the transition zone anomalies and the surface hotspot visible in the larger dataset used by Shen *et al.* (2002) suggest that the plume stem is tilted northward by about 9°, which could be explained by a background "mantle wind" either flowing northward in the upper mantle or southward in the uppermost lower mantle, or both. They suggest that a linear tilt is only an approximation and that the plume could rise more vertically in the upper 200 km; this could reconciliate the contradiction between the geophysical and geochemical anomalies especially of the Reykjanes Ridge, which are clearly more pronounced south of the plume, and the flow direction, which would let one expect such anomalies rather to the north of Iceland. A partially tilted, but elsewhere roughly upright plume is also in agreement with the flow deduced from the seismic anisotropy observations of Bjarnason *et al.* (2002) and with kinematic calculations of the buoyant rise of plumes by Steinberger (2000), which led to an estimated average tilt of 19° over the whole depth of the mantle. They propose that the flow velocity vs. depth profile is possibly constant over a considerable depth range above the 410 km discontinuity, but controlled by shear from the plates in its uppermost part, whereas the flow might decrease in the transition zone (Bjarnason, pers. comm., 2002). – Interestingly, in an along-ridge section of the older global tomography model of Zhang and Tanimoto (1993, fig. 23) as well as in the N–S section of Zhao (2001), it rather seems that the plume is tilted southward in the upper mantle, with its deeper parts being located beneath the Kolbeinsey Ridge; while the horizontal resolution of the study by Zhang and Tanimoto (1993) is much coarser than the plume diameter, and the along-ridge section will likely miss the structure seen by Shen *et al.* (2002), which does not lie beneath the Reykjanes Ridge, the resolution of the model by Zhao (2001) is said to be *ca.* 300 km below Iceland and should hence provide more detailed geometrical information.

However, the use of receiver functions for tracking the plume through the transition zone has also been criticized by some authors. Keller *et al.* (2000) note that the thinning effect was not found at several other hotspots and claim that it is thus not hard evidence for a lower-mantle origin of the plume; one should though rather state that it is not hard evidence for the lower-mantle origin of those other plumes. Similarly, Foulger and Pearson (2001) claim that the variations are "local" and can have several other causes such as "normal" lateral temperature variations or varying water content. Unfortunately, they do not describe in more detail what a local T variation is. The effect of varying water content would rather counteract the thermal deviation, if the anomaly is assumed to be associated with higher C_{H_2O} (*e.g.* Wood, 1995; Chen *et al.*, 2002b; Higo *et al.*, 2001), which is suggested by geochemical data and plausible for material derived *e.g.* from an eclogitic fragment of old crust; if the thinning of the transition zone is due to the thermal effect of a plume from the lower mantle, it must be quite hot to overcome the water effect on the phase boundary, but should then be visible beneath the 660 km discontinuity.

Using receiver functions from HOTSPOT registrations, Schlindwein (2001) found a low-velocity zone in the uppermost mantle beneath most of Iceland, except the West Fjords and east Iceland, which increases in strength and shallows towards the youngest parts of the island and ends abruptly at the east margin of the NVZ. She interpreted this LVZ as asthenosphere, and the difference between its top and the depth of the Moho may serve as a constraint on the thickness of the unmolten lithospheric lid, which would hence reach a maximum thickness of 8–10 km.

Controlled-source seismics — Under favourable circumstances, the uppermost mantle can also be probed seismically with controlled sources, if diving rays on long refraction profiles reach deep enough. This was in parts the case on several of the profiles discussed in section IV.1.2. From the SIST data, Bjarnason *et al.* (1993) estimated a mantle P-wave velocity of 7.6–7.7 km/s, based on observed apparent velocities occasionally as high as 7.74 km/s. For the mantle beneath northeastern Iceland, the FIRE data suggest 7.9 km/s under the NVZ and 8.1–8.3 km/s under the adjacent Tertiary areas; the corresponding estimates for the mantle temperature are $\sim 1500\,°C$ under the neovolcanic zone and $\sim 800\,°C$ in the lithospheric mantle (Staples *et al.*, 1997).

IV.2.2   Geochemistry

In this subsection, only a superficial and incomplete review of some more recent publications out of the wealth of geochemical investigations from the last three decades can be given. Central issues in these investigations are how many and what kind of mantle sources are involved in basalt production on and around Iceland, and to which extent these sources mix.

Number and origin of geochemical sources — On Iceland itself, one expects that the volcanism is characterized geochemically by contributions from the plume and from the normal oceanic asthenosphere; however, it is still subject to controversy whether the common MORB source really contributes to Icelandic volcanism and how many non-MORB sources are involved. There is agreement on the notion of a chemically heterogenous plume, but some workers postulate a major contribution from a depleted MORB source in addition to two other reservoirs like HIMU or EM-1 (Hanan and Schilling, 1997; Hanan *et al.*, 2000; Stracke *et al.*, 2003), whereas others find the influence of MORB on Iceland to be small to nil and regard the significant depleted component as a non-MORB one from the lower mantle. An old MORB-like mantle source might though have been picked up by the ascending plume at the 660 km discontinuity and carried upward without mixing, but rather as a "sheath" (Thirlwall, 1995; Fitton *et al.*, 1997; Hardarson *et al.*, 1997; Kempton *et al.*, 2000), which

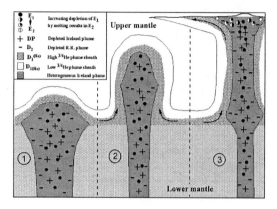

Figure IV.8: Schematic geochemical structure of the Iceland plume and its evolution (from Murton *et al.*, 2002, © 2002 Oxford University Press, reproduced with permission).

would influence more remote parts down the ridges (*e.g.* Murton *et al.*, 2002, see fig. IV.8); the latter view has been backed by numerical models by Farnetani *et al.* (2002). – Yet other workers prefer models with essentially two sources, whereby for the character of these sources suggestions range from MORB-source (Mertz and Haase, 1997) or another common north Atlantic endmember (Ellam and Stuart, 2000) together with a second, possibly regionally varying endmember to different types of sources derived from more local studies (*e.g.* Nicholson and Latin, 1992; Breddam, 2002). Nicholson and Latin (1992) explain the compositional variability they observe by imperfect mixing of melt from different depths, the deepest originating from about 140 km, from which they deduce a plume potential temperature as high as 1580 °C. Hanan and Schilling (1997) remark that the difference between binary and ternary mixing models cannot be resolved by REE analysis, because REE composition can also be modelled by mixing of pooled fractional melts from different depths in a homogeneous source, as was also shown by Slater *et al.* (2001).

A frequently found signature in many studies is that of recycled oceanic lithosphere, to which both the crustal and the lithospheric section contribute (Chauvel and Hémond, 2000; Skovgaard *et al.*, 2001; Breddam, 2002; Peate *et al.*, 2003), and it has been proposed as a candidate for the depleted non-MORB mantle source (*e.g.* Skovgaard *et al.*, 2001). However, opinions seem to differ on whether the entire composition can be explained by upper-mantle sources (Chauvel and Hémond, 2000) or whether certain undegassed lower-mantle components (*e.g.* FOZO) are also required (Stecher *et al.*, 1999; Hilton *et al.*, 1999; Breddam, 2002).

While the previous arguments are largely based on analyses of Pb, Sr, Nd, Hf, and REE, major element analysis also provides some information about the plume source. Scarrow *et al.* (2000) find that the source of the early Iceland plume was unusually Fe-rich (Mg#=0.855), similar to other flood basalts and comparable to basalts produced by anhydrous melting of the pyroxene-rich rock HK-66 (see section D.1.1). Similarly, Korenaga and Kelemen (2000) came to the conclusion that lavas from Þeistareykir and from southwestern Iceland and the Reykjanes Ridge must have different sources, the latter showing an unusually low Mg# of less than 0.88. The presence of old recycled oceanic crust, *i.e.* an eclogitic component in the plume source was proposed as a very probable explanation for this result, because it would provide the necessary Fe enrichment; such a component has also been postulated for other plumes, *e.g.* Hawaii (Feigenson, 1986).

On and beneath Iceland itself, Breddam *et al.* (2000) could show a strong correlation

Figure IV.9: Along-ridge positions of the ^{3}He/^{4}He, Bouguer, and seismic velocity anomalies (slightly modified from Breddam *et al.*, 2000, © 2000 Elsevier, reproduced with permission).

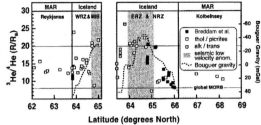

between the plateau-like maximum of the ^{3}He anomaly and the seismic velocity and gravity minima of the plume conduit (figure IV.9), which enabled them to deduce a width of not much more than 100 km for it; this is a very low value for r_P (see table IV.2), but it must be kept in mind that it might rather reflect the width of the damp melting zone, which is probably narrower than what would be considered as "the plume" with regard to the thermal or seismic anomaly. As they note, the strong incompatibility of He makes much of it leave the source already during the initial hydrous melting stage, *i.e.* before the plume has already undergone significant lateral spreading due to its interaction with the lithosphere or the melting-induced increase in viscosity; this could explain why the correlation with the geophysical signals from the plume stem is stronger for He than for other elements such as Pb. A certain fraction of the He is transported with the deflected plume material, though, which indicates that the viscosity effect is notable below the dry solidus already. – It should be noted that the distinct role of He as a tracer for the plume is only a proof for a plume origin in, or at least a contribution to it from, the lower mantle if the only possible source for a ^{3}He excess is indeed the lower mantle. Although a correlation between the plume and He is obvious, none of the mentioned studies yields a proof for an origin of the ^{3}He from the lower mantle. Similarly, an unusually high, primordial ratio of ^{22}Ne/^{21}Ne in some samples from Reykjanes and central Iceland has also been interpreted as possible evidence for a small undegassed lower-mantle component in the plume source (Dixon *et al.*, 2000; Moreira *et al.*, 2001); a corresponding solar ratio of ^{3}He/^{4}He was not observed in those samples, though, and was explained by decoupling of He and Ne *e.g.* during melting or mixing of different sources.

Water content — An issue of special interest with regard to melting, rheology and several geophysical observables is the water content of the mantle and the question whether the plume has a water content different from that of normal mantle (see section II.3.2). Recent investigations of basalt samples from the northern Atlantic (Jamtveit *et al.*, 2001), and especially from along the Reykjanes Ridge and Iceland itself (Nichols *et al.*, 2002) yield evidence that indeed the plume source has water concentrations of more than 300 ppm, probably as much as 620–920 ppm, whereas for the source of MORB at the southern Reykjanes Ridge a value of only 165 ppm has been found, which agrees well with independent estimates for the normal upper mantle (Wood, 1995; Saal *et al.*, 2002); the boundary between both sample sets is about 650 km from the plume center, thereby providing another constraint for the influence of the plume. Nichols *et al.* (2002) remark that there is considerable variability between samples from Iceland and suggest that the highest values might be partly due to the production of enriched melts under glaciated areas (Jull and McKenzie, 1996). Schilling *et al.* (1999) reaffirmed the high contents in water and other volatiles, especially CO$_2$, in the plume and their possible importance for the transport and distribution of He and other noble gases both to the south and the north of Iceland (see below).

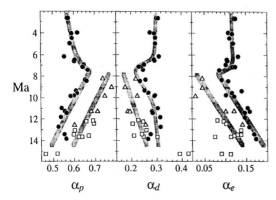

Figure IV.10: Temporal variations of the Pb mass fractions of the "depleted (d)", "plume (p)", and "enriched (e)" endmembers of Hanan and Schilling (1997); filled symbols are for eastern Iceland, open symbols for western Iceland (from Hanan and Schilling, 1997, © 1997 Elsevier, reproduced with permission).

Temporal variations — The mixing ratios of the three components of Hanan and Schilling (1997) vary spatially and also have not been constant over time: plots of mixing proportions resp. their ratios vs. eruption time (figure IV.10) show a decrease of the "enriched component (e)" contributions from 15 to 9 Ma and a pronounced pulse of the p component between 6 and 10 Ma, peaking at 7.5–8 Ma and coinciding with a boost of volcanic activity. This observation is in agreement with evidence for plume pulsing over the past 70 Ma with a periodicity of 3–8 Ma as recorded in several seamounts (*e.g.* O'Connor *et al.*, 2000), or more spectacularly in the V-shaped structures of the Reykjanes Ridge, and also with crustal thickness estimates from seismological observations (Allen *et al.*, 2002b). – However, temporal variability of the plume is not the only possible explanation for geochemical variations with time. As mentioned earlier, other authors (*e.g.* Hardarson *et al.*, 1997) suggest that the magmatism is controlled by an overriding lithosphere of varying thickness and by the opening and dying of rifts at different places. For instance, their analysis of basalts from northwestern Iceland shows a systematic shift of Pb isotope ratios between successive lava flows of the same location, but erupted from different rift zones.

Reykjanes and Kolbeinsey Ridges — The geochemical footprint of the plume is not restricted to Iceland, but extends along the Reykjanes Ridge (Schilling, 1973), as visible in REE, Pb, and ^{3}He anomalies, which seem to correlate well with *e.g.* the thermal anomaly over more than 1500 km (Taylor *et al.*, 1997); they claim that the plume actually contaminates a very large part of the North Atlantic asthenosphere and that the geochemical trends along the ridge can be explained best by mixing of MORB and plume melts in continuously varying proportions. The combined crustal thickness and geochemistry model of Maclennan *et al.* (2001a) suggests though that the region where melting is immediately affected by the upwelling of the plume is limited to a ∼150 km-radius zone beneath central Iceland and does not reach to the Þeistareykir volcano in the NVZ.

The presumed along-ridge flow of plume material can be expected to provide a radial, though distorted, geochemical cross-section through the plume and offer samples of melts from the marginal parts of the plume. In particular, material from the abovementioned sheath (Fitton *et al.*, 1997; Kempton *et al.*, 2000) does not enter the central part of the plume and is thus most likely to produce melts in ridge sections somewhat remote from Iceland (figure IV.8). A detailed analysis of the chemical abundances and Nd–Sr–He isotopy of a large set of samples from several hundred kilometers of the Reykjanes Ridge by Murton *et al.* (2002) displays not only the typical along-ridge gradients in several geochemical observables, but exposes also a chemical variability which lets these authors propose a total of six different sources of melt in Iceland and its surroundings, two of which are related with the sheath. In agreement with the link between certain sources and certain parts of the plume, not all of these sources are found everywhere in the region, but some are restricted to Iceland, whereas others only occur on the ridge.

As mentioned in section IV.1.1, the north–south asymmetry of the Icelandic crust is one of the most prominent tectonic characteristics and suggests that a corresponding asymmetry exists in the plume–MOR interaction, at least on the shallowest level, where the mantle melts. While a fairly gradual transition from the OIB-type, diverse Icelandic signature to MORB can be observed along the Reykjanes Ridge, the situation at the Kolbeinsey ridge is less clear. While Mertz *et al.* (1991) did not find plume–MORB mixing trends north of the TFZ especially in the Pb signature and defined the TFZ as a geochemical boundary, newer studies with a larger dataset yielded indication for influence of the plume north of Iceland, *e.g.* elevated ^{3}He/^{4}He (Taylor *et al.*, 1997; Botz *et al.*, 1999; Schilling *et al.*, 1999; Chauvel and Hémond, 2000). In a synoptical investigation of Pb, Nd, Sr, and He systematics, Schilling *et al.* (1999) emphasize the importance of He for the assessment of the influence of the Iceland plume and declare the Spar Fracture Zone, which is located half-way between Iceland and Jan Mayen resp. *ca.* 300 km north of the TFZ, as the approximate boundary, although high-^{3}He/^{4}He signature of Iceland extends even further northward. The signature of V-structures as far north as the northern Kolbeinsey Ridge and Jan Mayen pointing away from Iceland, which was reported by Jones *et al.* (2002) (see section IV.1.1), would also support the arguments in favour of a significant influence of the Iceland plume on the ridge in the north, and the synchronism of plume pulses on Iceland and planform anomalies around the Kolbeinsey Ridge observed by Abelson and Agnon (2001) let these authors suggest that during pulse episodes, a northern branch of the plume extending beyond the TFZ is transiently active; possible support of this notion can be taken from regional (*e.g.* ICEMELT) and global tomography (*e.g.* Zhang and Tanimoto, 1993; Bijwaard and Spakman, 1999; Zhao, 2001; Grand, 2002), where a northward continuation of the low-velocity anomaly is seen in the shallow upper mantle. Conversely, Mertz and Haase (1997) had argued that MORB source mixes with different enriched end-members, of which the southern is present in the plume; unfortunately, the boundary between both domains is not pinpointed, but one would suspect that it is related to the TFZ.

IV.2.3   Geodynamical modelling

Analogue and (semi-)analytical modelling — In their laboratory and numerical experiments on chemical plumes ascending beneath a spreading ridge, Feighner and Richards (1995) and Feighner *et al.* (1995) tried to apply their results for the shape of the steady-state plume head to the Iceland plume and came to the conclusion that it has now, *i.e.* after 60 Ma, become stationary, judging from the comparison of an observational value of the plume's waist width with their results of 850 and 900 km, respectively. Although it is possible that the Iceland plume has reached a quasi-stationary state now, it is still not clear how large the plume head actually is, given *e.g.* the unclear picture from seismic tomography at shallow mantle levels and the possible influence of the TFZ in northern Iceland, so that the quantitative assessment does not seem very reliable; strictly speaking, their implicit assumption that the plume does not pulsate is in contradiction with observations from gravity and geochemistry, and the relative motion of plume and ridge also is neglected. Another problem with applying these experiments to Iceland is the assumption of a (nearly) isoviscous rheology with $\eta_0 = 10^{21}$ Pa s, which is a strong simplification; it is difficult to assess whether thermal weakening and fusion-related dehydration stiffening would compensate to make this an acceptable approximation to the rheology of the upper, say, 150 km beneath Iceland.

On the basis of REE and isotope geochemistry, topography data, mass balance considerations (White, 1997), and ^{238}U–^{230}Th disequilibria (Elliott, 1997), the magnitude of active upwelling beneath Iceland was found to be rather moderate, and certainly much smaller than in Hawaii; White (1997) conjectures that much of the plume material is readily integrated into the diverging oceanic plates and follows its motion. From seismics-based crustal thickness measurements and REE concentrations the active upwelling was estimated to be 2–4 times (Allen *et al.*, 2002a; Holbrook *et al.*, 2001) or even up to ten times (Maclennan *et al.*, 2001b) larger than the mass flux due to rifting; higher values (5–20 cm/a in the center) are also supported by newer ^{238}U–^{230}Th data (Kokfelt *et al.*, 2003). The variations between these models might be attributable to different assumptions, *e.g.* concerning the potential temperature, melting entropy, or the depth for which the estimate holds, but the significant common conclusion is the importance of active flow for explaining the anomalies found beneath Iceland. In the widely cited study by Sleep (1990), a volume flux of 2.23 km³/a is given[2], which corresponds to a maximum upwelling velocity of about 4.6 cm/a according to eq. B.13, if one uses $R_P = 2r_P = 250$ km and assumes that the r_P values in table IV.2 represent the radius at which the anomaly has decayed to $1/e$ of its maximum. Schilling (1991) estimated a flux of 1.43 km³/a using eq. I.12, which would correspond to an upwelling at a maximum of about 3 cm/a for the same R_P; it should be noted that, although this agrees with the results supporting a stronger active upwelling in that it is three times the half-spreading rate, it does not imply that his high estimate of $\Delta T = 263$ K is correct, because the flux can also be produced in part by chemical buoyancy *e.g.* of depleted material, which he did not take into account in his model.

Numerical convection modelling — In the 1990s, a number of numerical experiments has been conducted on the interaction of plumes and ridges, many of them with special emphasis on Iceland. In general, they have matched some of the field observations while contradicting others.

Ribe *et al.* (1995) have used their numerical models of a buoyant plume (see section I.3.2) to estimate the dynamic and total topography of the Iceland plume and favour a cool, broad, and slow plume ($\Delta T = 93$ K, $r_P > 300$ km, $v_{zP} \approx 2$ cm/a; see table IV.2), which is in contradiction to geochemical results (cf. section IV.2.2), while a narrow, hot plume would produce far too much melt; it must be noted, though, that they assume ideal fractional melting, do not account for details of melt migration and assume a rather low mantle potential temperature of only 1324 °C, so that it is not clear if their melt production results are appropriate, although their values for normal oceanic crust are realistic. Ito *et al.* (1996) made a similar investigation, but assumed a slightly higher $\mathcal{T}$ of 1350 °C and included some effects related to melt dynamics, in particular buoyancy contributions from depletion and melt retention, although they did not explicitly model melt migration. Nonetheless, their observations are basically consistent with those of Ribe *et al.* (1995) in that their cool, broad plume ($\Delta T = 75$ K, $r_P = 300$ km) fits better the observed values of crustal thickness, topography, and gravity, while their hot, narrow plume ($\Delta T = 170$ K, $r_P = 60$ km) can better explain estimates of total crustal production rates, geochemical anomalies and the seismic

---

[2]In the paper of Sleep (1990), actually a buoyancy·flux of 1400 kg/s is given; this buoyancy flux is transformed to a volume flux estimate using the parameters mentioned in table IV.2 and eq. B.14 for better comparison. It should be noted that this estimate includes the possible flux of the weak plume postulated as the cause for the Jan Mayen hotspot. On the other hand, it is still questionable whether Jan Mayen is really the surface expression of a plume; *e.g.* it is far from being ranked primary in the Courtillot *et al.* (2003) compilation, because there is only little safe information.

tomography from ICEMELT. They find that the dynamically reinforcing depletion and melt retention terms and the damping effect of latent heat on temperature are secondary effects compared with thermal buoyancy and tend to cancel out; thus, the effect of melting on large-scale convection would be minor, as far as depletion, melt retention and latent heat are concerned, but its influence on dynamical topography, the mantle gravity signal or P-wave traveltimes is significant, especially for the hot, narrow plume. Both Ribe *et al.* (1995) and Ito *et al.* (1996) fail to reproduce channelling of plume material into the ridge, because the viscosity ranges accurately tractable with their methods do not allow for sufficiently large contrasts for low-viscosity jets being flushed into the ridge; however, Ito *et al.* (1996) note that their estimates for total excess crust production by a hot plume, which agree well with observations, suggest that along-axis melt generation is significantly higher than in their model.

A major progress was made by Ito *et al.* (1999) by also taking into account the effect of dehydration on viscosity (see section II.2.1). They found that the viscosity increase by loss of water to deep low-degree melts, which they assume to be of about two orders of magnitude, largely suppresses active buoyant upwelling, leading to a significantly lower melt production rate compared with earlier models without dehydration and subjecting control over upwelling in the melting zone mostly to the spreading rate of the plates; this is consistent with the results of Braun *et al.* (2000) for the dry melting domain and removes several problems with the hot, narrow plume from Ito *et al.* (1996). Modelling of incompatible element concentrations indicates an enriched source, but still does not match the field evidence for gradual mixing of plume and MORB source by along-axis melt migration. Melting is restricted to $z > 40\,\mathrm{km}$ in these models by assumptions about hydrothermal cooling at the ridge and application of a criterion for the exhaustion of clinopyroxene as a melting switch (see section II.3.2). It is actually possible that Ito *et al.* (1999) underestimate the effect of water and melting dehydration, because they do not assume an elevated water content for the plume (see section IV.2.2). However, while it is reasonable to assume stiffening of the mantle, a contradiction to viscosity estimates from postglacial rebound measurements (*e.g.* Sigmundsson, 1991) and postseismic/postdiking deformation observations (Pollitz and Sacks, 1996) emerges at this point, because these indicate values of $3 \cdot 10^{18}$–$10^{19}\,\mathrm{Pa\,s}$ for the uppermost mantle beneath Iceland. Apart from the influence of non-Newtonian behaviour, Ito *et al.* (1999) proposed that estimates from rebound models are essentially sensitive to the structure of the uppermost mantle, which can be controlled significantly by accumulated or retained melt, while its effect on larger-scale mantle convection would be marginal.

A very peculiar feature of the Iceland plume never reproduced by stationary plumes are the V-shaped patterns of the Reykjanes Ridge (see section IV.1.1). It has repeatedly been suggested that they were generated by pulsations of the plume and a concomitant enhancement of the injection of plume material into the ridge (Vogt, 1971, 1976). Recently, Albers and Christensen (2001) have succeeded in producing a very similar structure in numerical models with large viscosity contrasts between plume and lithosphere and low spreading rates; Yale and Phipps Morgan (1998) had already observed in models with a thermal and compositional lithosphere including a narrow asthenospheric channel beneath the ridge that excess flux from a plume would be strongly focussed and yield a dynamic topography in good agreement with observations if the plume is hot and narrow. However, the model of Albers and Christensen (2001) results in a very narrow ($r_{\mathrm{P}} \approx 40\,\mathrm{km}$), hot plume stem with a very low viscosity ($< 10^{17}\,\mathrm{Pa\,s}$), which seems to contradict seismic tomography models and estimates of viscosity and crust generation, whereas a broader plume would have too large a buoyancy flux (cf. table IV.2). – As an alternative, Ito (2001) proposed an extension of the model of Ito *et al.* (1999) with a plume of varying radius, but constant maximum ΔT and moderate viscosity contrast. The radius variations of the plume bring ring-like pulses of abnormally hot material towards the edge of the lithosphere, where they spread radially, conserving the higher temperatures preferentially in the along-axis portions while conductive cooling damps them below older lithosphere. The propagating pulse generates a V-shaped crustal thickness and gravity anomaly; this mechanism does only work if dehydration-enhanced viscosity is included in the model. Ito (2001) points out that in spite of the flow velocity decrease predicted for radially spreading pulses, the curvature of the V-shaped structures is only weak, so that the patterns appear almost as linear as pulses of along-axis injections would be. – While the models of Albers and Christensen (2001) and Ito (2001) attribute the V-shaped patterns to variations of flow in the plume conduit, one could also speculate that the combination of a plume–ridge setting with the mechanism of radially propagating waves in a plume head found by Bercovici (1992) (see section I.3.2) could lead to such an anomaly. It is unclear, though, if this idea can be brought in agreement with the results from geochemistry, seamount generation etc. – A somewhat different mechanism leading to V-shaped patterns was recently found by Rabinowicz and Briais (2002) in three-dimensional models of a system including a hotspot and segmented ridges which can adopt their position to the current stress field of the lithosphere. This model features a complicated polyhedral pattern of convection cells and small-scale convection rolls perpendicular to the ridge which results in significant along-axis variations in temperature and crustal thickness as well as in the formation of transform faults. A superimposed along-ridge large-scale convection current leads to a corresponding drift of the local anomalies beneath the spreading plates, resulting in a V-shaped structure which in this case is not controlled by temporal variations of the plume.

A special topic not included in these models is the initial phase of plume activity during arrival and spreading of the plume head. Similar to earlier studies, Nielsen *et al.* (2002) proposed that the plume accumulated some material beneath a thin spot of the lithosphere before rifting produced the actual break-up and allowed for fast and voluminous melting with a large amount of low-viscosity material travelling along the rift; they remarked that a viscosity increase due to strong melting and dehydration would have prevented the early Iceland plume from spreading over large distances along the opening ridge and suggest that the plume had only moderate excess temperatures of 100–200 K. Larsen *et al.* (1999) also have modelled the early plume with a model with non-Newtonian, p–T-dependent viscosity and propose that the nonlinear rheology of the plume enabled its fast spreading beneath the lithosphere and can thus explain the almost simultaneous appearance of flood basalts over a large area from western Greenland to the northern British Isles. On the other hand, as they remark themselves, plume branching beneath Greenland would be an alternative explanation, and is possibly better supported by the strong anomaly visible in seismic tomographic images of the upper mantle in that area (Bijwaard and Spakman, 1999). Keen and Boutilier (2000) modelled the rifting process with a similar rheology and emphasis on small-scale processes in the rift and found that the models which matched best east-Greenlandic crustal thicknesses require a plume head with a temperature contrast of only about 100 K and a thickness of only 50 km; as this value is not expected to increase in later stages of plume evolution, it might provide an explanation why the plume head is not visible in the ICEMELT upper-mantle tomography images, which do not resolve structures above 100 km depth.

IV.2.4 Synopsis

In the last years, the structure of the Iceland plume has become somewhat clearer thanks to the combined efforts of seismological surveys, geochemistry, and numerical modelling, especially as far as the upper mantle is concerned; the characterization of the plume in the lower mantle has evolved at least so far that Iceland is now viewed as one of the best-established examples for a plume which indeed traverses the whole mantle. Nonetheless, the number of geochemically distinct sources and the depth of the plume's root remain unresolved.

Seismic tomography and convection models have converged in the last years towards a model of a rather hot and moderately narrow plume in the upper mantle, which leads to a significant amount of active upwelling: the values in table IV.2 suggest that the maximum temperature anomaly at a depth of about 200–300 km lies between 150 and 200 K and that the radius of the plume is 100–150 km; the extreme r_P values from geochemistry and global tomography are less reliable due to the lack of resolution of these methods. It is, however, remarkable that the chemical ΔT_P estimates tend to higher values than those of other methods. This could be due to insufficient consideration of source heterogeneity, *e.g.* contributed by an eclogitic component, but it is also possible that the simplifications in $\Delta v_{P,S}-\Delta T_P$ conversion and the limited resolution in seismological models and the numerical limitations of convection models fail to reproduce a very narrow plume core with significantly higher temperature and lower viscosity. To resolve this issue, more models with the ability to handle large temperature and viscosity contrasts have to be calculated, and the resulting tomographic images have to be determined. Possible deficiencies of the numerical models to reproduce the flow field of the plume accurately are also indicated by the large discrepancy between volume flux estimates from the preferred convection models of Ito *et al.* (1999) and Ito (2001) and those from other methods; they could be due to an overestimate of the radius, but again, it is also possible that there are too strong simplifications in the models with a lower volume flux. The fact that the temperatures for the initial plume head have been found to be a bit lower than those of the stem (Keen and Boutilier, 2000; Nielsen *et al.*, 2002) is not at odds with the above estimates, though, because some cooling by conduction and melting is to be expected.

One of the fundamental questions concerning the Iceland plume is its depth of origin, and this is an issue which is closely linked to the debate whether plumes exist at all. Most researchers seem to support the notion that plumes from the CMB do indeed exist, and that Iceland is located above such a plume; as an example, in the assessment of the primary-plume character of several hotspots by Courtillot *et al.* (2003), Iceland matches four out of five criteria, the remaining one being the existence of a hotspot track, which was regarded as not definitely proven, because much of a possible track was suppressed or masked by the presence of the Greenland craton. Nonetheless, the location of the source zone of the Iceland plume has been regarded as unclear and not well constrained since the first geodynamic models which tried to relate the hotspot activity on Iceland to the plume concept of Morgan (1971), although it seems that preference has traditionally been given to an origin at the CMB (*e.g.* Vogt, 1974).

The picture resulting from seismological findings beneath the shallow mantle is particularly confusing, the deeper the level of the mantle under consideration is, the more. In general, it must be accepted that the ability especially of global tomography to resolve and image the shape and magnitude of deep low-velocity anomalies is still very limited compared to the imaging of fast anomalies, because they tend to be thinner, are often less well sampled and can be masked by effects such as wavefront healing (*e.g.* Grand *et al.*, 1997). In principle, the poorer visibility is also at least qualitatively expected from mineral physics, because the anharmonic variation of seismic velocities with temperature depends strongly on thermal expansivity for a given composition (see appendix D.1.8); as α of transition-zone and lower-mantle minerals is considerably smaller than α of upper-mantle minerals under the respective p–T conditions, the same holds for seismic velocity anomalies at different depths. The presence of a gap in the LVZ in the mid-mantle beneath Iceland (see section IV.2.1) is not necessarily evidence against the CMB origin of the plume either, because it is known from several independent investigations that plume conduits thin, possibly to a size beyond seismic resolution, or even disrupt (see p. 3). On the other hand, all tomographic studies agree that there is a pronounced anomaly in the upper mantle, and all yield some evidence of a low-velocity structure with a complicated shape in the lower mantle as well. Neither the ICEMELT nor the HOTSPOT data provide a reliable image of it beneath a depth of 400–450 km, but Foulger *et al.* (2001) remark that the HOTSPOT data do not show its bottom, which suggests that it reaches at least to 660 km depth; nonetheless, and in spite of a model fidelity test they present, which shows a fade-out of a cylindrical test anomaly beneath 413 km (their figure 12), they claim that the upwelling is confined to the upper mantle – a conclusion which is not warranted by their data, given the depth limit of resolution. Recently, Ritsema and Allen (2003) demonstrated that a regional network with an aperture of more than 1000 km and dense (50–100 km) station spacing would be necessary to track the anomaly down into the lower mantle; all regional tomography experiments on Iceland were far from reaching such a large aperture. To justify their notion of a shallow origin of the plume, Foulger *et al.* (2001) propose that the tabular shape of the anomaly in the lower part of the upper mantle, which they see in some of their sections, could be due to convection induced by the opening of the ocean and by lateral temperature gradients at the edges of the adjacent cratons; focussing of such tabular upwellings into cylindrical diapirs is corroborated by numerical and laboratory convection models. Even if one accepts that the deformation is indeed real – which is not clearly visible in the data, as *e.g.* comparison of the v_P and v_S anomaly shows (Foulger *et al.*, 2001; Allen *et al.*, 2002a) – there is though still a possible alternative notion that the low-viscosity conduit of the plume is deformed by the northward mantle flow, for whose existence evidence was presented on the basis of seismic anisotropy measurements; the effect would be strongest beneath the asthenosphere, while in the asthenosphere the lower viscosity would allow for a less disturbed cylindrical upwelling. This would also be in agreement with a tilted plume as proposed by Shen *et al.* (2002); as these authors noted, downward smearing of the anomaly of a tilted plume could also mimic a tabular shape at greater depths. As Ritsema and Allen (2003) made clear, the question of the depth of origin cannot be decided on the basis of seismological data at the present stage, but the preliminary evidence for a deeper origin from global tomography and receiver function analysis would justify a larger regional investigation.

If the plume stems from the lower mantle, one would expect this to be expressed in its geochemical signature; therefore, many studies have been conducted to clarify this in the context of the question how many and what sources are involved. Although there is a great many of different ideas concerning the number and character of sources, some points seem to be basically agreed upon. Obviously, the mantle beneath Iceland and its surroundings is quite heterogeneous, and especially analyses of noble gases provide strong evidence that there is a contribution from the lower mantle, although it seems that this contribution is not dominating. Another important result is that old recycled oceanic lithosphere – probably both the crustal and the mantle component – are significant sources of Icelandic melts. It would be interesting if evidence for recycled old lithosphere can also be derived from other than geochemical data. One could speculate that the weak high-velocity zone visible in some global tomography models in the transition zone or at the top of the lower mantle near Iceland is a candidate for such an old slab, but recycled material could as well stem from the lower mantle. In contrast, it is still controversial whether depleted mantle sources on Iceland are identical with the MORB source or not, while in more remote regions,

MORB mantle is certainly involved. Along the Reykjanes and Kolbeinsey Ridges, gradients are observed in several geochemical markers. They are a manifestation of the decreasing influence of the plume with distance due to mixing with other mantle sources, but could also partially represent a horizontal cross-section through the spreading plume head with different parts of a sheath previously wrapped around the ascending plume producing varying melts at different distances from the axis, as proposed *e.g.* by Murton *et al.* (2002) (also see figure IV.8).

Bathymetry, gravity, and crustal thickness measurements show though that the influence of the plume towards the south is more pronounced. The V-shaped pattern of these anomalies also yields evidence for temporal changes in the flux and melt production of the plume, which are also observed in the chemical signature. These changes could be caused by the relative motion of the plume and the mid-Atlantic ridge and concomitant ridge jumps or by irregularities in the plume itself, or by a combination of both. The passage of the plume beneath lithosphere of varying thickness would cause changes in melt production in the first case; an argument for the second case is the actual irregular form of the plume and the visibility of variations of the anomaly in tomographic images (cf. figure IV.5). The significance of the lithosphere-controlled mechanism has not yet been clarified in numerical models, which up to now have considered the current ridge-centered setting, but not the motion of a plume from beneath a craton to an occasionally shifted ridge, and the details of the mantle-controlled model are also still ambiguous, as several mechanisms have been proposed.

CHAPTER V

RESULTS OF NUMERICAL MODELLING

In this chapter, the results of numerical modelling of a plume–ridge system, usually a ridge-centered plume, performed in this study are presented. It is divided in two sections: the first shows and discusses some general effects related to melting, phase transitions, and material properties like those described in chapters I and II in models of simplified design; the second attempts to approach realistic mantle conditions more closely in order to provide results comparable to the situation of Iceland (chapter IV). However, in all models the choice of certain parameters is tied to those of the Iceland hotspot; in particular, the half-spreading rate is kept constant at about 1 $^{cm}/_{a}$, $i.e.$ the influence of changes in v_{r} will not be investigated.

Before going through the models, it is necessary to explain some general aspects of the default model design of this study apart from those specific to Iceland which will not, or only in exceptional cases, be changed. This concerns in particular the choices for the potential temperature and the melt extraction threshold.

As mentioned in sections II.3.1 and II.3.2, the potential temperature of the mantle is a fundamental quantity in melting models, yet it is not very well constrained (cf. table II.1). Many authors use a value of 1350 °C ($e.g.$ Ito $et\ al.$, 1999) or a bit less, but it seems that in those cases where the dynamics of fractional melts are considered explicitly ($e.g.$ Ghods and Arkani-Hamed, 2000), higher $\mathfrak{T}$ of 1400 °C or even more are preferred, because otherwise tighter constraints like the thickness of normal oceanic crust cannot be matched; this is also in agreement with geochemical investigations which predict a generally lower productivity for fractional melting compared to batch melting, on which the lower $\mathfrak{T}$ are usually based ($e.g.$ Langmuir $et\ al.$, 1992; Nicholson and Latin, 1992). This observation was also persistent in the models made during this study and has eventually resulted in a choice for $\mathfrak{T}$ of 1410 °C as a default. However, it should be borne in mind that usually anhydrous melting of a peridotite mantle is assumed; the anticipated increase of the amount of produced melt in the case of hydrous melting of peridotite, at least for passive upwelling, or of mantle with a small eclogitic component might allow for slightly lower potential temperatures.

Another crucial parameter is the melt extraction threshold φ_{max}. As detailed in section III.2, the treatment of melt migration is ruled by the principle that melt in excess of a certain threshold will not remain in the mantle and percolate slowly through the matrix, but will be flushed from the melting region to the eruption site at the spreading center via a fast transport path. This is a simple, approximate way of representing the bimodal transport of melt in the mantle by channels, dikes or whatever similar structure described in section II.4.2. In most cases where melt segregation was modelled, it was decided to keep the threshold at quite low values, usually at 1 %. The rationale is firstly, that several independent pieces of evidence from rock physics and geochemistry indicate that especially in deforming mantle, melt connectivity is established at very low porosities already ($e.g.$ McKenzie, 1985b; Zimmerman $et\ al.$, 1999), as discussed at length in sections II.1.5 and II.2.5, and secondly, that normal oceanic crustal thicknesses from models with a higher threshold were notoriously too small compared with observations, unless potential temperatures were assumed which were regarded as unreasonably high. Postponing a discussion of the trade-off between $\mathfrak{T}$ and φ_{max} to section V.2, where it is considered in the framework of Iceland-like models, and chapter VI, it is only noted here that such a trade-off exists and would allow for lower $\mathfrak{T}$ in combination with lower φ_{max}.

The crustal transport mechanism has been chosen to realize the most simple constellation of the possibilities analyzed in section II.5.2, namely that the crust formation zone and the spreading zone have the same width and that the width does not change along the ridge, *i.e.* it is the same above the plume and at normal MOR; material is added uniformly over the whole x extent of the crust formation zone. This choice causes the crustal thickness to be approximately constant along x profiles across normal ridge; it seems to be the most natural choice and avoids the formation of crustal features near the spreading center which are not well constrained or would have to be regarded as artefacts, because the corresponding structures in nature cannot be resolved well by the numerical grid.

V.1 GENERAL EFFECTS RELATED TO MELTING, PHASE TRANSITIONS, AND MATERIAL PROPERTIES

V.1.1 Mantle buoyancy

As discussed in sections I.1.2, I.1.4, I.2.1 and I.2.2, several factors influence the buoyancy of upwelling mantle material: variations in temperature, chemical composition resp. depletion, melt content or phase transitions. As several aspects of these factors have already been investigated, albeit mostly with models without plumes (Sotin and Parmentier, 1989; Niu and Batiza, 1991; Scott and Stevenson, 1989; Su and Buck, 1993; Cordery and Phipps Morgan, 1993; Barnouin-Jha *et al.*, 1997), it is not attempted to repeat these studies and explore the related effects in a large parameter space. However, as these effects are of substantial importance for plume–ridge systems as well, four simple models have been run to test the program for reasonable results and probe its reaction on effects which will play a role in more complex models.

In order to isolate the effects of buoyancy, the models have been kept very simple: all models are isoviscous, start with an adiabatic geotherm combined with the $T(x, z)$ distribution for a spreading mid-ocean ridge, and no phase changes have been included. Melting is included to provide depleted material and melt, but no temperature changes due to melting enthalpy are calculated ($L_m = 0$); migration of melt is not considered. The essential model parameters are listed in table V.2. All models cover a timespan of 50 Ma in order to let the plume evolve properly and reach a state where it does not change appreciably anymore, at least in its central parts. In the first model (SB1), only the buoyancy due to the excess temperature of the plume is accounted for; in the second model (SB2), the effect of density reduction due to depletion is added, but the melt is assumed to have no effect on the density; in the third model (SB3), the depleted material is assumed to have the same density as the unmolten mantle, but the density effect of retained, incompressible melt with a density and a thermal expansivity different from that of the solid is included. To emphasize the effect of retained melt, a porosity up to 10 % is allowed for; in the mantle, however, such high values are in general not to be expected, as discussed in sections II.1.5 and II.2.5. The fourth model (SB4) features a chemical instead of a thermal plume: a material with a density lower by $-30.195\,\mathrm{kg/m^3}$ is pumped through the circular influx zone of the model bottom with the usual

SB1	thermal buoyancy only
SB2	thermal and depletional buoyancy
SB3	thermal and porosity-related buoyancy
SB4	chemical buoyancy only (model with two sources)

Table V.1: Model runs of series SB.

Convection grid		
(x_m, y_m, z_m)	grid dimensions	(1500 km, 1000 km, 660 km)
		108×72×100 pts.
η_0	viscosity	10^{21} Pa s
g	gravity acceleration	9.9 m/s²
ϱ_0	mantle reference density	3660 kg/m³
c_p	isobaric specific heat	1350 J/kg K
$\varkappa$	heat conductivity	3 W/m K
α	mantle thermal expansivity	$3.3 \cdot 10^{-5}$ 1/K
$\mathcal{T}$	mantle potential temperature	1410 °C
v_r	half spreading rate	1 cm/a
ΔT_P	maximum plume excess temperature	250 K
r_P	plume radius	125 km
v_P	influx velocity at plume center	3 cm/a
$\Delta \varrho_C$	density difference mantle/chemical plume (SB4)	-30.195 kg/m³
Melt grid		
(x_f, y_f, z_f)	grid dimensions	(611 km, 1000 km, 153 km)
		133×214×24 pts.
$\Delta \varrho_{dp}$	density difference fertile mantle/residue (SB2)	-150 kg/m³
ϱ_{f0}	reference melt density (SB3)	3000 kg/m³
α_f	melt thermal expansivity (SB3)	$6.3 \cdot 10^{-5}$ 1/K
φ_{max}	threshold porosity (SB3)	0.1

Table V.2: Model parameters for the simple buoyancy (SB) models.

v_z profile, but without the corresponding profile on the concentration, because the chemical boundary of a plume is expected to be much sharper than the thermal; this density difference corresponds to the density decrease $\varrho_0 \alpha \Delta T_P$ in the center of the thermal plume source in models SB1–SB3.

In the earlier stages of the runs, the background mantle beneath the axis rises at about the velocity imposed by the spreading ridge. Due to the relatively high viscosity, the plumes are also relatively slow, compared with later models with temperature-dependent viscosity: their maximum vertical velocities, which are observed at mid-depths, lie around 5 cm/a for the thermal plumes and at *ca.* 7.5 cm/a for the chemical plume, and they reach the lithosphere base after 23–25 Ma in SB1–SB3 and some 2 Ma earlier in SB4; plume centerline v_z does not change much after that. The stronger buoyancy of the chemical plume might be due to the fact that its density contrast does not shrink during ascent, because it does not experience diffusion as do the thermal plumes; moreover, the chemical plume is a sharp, massive structure whose density contrast is at its maximum throughout the whole volume, whereas the thermal plumes are fuzzy and have a smaller density contrast at their outer parts.

The buoyancy differences between the thermal plumes are smaller and more restricted to the upper, near-ridge regions of the model, because their cause has its origin in the shallow melting processes there. For SB2, given a melting degree of about 0.28 under normal ridge and of up to more than 0.4 in the plume head, one reaches density differences of -42 kg/m³ and up to more -60 kg/m³, corresponding to temperature contrasts of about 370 K and 540 K, respectively; such high values would not be reached in nature, of course, because enthalpic

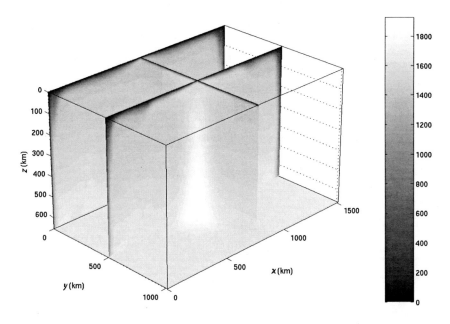

Figure V.1: Temperature field (in °C) of run SB2 at $t = 34\,\text{Ma}$.

cooling will lower the temperature of the melting mantle, and phase exhaustion will effectively shut down melting at significantly lower f (see section II.3.2). The density contrast in SB3 is even larger ($-77\,\text{kg}/\text{m}^3 \mathrel{\hat{=}} 672\,\text{K}$), but is equally far from what is expected in nature, because a porosity of 0.1 will not be reached in the whole mantle region (see section II.1.5); furthermore, the compressibility of the melt was not included here. – The $v_z(z)$ profiles in figure V.2 illustrate the differences between mostly passive upwelling at normal ridges and active upwelling within a well-evolved plume. The plume profiles show how the buoyant effect of the melt-related processes is restricted to the upper part of the model, mainly to the depth range of the melting region, whereas thermal and source-determined chemical buoyancy affects the large-scale dynamics of the upwelling. Differences between depletion-driven and melt-driven buoyancy are probably related to the different spatial distributions of the buoyant masses: the buoyant melt is spatially more confined.

As all models start with completely unmolten and undepleted mantle, a layer of light depleted and melt-bearing material develops beneath the lithosphere and moves outward. This leads to a gravitational instability of the adjacent mantle, especially of the cold lithosphere (figure V.1). It shows a tendency to form drops, and at the boundary between depleted and undepleted mantle, where the contrast is largest, descending lobes of colder lithospheric material are a common feature. Obviously, they will modify the normal flow field and are considered to be a consequence of the initial conditions, because in nature the first stages of the formation of the depleted layer take place in a lithospheric framework different from the one imposed as a starting condition in these models. However, in subsequent models in general a p–T-dependent rheology is used, and the density contrast is less than here; therefore, the effect is smaller, although it is still present.

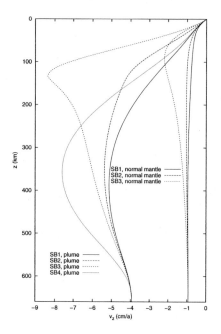

Figure V.2: $v_z(z)$ profiles at the spreading ridge for models SB1–4. The normal ridge profiles are taken at $t = 2$ Ma, the plume profiles at $t = 34$ Ma; the normal ridge profile of SB4 is virtually identical to that of SB1, because the difference between the models is restricted to the plume. v_z is negative, because the material flows upward, while by convention the z unit vector points downward.

V.1.2   Phase transitions

As mentioned in section I.2.2, several phase transitions occur in the upper mantle and have an effect on its thermal and density structure which is, or might be, significant. Interest has focussed on the phase transitions of olivine because of their magnitude and their relation to seismic discontinuities, but also on the transitions of the Al-bearing phase, because they lie within or near the melting zone beneath MOR and are important for melt chemistry and possibly also for the thermal structure. Therefore, the algorithm for handling the olivine phase transitions used by Marquart *et al.* (2000) has been extended and generalized to include both types of phase changes (see section III.4.3). To illustrate the effects, four models have been calculated: PT1 has no phase changes and is practically identical to SB1, PT2 contains a linear, continuous transition from olivine to ringwoodite as described in section III.4.3, PT3 and PT4 include the garnet–spinel and both the garnet–spinel and the spinel–plagioclase transition, respectively. The characteristic model parameters are listed in table V.4, and characteristic geotherms are shown along with phase boundaries in figure V.3.

The olivine–ringwoodite transition is visible as a linear superadiabatic decrease of temperature in $T(z)$ profiles in PT2–4 (see figure V.3); the temperature drop in excess of adiabatic cooling is 46 K beneath normal ridges and 52 K in the plume, in excellent agreement with the

PT1	no phase transitions
PT2	olivine–ringwoodite transition
PT3	olivine–ringwoodite and garnet–spinel transitions
PT4	olivine–ringwoodite, garnet–spinel, and spinel–plagioclase transitions

Table V.3: Model runs of series PT.

X_{ol}, X_{gt}	reference olivine and garnet contents	60.1 %; 7.3 %
Olivine–ringwoodite (PT2–4)		
$\Delta\varrho$	density contrast	$302\,^{kg}/_{m^3}$
ΔS	entropy change	$53.779\,^{J}/_{kg\,K}$
ϑ	Clapeyron slope	$2.863\,^{MPa}/_K$
$T_{\vartheta 1,2}$	p intercepts of Clausius–Clapeyron curves	$-3500\,°C$; $-4700\,°C$
Garnet–spinel (PT3–4)		
$\Delta\varrho$	density contrast	$-51\,^{kg}/_{m^3}$
ΔS	entropy change	$-73.007\,^{J}/_{kg\,K}$
ϑ	Clapeyron slope	$\frac{p-1.5}{0.01p+0.305}\,^{MPa}/_K$
ΔT_{ϑ}	T shift of Clapeyron curve	$1400\,°C$
$p_{\vartheta 1,2}$	poles of Clausius–Clapeyron curves	$1.4\,GPa$; $1.6\,GPa$
Spinel–plagioclase (PT4)		
$\Delta\varrho$	density contrast	$-70.3\,^{kg}/_{m^3}$
ΔS	entropy change	$-130.269\,^{J}/_{kg\,K}$
ϑ	Clapeyron slope	$0.8\,^{MPa}/_K$
$T_{\vartheta 1,2}$	p intercepts of Clausius–Clapeyron curves	$350\,°C$; $300\,°C$

Table V.4: Model parameters for the phase transition models; other parameters are as in table V.2. Variations are relative to a garnet lherzolite in the olivine stability field, which is the default reference mantle material. The Clapeyron slope of the garnet–spinel transition is the inverse of the derivative of eq. D.14, with p in GPa. The $\Delta\varrho$ and ΔS values for the olivine and the Al-phase transitions have been taken from, or determined after, Marquart *et al.* (2000) and Phipps Morgan (1997), respectively.

theoretically expected values. As expected, the transition interval from the α phase to the γ phase is shifted to greater depth within the plume, by about 20 km from 420 km in normal mantle; noting that this value is slightly larger than the seismically determined depth of 410 km, it turns out that the position of the boundary can be used to calibrate the reference density ϱ_0 of the model, at least to within the resolution of the model discretization. – The $v_z(z)$ profile in figure V.3 shows its effect on the buoyancy of a plume: while both the PT1 and the PT4 plume exhibit some velocity variations in the deep upper mantle, possibly when adjusting their thermal buoyancy to the imposed influx velocity, the PT4 plume experiences an additional acceleration in the depth range of the ol–rw transitional regime, which also results in a slight increase of the Nusselt number. As expected, such an effect is not observed in normal mantle, where the velocity is essentially controlled by passive upwelling due to plate spreading and no lateral density variations result in additional buoyancy.

The garnet–spinel in PT3–4 and the spinel–plagioclase transitions in PT4 have a smaller, but nonetheless visible and in certain situations non-negligible magnitude; the temperature shifts are 6.5–7 K and about 12 K, respectively, and are apparently both well reproduced by the numerical model, under normal MOR as well as in the plume; however, the latter transition lies at the edge of, or within, the thermal boundary layer and cannot be seen itself as a kink as can the other two. With respect to geochemical considerations it should be noted that if melting had been included, it would have begun already in the uppermost part of the garnet stability field even at normal ridge and could have generated the garnet signature reported in MORB, as mentioned in sections II.2.5 and II.3.2. The plume, where the transition is shifted to greater depth by more than 30 km, would definitely have begun to melt

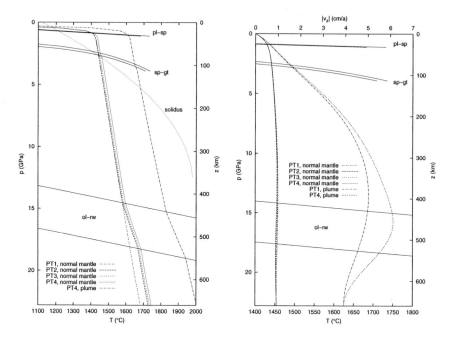

Figure V.3: Some spreading center geotherms (left) and $v_z(p, z)$ profiles (right) from models PT1–4, and phase transitions; the curves labelled as "normal mantle" show $T(p, z)$ resp. $v_z(p, z)$ beneath the axis of a ridge location remote from the plume, at 6 Ma model time, whereas the plume curves go through the centerline of the full-grown plume at 36 Ma model time. In the left plot, the solidus of lherzolite (eq. II.25c, Hirschmann, 2000) is drawn for convenience, although melting was not included in these models. The phase boundaries are drawn far into the supersolidus region for better orientation only; they should rather end near the solidus if melting occurred.

deeper in the garnet stability field (figure V.3), probably to the point of garnet exhaustion, so that the kink due to the transition would be reduced, if not absent (cf. section V.1.6). Of course, the geotherms would be refracted at the solidus if melting took place. – The phase transitions related to the Al-bearing phases have no appreciable effect on upwelling velocity, neither in the plume, nor in normal mantle. This confirms the finding of Scott and Stevenson (1989) (see section I.2.2).

In this context a remark concerning the potential temperature is advisable. The potential temperature is defined here as the temperature the mantle would have if it were transported adiabatically to the surface without melting (see p. 35); hence, if all phase transitions are included, the geotherm is higher at greater depth than in a mantle model with less or without phase changes. In this model series, the temperature in the depth range of the spinel stability field, where most of the normal mantle melting would occur, is higher by 12 K in PT4, compared with PT1–3, and $T(z = 660 \, \text{km})$ lies 46, 53, and 65 K above the transition-free PT1 in runs PT2, PT3, and PT4, respectively.

Figure V.4 shows a snapshot of the temperature and phase fields of run PT4 at 20 Ma model time, *i.e.* when the plume is about to reach the lithosphere. As in former models, the widening of the plume head is well visible in the T field. The acceleration at mid-mantle

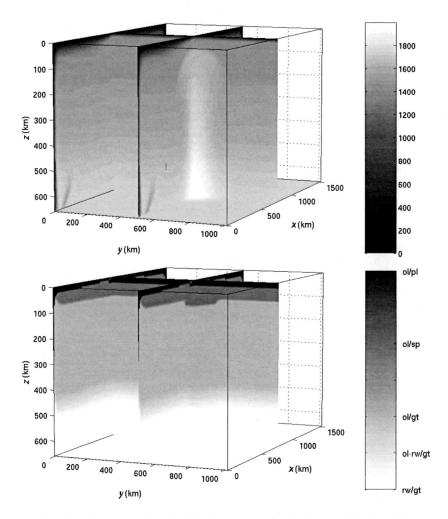

Figure V.4: Temperature (top, in °C) and phase fields (bottom) of run PT4 at $t = 20$ Ma.

depths leads to some thinning, and conductive heat loss diminishes the magnitude of the anomaly; nonetheless, the $T(z)$ profiles in figure V.3 show that it reaches the lithosphere with still considerable strength. The phase field shows two important features apart from the essentially horizontal layering of the phase stability fields in undisturbed mantle, namely, a downward shift of the lower boundary of the spinel stability field in the plume head and a depression of the olivine–ringwoodite transition interval, which are both caused by the higher temperatures in the plume. The effect on the spinel–garnet transition is likely to be smaller if melting is included, but the effect on the 410 km discontinuity is of great interest with respect to the seismological detection of plumes and has been applied to the Iceland plume by Shen *et al.* (1998, 2002) (see sections IV.2.1 and VI.2). – It should be noted that the stability field of plagioclase is not determined very accurately due to the lack of constraints on the phase boundary at low p–T conditions; it is, however, in qualitative agreement with expectations in that it reaches to greater depth in the hot, shallow mantle at the spreading center, compared to the adjacent cooler lithosphere.

An interesting detail of phase transitions is the thermal precursor in the upstream direction of the flow (see section I.2.2), which in principle should be visible in the models as well. However, for the model parameters for the olivine transition used here, both the univariant and the divariant analytical model of Schubert *et al.* (2001) yield a critical velocity of about $5 \, \mathrm{mm/a}$ and a precursor amplitude of less than $1 \, \mathrm{K}$, decaying over a distance shorter than the vertical resolution of the model; therefore the absence of the precursor in the model is in agreement with theory. For the garnet–spinel transition, the critical velocity is about $2.6 \, \mathrm{cm/a}$, which is more than the upwelling velocity in normal mantle; however, the precursor again is too small to be resolved[1], and is not significant anyway.

V.1.3   Variable thermal expansivity

In the frame of this study, the possibility of accounting for the p–T dependence of α had also been considered (see section I.2.3). While fundamental aspects of the effect have been investigated by Schmeling *et al.* (2003), the extension of the method as detailed in section III.4.4 will be applied here to the particular framework of a ridge-centered plume. Four different models have been run, whose essential common features are the inclusion of all phase transitions present in model PT4 of section V.1.2 and a potential temperature of $1410 \, °\mathrm{C}$: Model VA1 is identical to PT4; model VA2 uses a depth-dependent α given by

$$\alpha(z) = \frac{\gamma c_V \varrho_0}{K_{T0}(1 + c_1 z + c_2 z^2)}, \tag{V.1}$$

where $\gamma = 1.26$, $\varrho_0 = 3500 \, \mathrm{kg/m^3}$, $K_{T0} = 129 \, \mathrm{GPa}$, and $c_1 = 4.133 \cdot 10^{-7} \, \mathrm{m^{-1}}$, $c_2 = 1.18 \cdot 10^{-12} \, \mathrm{m^{-2}}$ are parameters derived from fitting a second-order polynomial of K_T/ϱ to PREM for $25 \, \mathrm{km} \leq z \leq 660 \, \mathrm{km}$ (Schmeling, pers. comm., 2002; Kreutzmann *et al.*, 2004); model VA3 has p–T-dependent α, but assumes a rather simple harzburgitic mineralogy without phase transitions for determining α, with X_{ol} as in table V.4 and olivine and orthopyroxene being solid solutions with Mg#=0.9 (figure V.5). Model VA4 works the same way, but is based on a mantle composition approximating more closely the full complexity of major mineral phases, including their phase transitions; the details of this model are described in the appendix on the application of the method for calculating thermoelastic properties, appendix E. With this set of models, it should be possible to assess which degree of simplification in the choice of α yields a good compromise between realistic physics and numerical efficiency. It is important to note that the adiabatic gradient in the simple model VA1 is assumed constant (except the shifts at the phase boundaries), *i.e.* the geotherm is a linearized solution of eq. II.18b, whereas the starting adiabats of VA2–4 are based on a full exponential solution; this results in notably different mantle temperatures at the bottom of the model.

The $T(z)$ profiles in figure V.6a show that the normal mantle geotherms hardly differ in the upper *ca.* $250 \, \mathrm{km}$ of the mantle, but diverge especially in the transition zone, so that VA1 is about $21 \, \mathrm{K}$ hotter at the model bottom than the PREM-derived VA2. VA3 and VA4 are nearly identical and give a base temperature *ca.* $40 \, \mathrm{K}$ below that of VA2; as the difference between VA1/2 and VA3/4 at the top of the transition zone is about $20 \, \mathrm{K}$, it is justified to regard both petrological α determinations as good approximations to reality, at

---

[1] A model with doubled vertical resolution gave results differing by only 1 K and did not permit to pinpoint the precursor convincingly either. This indicates further that the resolution for these runs was adequate.

VA1	constant α, linear adiabat
VA2	depth-dependent α, adiabat after eq. II.18
VA3	p–T-dependent α for fo–fa–oen–ofs mineralogy, adiabat after eq. II.18
VA4	p–T-dependent α for 18-endmember mineralogy of app. E, adiabat after eq. II.18

Table V.5: Model runs of series VA.

least for the upper mantle above the transition zone. In the plume center, for which the profiles have been taken at $t = 32$ Ma, when the T field did not change appreciably anymore there, the differences between the models are a bit stronger in the upper mantle as well, but they are still remarkably similar and do not differ by more than *ca.* 25 K; this is true even for VA3 and VA4 above the transition zone, although they (and also VA2) start at lower temperatures than VA1, because their adiabatic gradients, which are proportional to α, are significantly lower in the transition zone. The differences in the plume geotherms are important because they control the temperatures of the plume in the melting region and the depths at which the onset of melting and the spinel–garnet transition occur, as can be seen from the position of the corresponding kinks in the plume geotherms; this also affects the chemistry of produced melts. The $v_z(z)$ plots in figure V.6b do not display very large differences in the ascent velocities, but in the depth range of initial melting in the plume, *i.e.* around 100 km, they can become as big as 17 %, which would result in somewhat enhanced melt productivity in the models with z-dependent or even p–T-dependent α. The differences are at least in part due to subtle differences in α. As one would expect, the VA1 plume is the fastest one in the transition zone, but the slowest in the shallow upper mantle, whereas VA3 and VA4 are slower in the transition zone and faster at small depths, because their thermal expansivities, and therefore their buoyancies, increase with decreasing z, in agreement with the conclusions of previous studies. As can be seen in figure V.6c, α reaches the highest values in the hottest, shallowest parts directly beneath the spreading center, especially in the plume head. However, it must be recalled that these models do not include melting; therefore, the absence of enthalpic cooling might lead to an overestimate of the solid rock α, unless the residue has a higher bulk expansivity. – It should be emphasized here that the prescription of a vertical velocity at the model bottom as a boundary condition is a

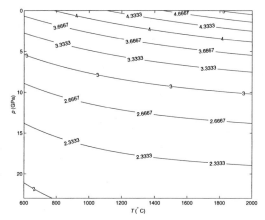

Figure V.5: $\alpha(p,T)$ for the simplified fo–fa–oen–ofs system of VA3. Comparison of this map with the $\alpha(p,T)$ plot for a more complex mineralogy in figure E.2 shows a close agreement for p–T conditions of the olivine stability field, but a significantly higher expansivity in this system in the p range of the transition zone.

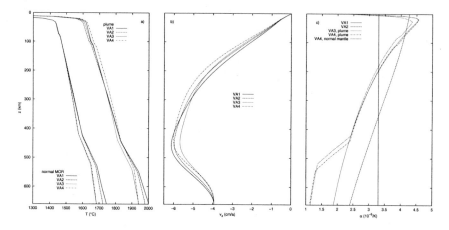

Figure V.6: Some spreading center geotherms (a), $v_z(z)$ profiles (b) and $\alpha(z)$ profiles (c) from models VA1–4; the curves labelled as "normal MOR" or "normal mantle" show profiles beneath the axis of a ridge location remote from the plume, at 4 Ma model time, whereas the plume curves go through the centerline of the full-grown plume at 32 Ma model time.

strong restriction, which largely reduces the dynamic effect of α in the lowermost parts of the model: the large differences in α in the transition zone between the four models would result in greater variability in the upwelling patterns than is observed here; the more complex models would show a weaker upwelling at great depth due to their smaller expansivity, if v_z were not imposed at z_m.

Two details related to the full three-dimensional variability of α in models VA3 and VA4 are worth some further consideration, although their physical effect is only secondary. First, the generally α-increasing effect of the temperature leads to a local enhancement of the buoyancy in the plume in addition to the obvious thermal buoyancy; this effect would not appear in a model with only p-dependent α as VA2, where all mass elements at the same depth level have the same expansivity. Second, the differences in the depth where a phase transition is crossed, also give rise to a contrast in α, as can be seen in the "plume" and "normal mantle" geotherms in figure V.6c and in figure V.7. As both the ol–rw transition and the gt–sp transition occur at greater depths in the hot plume and in both cases the lower-p phases have a larger expansivity, the plume experiences a stronger buoyancy in the transitional regime than normal mantle in model VA4, again in addition to thermal buoyancy and buoyancy due to chemical and mineralogical differences which are present anyway in all models. Figure V.7 shows the relative α contrast of the plume with respect to the background mantle and reveals that this effect is greatest in the ol–rw transition, where it reaches about 10 % of the background value. Comparing the $v_z(z)$ profiles of VA3, where the mineralogical change related to the phase transition was not included in the computation of α, and VA4, where it was included, one notes that both profiles begin to differ visibly at *ca.* 530 km, near the lower boundary of the phase transition, and that the VA4 plume indeed rises a bit faster; above the phase transition, both profiles are approximately parallel up to a depth of *ca.* 250 km, possibly because α is practically identical there in both models. The difference in the upwelling velocities between both models reaches some 6–7 %, which suggests that it is related to this effect.

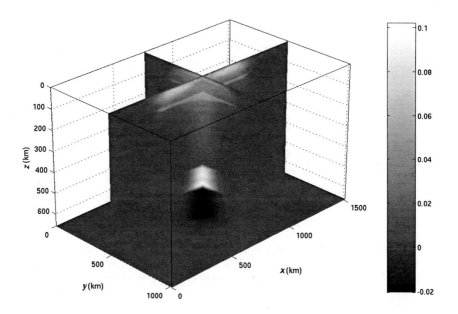

Figure V.7: $(\alpha_{\mathrm{pl}} - \alpha_{\mathrm{bg}})/\alpha_{\mathrm{bg}}$ in model VA4 at $t = 32\,\mathrm{Ma}$; α_{bg} is the expansivity of the background mantle, taken at $y = 0$.

V.1.4   Variable viscosity

In order to make the effect of other factors controlling convection more easily visible, the models shown so far have used a constant viscosity. For later models, which strive for a higher degree of realism, this simplification is dropped, and the effects of pressure resp. depth, temperature, melt content and dehydration will be considered according to eq. III.4.1. In a first series of models, individual effects are highlighted separately; to keep these models simple, phase transitions are not considered, and α is assumed constant again. It should be noted that now, as the model setup comes closer to that of the more complex models in terms of rheology, which attempt to reproduce the possible situation beneath Iceland, the treatment of the side boundaries is changed, and the convection box is made twice as large in the x direction, according to the remarks on p. 76.

First, the immediate effects of pressure and temperature are considered: model VV1a has a weakly ($b_T = \ln 10 = 2.303$), model VV1b a strongly ($b_T = \ln 1000 = 6.908$) temperature-dependent rheology; model VV2 combines the T dependence of VV1b with a pressure dependence using $b_z = 2.303$; model VV2a additionally assumes a constant melting enthalpy of $550\,\mathrm{kJ/kg}$, whereas in the other models melting was not included, but the other melt-related parameters were chosen in such a way that melt would have no other effect on buoyancy or rheology. A model with only z-dependent rheology, actually VV2 minus the T dependence, resulted in a very thick, slow plume greatly different from the other examples; a comparison would not give relevant insights, so that this case is not considered further. In all these models a scaling viscosity η_0 of $5 \cdot 10^{22}\,\mathrm{Pa\,s}$ was used.

The temperature term in the rheological law causes the cold boundary layer at the top to be stiff, with the maximum viscosity being η_0 at the surface, whereas the depth term,

VV1a	weakly T-dependent rheology, $b_T = 2.303$
VV1b	strongly T-dependent rheology, $b_T = 6.908$
VV2	p–T-dep. rheology, $b_T = 6.908$, $b_z = 2.303$; no enthalpic cooling during melting
VV2a	like VV2, but with a melting enthalpy of $550\,\mathrm{kJ/kg}$
VV3a	isoviscous
VV3b	φ-dependent rheology, with $\leq 1\,\%$ retained melt; $s = 28$
VV3c	like VV3b, but with $\leq 3\,\%$ retained melt
VV4	p–T–φ–C_{H_2O}-dep. rheology, $b_T = 6.908$, $b_z = 2.303$, $s = 28$; no enthalpic cooling
VV4a	like VV4, but with a melting enthalpy of $550\,\mathrm{kJ/kg}$; chemically homogeneous
VV4b	like VV4a, but with a more hydrous plume

Table V.6: Model runs of series VV.

if present, leads to a relatively moderate increase of the viscosity with depth; hence, the combination of both effects results in a quasi-asthenospheric low-viscosity zone while also enabling the inclusion of local lateral variations and is therefore a better model than the use of either temperature- or depth-dependent viscosity of whatever kind. In model VV1a, the T dependence is rather weak, so that even the hot parts of the mantle have a rather high viscosity: even in the hottest, and hence least viscous region, the source of the plume, η is as high as *ca.* $0.1\eta_0$. As a result, the plume ascends slowly and remains quite broad, and even after 25 Ma model time, it is still at a depth of *ca.* 200 km; with regard to the ascent velocity, it should be noted though that the calculated velocity field of this model is maybe not very precisely determined, because the iteration is strongly relaxated. Model VV1b is in stark contrast to VV1a, because the strong T dependence makes the lower part of the box, and especially the plume, quite fluid, the latter reaching viscosities between *ca.* $10^{-3}\eta_0$ and $10^{-4}\eta_0$. The plume is much narrower and faster, so that its head reaches the base of the lithosphere after little more than 5 Ma and then spreads out preferentially in the direction of plate motion, having formed a 1000 km-wide head in the upper 100 km of the mantle after 25 Ma. As the viscosity does not change significantly beneath the lithosphere, the diameter of the plume stem is roughly constant throughout the model, after shrinking to about one quarter of the source region diameter in the first few grid planes above the bottom. The plumes of models VV2 and VV2a, where depth dependence was included, do not differ substantially from VV1b; basically, the higher viscosity in the lower parts of those models slows down the plumes to vertical velocities of about 20 $^{cm}/_a$, *i.e.* less than half that of VV1b, so that the stem is also a bit thicker. The inclusion of enthalpic cooling in model VV2a affects plume dynamics in the shallow mantle only: compared with VV2, the temperature in the melting region is up to ~ 100 K lower in the plume head and 40 K under normal ridge, which results in a viscosity increase by a factor 1.3–1.7 in the plume; in the real mantle, such a temperature difference would probably enhance the viscosity more. The currently or previously molten and accordingly cooled layer of higher-viscosity material also tends to erode the base of the lithosphere a bit, but the effect is only small here. – It should be noted that the anhydrous melting function by Katz *et al.* (2003) was used here for better consistency with forthcoming hydrous melting models; as that parameterization includes a cpx-out criterion, the $T(z)$ and $\eta(z)$ profiles show kinks at the depth where the corresponding melting degree is reached, so that further melting is strongly reduced.

The next set of models focusses on the effect of retained melt: VV3a is actually a constant-viscosity model, VV3b has a maximum of 1 % retained melt, and VV3c has a maximum porosity of 3%; $s = 28$ is used in this and in the following model series. As these models use the McKenzie and Bickle (1988) parameterization, an additional model with $\varphi_c = 0.01$ and the Katz *et al.* (2003) parameterization was computed, but no appreciable difference was observed, although this would probably be different at significantly higher porosities, when a wider range of differences in local melt content is possible. To prevent plume ascent from being too sluggish, a reference viscosity of 10^{21} Pa s has been used, and b_z and b_T were set to zero; melt segregation was not included, which is probably one reason why the maximum effect of melt on viscosity is reached almost everywhere. The effect itself, though, is only minor at these relatively low melt contents, as expected from laboratory experiments (see section II.2.1): η decreases by only 25% in VV3b and by about 50% in VV3c. The differences in the dynamics and the temperature field are stronger: while plume convection in VV3b is only slightly stronger than in the isoviscous model VV3a, the plume head of VV3c spreads more strongly against the top thermal boundary layer, which forms instabilities; this is probably also a consequence of the rather low reference viscosity and probably not important in view of the more realistic rheologies of later models.

Strictly speaking, a model which would only take into account dehydration stiffening would be necessary to complete this collection of models. However, tests have confirmed the expectation that the strong incompatibility of water would result in a strong increase in viscosity and lead to the formation of a rather sharply bounded high-viscosity body evolving into a dry, stiff layer upon plate spreading. As the large viscosity step is difficult to handle and, viewed separately from the other factors, very unrealistic, this case is not shown here. It is noteworthy that the viscosity increase due to dehydration led to a strong local increase of shear in the very region where the flowlines are most sharply bent. This shear resulted in frictional heating by up to some tens of kelvins, but in more realistic models, the lower viscosities are expected to reduce the effect.

The final set of models, VV4, VV4a, and VV4b combines all effects, whereby a strong T dependence was chosen; water is handled according to the method outlined in section III.4.2. The additional feature in VV4a and VV4b compared to VV4 is the inclusion of melting enthalpy as in models VV2 and VV2a; in VV4b, a passive field has been used to mark the plume as chemically different in that it has a higher, but uniform, water content. For these models, the reference viscosity has been set to $5 \cdot 10^{22}$ Pa s again for better comparison with the model subset VV2/VV2a; this η_0 refers to the normal hydrous mantle rock. The strong viscosity increase due to dehydration made it also necessary for numerical reasons to limit the maximum viscosity to η_0, because this value would otherwise be exceeded at least in the upper thermal boundary layer, which initially is not depleted and hence contains water; apart from this, the viscosity increase due to dehydration was limited to a value of 100 times the value without dehydration. For the water content of undepleted mantle in VV4 and VV4a and for the background mantle in VV4b, the uniform value of 140 ppm after Saal *et al.* (2002) has been used, whereas 500 ppm was chosen for the wet plume in VV4b as representative for wetspot source material, in particular for Iceland (see sections II.3.2 and IV.2.2); the exponent b_C in eq. III.16 was set to 1. The partition coefficients are $D_{ol}^{H_2O} = 0.003$, $D_{opx}^{H_2O} = 0.015$, $D_{cpx}^{H_2O} = 0.03$, and $D_{gt}^{H_2O} = 0.003$ after Hirth and Kohlstedt (1996), and the melting proportions are $\mathcal{M}_{ol} = 0.013$, $\mathcal{M}_{opx} = 0.087$, $\mathcal{M}_{cpx} = 0.36$ and $\mathcal{M}_{gt} = 0.54$ after Kostopoulos and James (1992) (also see appendix C.4); the restriction to garnet lherzolite is appropriate, because no phase changes are considered in these models and because dehydration would essentially occur in the garnet stability field even if the spinel–garnet transition had been included. The presence of up to 1% retained melt was taken into account in the manner described on p. 41. As usual in the case of hydrous melting, the parameterization of Katz *et al.* (2003) has been used.

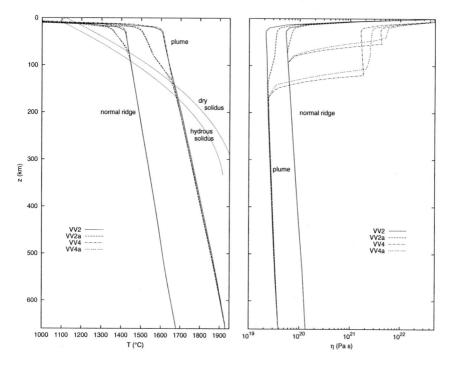

Figure V.8: Summary view of temperature (left) and viscosity (right) depth profiles at the spreading centers of runs VV2, VV2a, VV4, and VV4a at $t = 25$ Ma, showing the effect of latent heat consumption and dehydration on viscosity. The dry solidus is the curve from Hirschmann (2000), the hydrous solidus was calculated from it according to the formalism described in section II.3.3; the increase of the latter in the uppermost kilometers is due to water saturation.

Figure V.8 shows a summary of plume center and normal ridge $T(z)$ and $\eta(z)$ profiles of the p–T-dependent models VV2 and VV2a and the p–T–φ–C_{H_2O}-dependent models VV4 and VV4a at 25 Ma model time. Obviously, the depth profiles of all models are virtually identical outside the melting region. Within the melting region, it is essentially the latent heat which makes a difference between the temperature profiles of each series; furthermore, as melting starts about 30 km deeper in the plume and about 20 km deeper below normal ridge, enthalpic cooling occurs already at slightly greater depths in the hydrous model VV4a than in the dry VV2a, but the low productivity prevents it from becoming important for T or η at that depth. At shallower levels though, the continued melt production leads to considerable cooling and results in a viscosity increase by a factor of *ca.* 1.5.

For models VV2a, VV4, VV4a, and VV4b, figure V.9 shows some more details of T, f, and η in the melting domain. In the $\eta(z)$ profiles, it can be seen that for hydrous melting, there is a steep increase over a few tens of kilometers, until the maximum allowed viscosity is reached after 6–7 % melting; from then on, the rock is considered dry, which results in a constant viscosity. Notably, the reduced melt production when enthalpic cooling is accounted for broadens the hydrous melting interval and results in a smoother viscosity profile for models VV4a and VV4b compared to VV4. Furthermore, the higher temperature

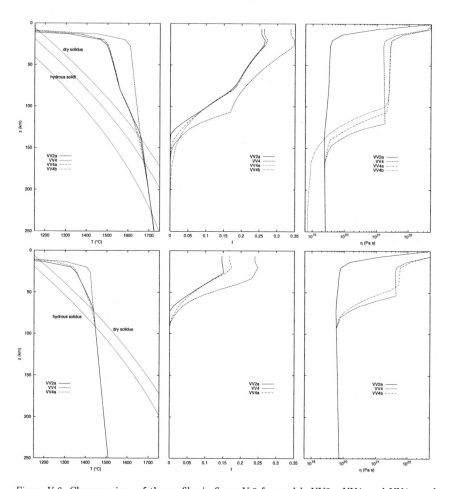

Figure V.9: Close-up views of the profiles in figure V.8 for models VV2a, VV4, and VV4a, and similar profiles for VV4b, showing the effects of latent heat consumption and dehydration on temperature (left), melting degree (center) and viscosity (right) for different dry and hydrous models; upper row: plume, lower row: normal ridge. The hydrous solidus in the $T(z)$ plot for the ridge is that of normal MORB source mantle. The low viscosity of the plume stem in VV4b in and below the damp melting zone is caused by the enhanced water content of the unmolten plume.

of model VV4 results in a much higher melting degree of up to 35 % in the plume, which is certainly unrealistic. By contrast, enthalpic cooling also retards the increase of f, so that the maximum melting degree is about 10 % lower. The cpx-out criterion included in the f parameterization is reached at shallower depths and results in a kink of the geotherm in VV2a, VV4a, and VV4b due to the reduction of melting and cooling; in VV4, there is no kink, because variations in melt production do not affect the T field here. The reduced cooling rate in VV2a, VV4a and VV4b also leads to a very slight decrease of the slope of $\eta(z)$. The reducing effect of φ on viscosity is hardly directly visible in the profiles; it is

152   *Results of numerical modelling*

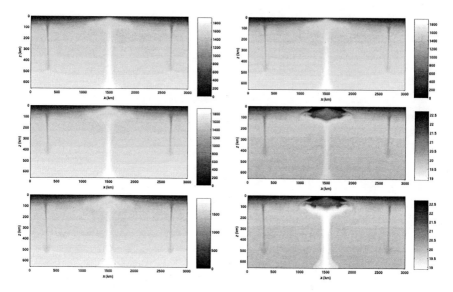

Figure V.10: $x-z$ (*i.e.* perpendicular to the ridge) slices through the plume of T (left column, in °C) and $\log \eta$ (right column, η in Pa s) for models VV2a (top), VV4a (center), and VV4b (bottom) at 25 Ma model time.

masked by the concomitant dehydration stiffening, because the porosity is always very low, but its effect is essentially to flatten the slope of the viscosity increase at the base of the melting region a bit. – Essentially the same effects are observed beneath normal ridge, albeit in a weaker form; in particular, the cpx-out limit is hardly reached, which is in agreement with field observations and melting models which argue for the existence of four mineral phases in the lherzolite throughout the MOR melting zone (see sect. II.3.2; Dick *et al.*, 1984; Kostopoulos, 1991). At the top, the slight reduction of f coinciding with the sharp decrease of T and the increase of η marks the position of the recrystallization zone, where the $\sim 1\%$ retained melt solidifies.

A more comprehensive insight of the effect of different rheologies is provided by cross-sections of the T and η distributions, in particular those through the plume (figure V.10). The temperature fields of VV2a, VV4a and VV4b plotted in this figure seem to be quite similar at first sight, but closer inspection reveals that the ascent of the plume had encountered some hindrance in VV4a and VV4b which led to flattening and spreading of a part of the head at a greater depth than in VV2a: although the plume heads have a similar width, the VV2a head does not reach deeper than *ca.* 100 km, whereas the VV4a and VV4b heads grow out of the stem smoothly at about 150–170 km depth, which corresponds to the onset of hydrous melting. A more spectacular difference appears in the viscosity field: while there is the expected quite close correspondence between η and T in the p–T-dependent model VV2a, the patterns are vastly different in the melting region of VV4a and, even more so, of VV4b, where the additional dependences on φ and, most importantly, on C_{H_2O} are included. In the center of the plume head, a high-viscosity plug has formed due to dehydration, with values of the order of the base of the lithosphere, while still unmolten or hardly molten plume material with a low viscosity has spread out at the bottom of the plug in a bowl-like shape.

Figure V.11: x–z slice through the plume of $\log \eta$
(η in Pa s) for model VV4a at 10 Ma model time.

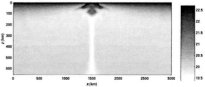

In fact, the steep viscosity increase coincides with a drop of the upwelling velocity to nearly passive upwelling, which is not the case in models without dehydration-dependent viscosity; the contrast is particularly strong in model VV4b, where the plume ascends faster than usual due to the combined effect of temperature and higher water content in the unmolten stem. – The maximum-viscosity cap on top of the plume head is a remnant from the initial melting event in the first timestep. The roof-like feature between it and the plume head is material from normal ridge-melting pushed aside by the upwelling of the plume (also see figure V.11); it has a slightly higher viscosity, because it is a bit cooler than the plume.

As a result of reduced upwelling, the melt production rate is reduced in dehydrating plumes by 10–20 % compared with VV2-type ones (after subtracting productivity of the respective ridges), as can be seen in figure V.12; this is expected to result in a thinner crust in models with dehydration. When water is included, the melting zone reaches deeper, as expected, but most of it does not spread as far away from the ridge as in the VV2-type models as a consequence of the high viscosity of its shallower parts; only the damp melting zone, whose viscosity is still relatively low, extends far off the center. Figure V.12 also shows the weak-melting zone of the stem of model VV4b, which is visible as a narrow, conic structure beneath the main melting regime. Furthermore, a lens-shaped zone with strongly reduced melt production is present in both models in the top of the plume head, but virtually absent under normal MOR; this is an expression of clinopyroxene exhaustion. As an important consequence of the existence of this low-productivity lens, it can be postulated that most of the melts erupted from a ridge-centered plume stem from deeper parts of the plume, so that a large fraction of the melt present in the upper part of the plume head also must have been created at those deeper levels. This effect should also make the productivity of

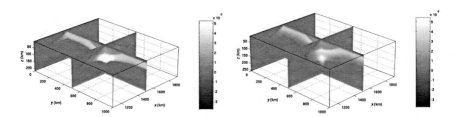

Figure V.12: Melt production rates for a model with p–T–φ-dependent viscosity similar to VV2a (left) and for model VV4b at model time $t = 25$ Ma (right). The high production rates of the left model have been clipped at the maximum of the right one for better comparison; they are up to more than five times as high in a small part of the high-productivity region at the plume center. Another feature of these images to be noted is the dark-coloured shell with negative values surrounding the top of the molten zone; at this boundary, all of the remaining retained melt recrystallizes. Note that the box of the hydrous model reaches *ca.* 40 km deeper than the other one.

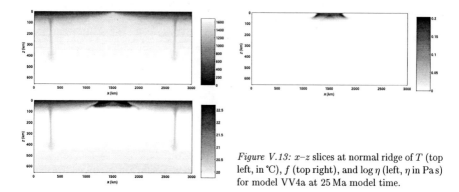

Figure V.13: x–z slices at normal ridge of T (top left, in °C), f (top right), and $\log \eta$ (left, η in Pa s) for model VV4a at 25 Ma model time.

a plume insensitive to the thickness of the overlying lithosphere, although only to a limited extent, which would depend on factors like the temperature anomaly and the water content; a thick crustal root has also been expected to exert some control on the final depth of melting (see section II.3.2), but it would also lie in the region of reduced melting, so that its impact might not be that large. Altogether, three local maxima of melt production can be seen at the plume axis: the deepest one lies in the region where the upwelling rate is still significantly above the passive rate and the productivity given by the concave-up $f(\tilde{T})$ curve already increased, whereas the two shallower ones are located in the depth interval where dehydration has forced v_z to the passive rate; the middle maximum is due to the maximum slope of $f(\tilde{T})$ before reaching the cpx-out point, and the weak shallow one is caused by the late-stage increase of $f(\tilde{T})$ in the cpx-free interval. – Under normal MOR, the melt production rate is far less reduced if dehydration is taken into account, but on the other hand, the total volume containing partial melts is also larger. In particular, the onset of melting occurs at greater depth, namely in the garnet stability field; this has the important geochemical implication of ensuring the presence of a garnet signature in MORB (see section II.3.2).

Under undisturbed ridge, the smaller extent of melting leads to weaker effects, but compositional changes also result in a significant difference in the rheological structure when comparing models VV2a and VV4a. In VV2a, rheology is essentially tied to the thermal structure, which produces a zone of particularly low viscosity in the "melting triangle" directly beneath the spreading center; the p dependence is irrelevant with regard to lateral variations, and the φ dependence can also be neglected at the low retention porosities applied here, for that matter. By contrast, inclusion of dehydration would result in an approximately horizontal rheological structure in the long run in a model like VV4a (figure V.13), similar to the compositional layering, of which it is a result, and quite different from the thermal structure of the lithosphere; although in reality the absence of imposed maximum viscosities of the kind used in these models would still yield some lateral variations, it remains true that the "melting triangle" under the ridge would not have such a low viscosity as the asthenosphere, as it was the case in models with only p–T-dependent viscosity. This supports the earlier findings *e.g.* of Braun *et al.* (2000) and has important implications for melt focussing, as pointed out in section II.4.2.

Although these models have not quite been tuned to reproduce crust formation in a real plume–ridge system and melt segregation was not included, the outstanding contrasts in plume behaviour between anhydrous and hydrous models raise the question how this difference is expressed at the surface. From the mantle picture, one would expect a broader

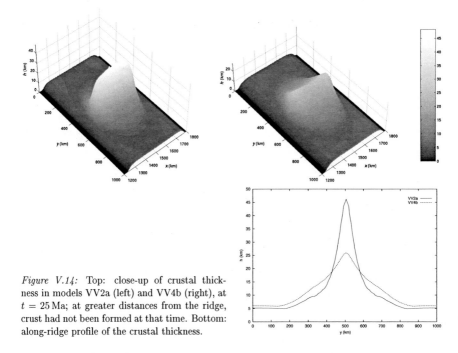

Figure V.14: Top: close-up of crustal thickness in models VV2a (left) and VV4b (right), at $t = 25\,\mathrm{Ma}$; at greater distances from the ridge, crust had not been formed at that time. Bottom: along-ridge profile of the crustal thickness.

region with plume-affected crust in the hydrous models, in addition to a possible slight increase in crustal thickness at normal ridge due to the deeper onset of melting. Figure V.14 shows that this is indeed the case. At normal crust, the thickness is greater by a few hundred meters in model VV4b, compared to model VV2a, because melting begins at greater depth; the crust production above the plume, however, is greatly reduced, with the peak values differing by a factor 1.7, which highlights the importance of active upwelling for melt generation in the plume, which is mostly suppressed in the main melting zone of the hydrous model. On the other hand, the hydrous plume crust covers a broader segment along the ridge, as can be seen from the $h(y)$ profile. The total melt resp. crust production of the plume is nonetheless notably reduced, as forthcoming models suggest (see series ICE3, section V.2.3).

V.1.5   Effect of melting enthalpy

The effect of the consumption of latent heat during melting was already highlighted to some extent by the contrast between models VV2 and VV2a, and VV4 and VV4a/b in the previous section, but only in the sense that this effect is of some significance at all in that it finds its expression in the geotherm as a kink at the bottom of the melting region and a depression of the geotherm over the whole melting interval; furthermore, a constant value of $550\,\mathrm{kJ/kg}$ had been assumed for the melting enthalpy there. As pointed out in section II.1.3, the value of the entropy of melting is not known very well, with the most commonly used values lying in the range from 0.3 to $0.4\,\mathrm{J/g\,K}$ (Hirschmann *et al.*, 1999; McKenzie and O'Nions, 1991; Kojitani and Akaogi, 1997); the constant enthalpy used in the previous models corresponds

ME1	constant melting entropy, $\Delta S_\mathrm{m} = 0.3\,\mathrm{J/_{g\,K}}$
ME2	constant melting entropy, $\Delta S_\mathrm{m} = 0.3268\,\mathrm{J/_{g\,K}}$
ME3	constant melting entropy, $\Delta S_\mathrm{m} = 0.4\,\mathrm{J/_{g\,K}}$
ME2C	constant melting enthalpy, $L_\mathrm{m} = 550\,\mathrm{kJ/_{kg}}$

Table V.7: Model runs of series ME.

to $0.3268\,\mathrm{J/_{g\,K}}$ at the potential temperature of these models. Obviously, prescribing the entropy instead of the enthalpy in the model makes the enthalpy T-dependent, because $L_\mathrm{m} = \underline{T}\Delta S_\mathrm{m}$. To assess the difference in mantle temperature and melt production between these bounds and the importance of the T dependence of enthalpy, four models are compared: models ME1, ME2, and ME3 assume entropies of 0.3, 0.3268, and $0.4\,\mathrm{J/_{g\,K}}$, respectively, whereas model ME2C is the same as VV2a in many respects and features the usual constant latent heat value. – In this and the following model series, segregation of melt by porous flow is included, which was not the case in the previous models.

The model setup for these models and those in the following sections is in general quite similar to the one of model VV2a, but the convection grid was chosen to reach down to $410\,\mathrm{km}$ only, because the transition zone is not relevant to the questions addressed here; accordingly, the temperature anomaly was reduced to a peak value of $180\,\mathrm{K}$ at that depth, which is though a bit less than the temperature difference between the plume and the background in model VV2a. The viscosity in the following models is p–T–φ dependent. In contrast to the VV series, the melting parameterization of McKenzie and Bickle (1988) has been used here.

As melting is a rather localized process and the variations considered in this section are even more so, one would not expect substantial differences in the general characteristics of convection between the three runs. This is confirmed by comparing $v_\mathrm{rms}(t)$ and $Nu(t)$, which are very similar for all models: the $v_\mathrm{rms}(t)$ values differ by less than $2\,\%$ for most of the time, and $Nu(t)$ by less than $1\,\%$, whereby Nu decreases with increasing ΔS_m, because the stronger cooling reduces the vertical temperature gradient and hence the heat flux at the top. In contrast, substantial differences appear in variables which characterize processes in and near the melting zone, *e.g.* T, which is affected by the consumption of latent heat of melting, and possibly also of phase transitions, or f (figure V.15).

The mentioned small differences in Nu are due to different temperatures of the mantle at the top of the melting zone as a consequence of the different melting entropies: the most strongly depressed geotherm corresponds to the smallest Nu. The effect of cooling is cumulative, so that the gap between the model with the lowest ΔS_m, ME1, and the one with the largest, ME3, is larger at the top of the plume axis ($29\,\mathrm{K}$ at $27\,\mathrm{km}$ depth) than at the top of the MOR melting region ($14\,\mathrm{K}$), as can be seen in figure V.15. This looks like an unimportant difference, especially with respect to the much larger uncertainty in mantle potential temperatures – and with respect to several variables, *e.g.* the viscosity, this view is justified –, but nonetheless it results in differences in f of 2.5–$3\,\%$, which are possibly significant for geochemical or petrological conclusions, especially if the degree of depletion comes close to the exhaustion of a mineral phase, which was not explicitly included here. Furthermore, it also yields notable differences in crustal thickness, especially in the case of the plume: the steady-state values of the normal oceanic crust differ by *ca.* $1\,\mathrm{km}$ in models ME1 and ME3, those of the plume by about $7\,\mathrm{km}$ (figure V.16).

The other topic touched in this model series is the question whether there is a significant difference between the use of a constant and a T-dependent latent heat of melting, respectively, whether it matters if a constant enthalpy or a constant entropy is prescribed as an

Figure V.15: Temperature (left) and melting degree (right) for the models of series ME in the depth range of the melting region; upper row: plume, lower row: normal ridge.

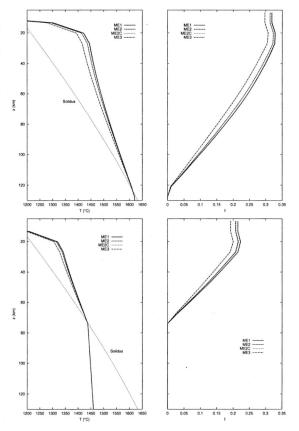

input parameter. Curiously, the choice of model parameters used here turned out to be such that the $T(z)$ and $f(z)$ profiles of model ME2C at the plume axis virtually coincide with those of model ME1, while those at normal ridge are practically congruent with those of model ME2 on the plots in figure V.15. To understand this, one has to compare the melting enthalpies and corresponding temperature changes at different depth levels, *e.g.* for simplicity near the top at 27 km and at the bottom of the melting region of the plume and the ridge at *ca.* 127 km and 74 km, respectively; in model ME2C, the enthalpy is 550 kJ/kg everywhere. In the plume of ME1, L_m is 515 kJ/kg and 568 kJ/kg at the top and bottom, respectively, whereas it is 557 kJ/kg and 617 kJ/kg in the ME2 plume; hence, the top and bottom values of model ME1 bracket the constant ME2C value quite closely, but those of model ME2 are both higher. Under normal ridge, the top and bottom enthalpies are 484 kJ/kg and 513 kJ/kg in model ME1 and 525 kJ/kg and 558 kJ/kg in model ME2, respectively; thus, the ME2C value lies within the ME2 range here, whereas the ME1 values are lower. Applying eq. II.21 with $c_p = 1350$ J/kg K and $\bar{\varphi}$ of less than 1 %, one can estimate that the difference in the temperature drop between model ME2C and its variable counterpart is at most of the order of 0.1 K. The T difference at the top of the melting region is only 10 K in the plume and about the half under normal ridge, thus the difference is indeed minor. However, it should be recalled

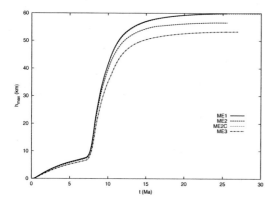

Figure V.16: Maximum crustal thickness as a function of model time, for model series ME. The strong increase between 5 and 10 Ma is closely related to the beginning of plume melting; earlier times give an idea of steady-state MOR crust thicknesses, later times of plume crust thicknesses.

that the total range of proposed values of ΔS_m is larger, so that it is desirable to constrain the enthalpies better.

Remarkably, the ME1, ME2, and ME3 curves in figure V.16 are roughly equidistant in the high-h_{max} segments corresponding to the plume, although the melting entropies are not; intuitively, one would expect the ME3 curve to lie deeper, because the difference between ΔS_m in that model to that of model ME2 is about three times as large as the one between ME1 and ME2, and this is indeed observed in the $t \lesssim 7$ Ma segment of the plot corresponding to MOR crust. The $h_{max}(\Delta S_m)$ plot in figure V.17 illustrates the problem in more detail: while the data for the normal ridge can be fitted very well with a straight line (standard deviation $2.7 \cdot 10^{-2}$ km), the linear fit for the plume data is poor (standard deviation 1.38 km). The most likely explanation is that the models with a smaller entropy resp. enthalpy remain hotter and reach a higher depletion, so that they come into the part of the $f(\tilde{T})$ parameterization eq. II.28a where the curve flattens, *i.e.* the productivity is reduced, whereas this effect is not as strong for the high-ΔS_m model ME3; hence, the latter produces relatively more crust. Under the ridge, all models remain in the almost linear segment of $f(\tilde{T})$, so that the linear fit works well. – The data points and entropy ranges for model ME2C which are also plotted show why ME2C is so similar to ME1 in the plume, but similar to ME2 at the ridge.

It must be noted that the assumptions on entropy are still simplistic here. In particular,

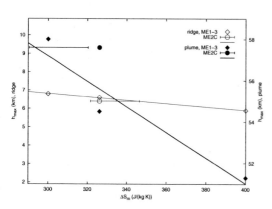

Figure V.17: Maximum crustal thickness as a function of melting entropy, for model series ME. The regression lines are for the models with T-dependent L_m only; the functions are $h_{max} = -0.0091\Delta S_m + 9.54$ for the ridge and $h_{max} = -0.062\Delta S_m + 75.73$ for the plume, with h_{max} in km and ΔS_m in J/kg K. Additionally, the error bars indicate the melting entropy ranges covered by the plume and ridge in model ME2C, the circles mark its entropy at the potential temperature.

PM1	no phase transitions; identical to model ME2
PM2	with garnet–spinel transition
PM2a	with garnet–spinel transition, garnet exhaustion during melting disabled
PM3	with garnet–spinel and spinel–plagioclase transition

Table V.8: Model runs of series PM.

compositional changes will cause the melting entropy to change over the range of the melting region, in particular in the plume. As mentioned in section II.3.1, it has been proposed that L_m will increase with f because of the higher L_m of forsterite (Johnson *et al.*, 1990); if this is true, the decrease of L_m with T would be counteracted, but further aspects such as the exhaustion of garnet, which also has a high L_m (see table D.11), might complicate the issue. Furthermore, the enthalpy and entropy also depend on pressure and are expected to increase with depth. If L_m is taken constant, so that ΔS_m is inversely proportional to T, the entropy would decrease towards greater depth, whereas for $\Delta S_m = const.$, both quantities increase, which is closer to what is expected in the real mantle. However, while the use of a constant entropy seems to fit the needs better, this series shows that both approaches can be tuned within the uncertainty range of the data to give very similar results, so that the foremost concern remains the improvement of the underlying thermochemical model.

Note on $h_{max}(t)$ plots — As in most models, a melt extraction threshold of 1 % has been assumed, and the extracted melt was added to the crust at the spreading zone. Plotting the maximum crustal thickness h_{max} present at a given time versus model time t as in figure V.16 gives a useful, albeit not completely precise impression of the development of the crust during a model run, and will therefore be used frequently in subsequent model sets. These $h_{max}(t)$ curves are usually divided into two segments: the first one, with a slow increase and low-to-moderate thicknesses, corresponds to the phase of model evolution when melting occurs beneath MOR only, while the plume ascends through the deeper mantle, but is still unmolten, or has a melt content below the threshold; the second segment, with a steep increase and subsequent flattening towards a usually much larger value, marks the onset of plume crust formation and its convergence towards a stationary value. The asymptotic value of the first segment serves as an approximation of the thickness of normal oceanic crust in a model and can be used for validation by comparing it to real-world data; the value of the second segment yields the maximum crustal thickness reached above the plume.

V.1.6 Effect of transitions of the Al-bearing phases on melting

Solid-state phase transitions are the other potentially important occasion where entropy changes influence the shape of the geotherm and the melting process. The effect of such transitions on convection dynamics has already been considered in the PT model series in section V.1.2, but that series was limited to isoviscous models and did not include melting. In this section, the effect of the transitions of the aluminous phases is revisited with emphasis on melting in the framework of a more realistic convection model with variable (p–T–φ-dependent) viscosity. Model PM1 has no phase boundary and is identical to model ME2 of the previous section; model PM2 includes the spinel–garnet transition, which lies near the bottom of the melting zone under normal MOR and well within the melting zone in the plume; in model PM3, the plagioclase–spinel transition, which lies at the top of the melting zone, is also included.

In sections II.3.2 and III.4.3, it had been proposed that the effect of the sp–gt transition

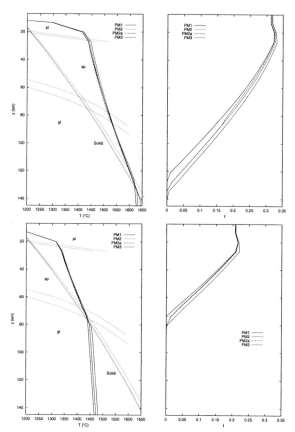

Figure V.18: Temperature (left) and melting degree (right) for the models of series PM in the depth range of the melting region; upper row: plume, lower row: normal ridge. The solidi and the phase transitions for the models are also drawn; the shallowest solidus corresponds to model PM1, the two deeper (barely separable) to PM2/2a and PM3, respectively.

could be reduced or even absent in a plume, because the deep onset of melting would exhaust garnet, which is a major contributor to the melt (*e.g.* Walter, 1998, see tab. C.6), and leave nothing of it which could be transformed; according to eq. II.33b, this would be the case if $f \geq \ln \left(\mathcal{M}_{\mathrm{gt}} / \mathcal{M}_{\mathrm{gt}} - X_{\mathrm{s,gt0}} \right)$, *i.e.* for $f \geq 0.145$, with $X_{\mathrm{s,gt0}} = 0.073$ and $\mathcal{M}_{\mathrm{gt}} = 0.54$ as in these models. To probe the effect of garnet consumption, an additional model PM2a with the gt–sp transition included, but the effect of melting on garnet content switched off was run, *i.e.*, in this model the fraction of garnet is kept constant.

All models were chosen to have the same potential temperature of the background mantle of 1410 °C. By the convention used here, this includes the geotherm shifts in the phase transitions, so that all models would have a temperature of 1410 °C at the top if no melting occurred and no thermal boundary layer existed, but a model with a phase transition with a positive Clapeyron slope will have a higher temperature below it; the alternative would have been to choose a different $\mathcal{T}$ for each model such that the geotherms in the deeper mantle are the same. For this reason, the deep geotherm of model PM1 is the coolest, those of PM2 and PM2a is higher by *ca.* 7 K, and that of PM3 is higher by another 9 K (figure V.18).

Figure V.18 shows the geotherms of all models along with the corresponding solidi. The solidi are different, because they are calculated in terms of T and p (see section III.4.3 and

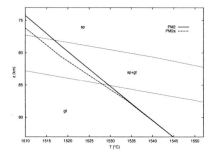 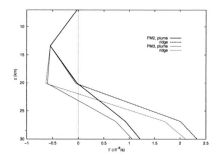

Figure V.19: Close-up of the plume axis geotherms of models PM2 and PM2a, and boundaries of the garnet–spinel transition region.

Figure V.20: Close-up of the melt production rate, Γ, of models PM2 and PM3.

appendix D.1.6), and not of T and z, which would be an unphysical approximation if it were used to force them to coincide. The reason for the shift is that the reference mantle is garnet lherzolite (with α-olivine), but if the shallow phase transitions are included, material with a lower density takes the place of garnet lherzolite in the uppermost mantle, thus reducing the lithostatic pressure, so that a given p is reached at greater depth; hence the solidus of PM1 is the shallowest, whereas those of PM2/2a and PM3 lie a bit deeper, although these latter nearly coincide, because there is only little plagioclase lherzolite in model PM3. This shift combines with the slightly higher temperatures of the models with phase transitions to move the onset of melting to greater depths; if the deep geotherms were identical, the $T(z)$ profiles of models PM2 and PM3 would be almost indistinguishable, but the p dependence of the solidus would still let melting begin a few kilometers deeper. It should though be noted that the fine structure is difficult to resolve, because the differences are small and only of the order of the vertical grid point spacing (~ 6.7 km) or less.

The common feature of ridge and plume geotherms is that they lie closer to each other in the melting zone than below, because the deeper onset of melting in the models with phase changes leads to earlier cooling by melting; hence, taking into account the phase changes results in relatively lower temperatures in the melting region. However, the differences below the ridge are very small. Strong depletion as in the plume can though make the differences larger again and reestablish the order of the geotherms from the deeper mantle in the shallower parts of the melting region, because the decrease of productivity leads to less cooling, the stronger the depletion is, the more; however, this also depends on the f parameterization used, and may be different for *e.g.* a concave-up $f(\tilde{T})$ (see section V.1.7). In agreement with the deeper onset of melting, models with phase transitions reach higher melting degrees, as shown in figure V.18, but they converge to some extent according to the concave-down shape of the melting function; however, it must be emphasized that this result is a rough approximation, because it does not consider any possible changes in melt productivity caused by the change in modal composition due to the gt–sp transition.

To assess the effect of the latent heat of the phase transformation on melting and the possible role of garnet exhaustion in the deep melting region, one has to compare model PM2 with model PM2a, where garnet consumption has been suppressed. As normal MOR melting occurs practically completely in the spinel stability field, the effect can only be observed along plume geotherms. Figure V.19 shows a close-up view of the plume axis geotherms of models PM2 and PM2a in the transition zone from the garnet stability field to the spinel stability field. As can be seen from the $f(z)$ profile in figure V.18, garnet is practically exhausted in model PM2 according to the gt-out calculation above just when the plume reaches the transition, so that actually no phase change occurs, and the PM2 geotherm crosses the boundary without being deflected. In contrast, the PM2a geotherm is visibly shifted to slightly lower temperatures, because the effect of garnet consumption on the latent heat of the transition was artificially suppressed here; this also results in a slightly lower depletion above that depth level (figure V.18). The offset between the boundaries and the actual kinks in the geotherm in figure V.19 is very probably due to errors introduced by the coarse sampling: the width of the transition corresponds approximately to the grid point spacing, but its boundaries do not coincide with grid points. Anyhow, the effect is reproduced as predicted, but it is obviously of minor importance. – In principle, the same applies to the spinel–plagioclase transition, which was included in model PM3; however, spinel exhaustion has not been implemented yet and is therefore not included in that model, so that the sp–pl transition is expected to affect all geotherms in this model. Indeed, one observes that T and f of model PM3 decreases more strongly and over a broader depth interval than in the other cases at the top of the melting region (at $z < 30\,\mathrm{km}$), where this phase change occurs, and is practically identical to the values of model PM2 above that level. Figure V.20 shows a close-up view of Γ near the top of the melting region for models PM2 and PM3 which shows that the melting rate is diminished on a larger depth interval in model PM3. This supports the idea that the plagioclase–spinel transition can contribute to the shutdown of melting at the top of the melting region, as suggested by Asimow *et al.* (1995), unless the ingredients of the reaction are not exhausted before, which might happen in plumes; on the other hand, it is questionable in view of the situation in Iceland (see section IV.1) if a peridotitic lithology would prevail at this depth above a plume instead of a gabbroic thick crustal root, where other phase relations would be applicable and no melting would occur anyway.

For completeness, $h_{\max}(t)$ is shown in figure V.21. It shows differences in maximum crustal thickness which are significantly larger than one would probably expect in view of the relatively small variations in T and f. However, these differences are rather to be regarded as an indirect effect of the different solidus positions and the concomitant variability in the size of the melting region, which is disproportionately large, because the solidus shifts partly affect the very depth level where the melting zone is broadest and where the productivity is highest. There are also differences of a few percent in the upwelling velocities which might contribute to the variability, although they are largest at levels deeper than the melting zone; the difference in the velocities might partly be related to small lithosphere instabilities in model PM3, which could be due to lateral density contrasts related to the sp–pl transition near the lower edge of the thermal boundary layer, but this is difficult to verify. Although these cases are not further considered here, it is expected that the alternative choice of potential temperatures mentioned at the beginning of this section and/or a different $f(\tilde{T})$ parameterization would both reduce the spread in the $h_{\max}(t)$ curves. Thus, the only effect in this picture which can be directly related to the thermodynamic effect of the phase transition is the rather small reduction of the crustal thickness of the plume in model PM2a compared to that of PM2.

V.1.7 *Effects of the choice of solidus and melting degree parameterizations*

In section II.3.3, the need of a proper definition of the rock solidus was discussed, and several parameterizations of T_s as an explicit function of pressure were summarized. To emphasize and quantify the effect of the solidus used in the melting model, two anhydrous solidi,

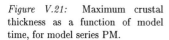

Figure V.21: Maximum crustal thickness as a function of model time, for model series PM.

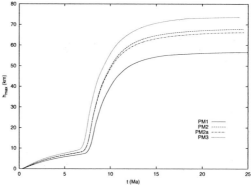

namely the default solidus function of this study, eq. II.25c (Hirschmann, 2000) (model SF1), and an own parameterization of the solidus of KLB-1 (model SF2) (see section II.3.3 and appendix C.2) are compared with respect to the resulting amount of melt and the thickness of the crust, the solidus depth, and the size of the melting region; hydrous melting will not be considered in this context, as it is a substantially different issue whose effect has already been investigated previously and will be addressed again in section V.2.3. Furthermore, a variant of model SF1 with the same solidus lowered by 35 K is also tested following the proposition by Katz *et al.* (2003) that the solidus was overestimated in the original parameterization due to difficulties in detecting very small initial melt contents (SF1a); lowering the solidus has the same effect on the depth of initial melting as raising the reference potential temperature, but it should be noted that it is not entirely the same with regard to viscosity, Rayleigh number, the thickness of the lithosphere etc. The melting degree parameterization by McKenzie and Bickle (1988) is used for this comparison; this is to make the three models comparable, although it contradicts the reasoning of Katz *et al.* (2003) with regard to the initial low-melting tail. It should be noted that in model SF2, the latent heat of phase transitions of the Al-bearing phases has not been included, although the KLB-1 solidus features the cusps associated with them in the present parameterization; this might seem inconsistent, but is necessary to ensure comparability of the models in this run, and the additional effect of phase change enthalpy can be assessed from the PM series. – For all models, the same liquidus of forsterite, eq. II.27, has been used as the upper bound of the melting interval.

While the size and maximum depth extent of the melting region is largely controlled by the solidus, the spatial distribution of melt productivity and hence possibly also the amount of melt produced depend crucially on the form of f along the p–T path of the ascending rock. As mentioned in sections II.3.2 and II.3.3, melting experiments on various peridotites of different composition gave both concave-down and concave-up $f(\tilde{T})$ functions, represented by the parameterizations of McKenzie and Bickle (1988), eq. II.28a, and Katz *et al.* (2003); the

SF1	T_s of Hirschmann (2000), $f(\tilde{T})$ of McKenzie and Bickle (1988); identical to ME2C
SF1a	like SF1, with T_s lowered by 35 K
SF2	KLB-1 solidus, $f(\tilde{T})$ of McKenzie and Bickle (1988)
SF3	solidus of Hirschmann (2000), $f(\tilde{T})$ of Katz *et al.* (2003)

Table V.9: Model runs of series SF.

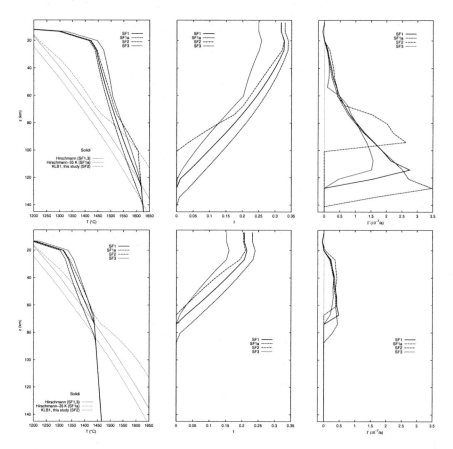

Figure V.22: Temperature (left), melting degree (center) and melting rate (right) for the models of series SF in the depth range of the melting region; upper row: plume, lower row: normal ridge.

quasi-eutectic parameterization of McKenzie and Bickle (1990) was not considered further, because earlier tests not shown here indicated that the associated step-like increase in melt content can lead to numerical problems. Complementary to models SF1, SF1a, and SF2, the differences introduced by the low-melting tail, the concave-up shape of $f(\tilde{T})$, and the explicit cpx-out criterion in the modified anhydrous model after Katz *et al.* (2003) (see section II.3.3), which distinguish that parameterization from that of McKenzie and Bickle (1988), are investigated by a further model, SF3; it uses the original solidus from Hirschmann (2000) in SF1, because that is the standard, but applies the other f parameterization. – All models can be related to previous series, because SF1 is identical to ME2C, *i.e.* a constant latent heat of 550 kJ/kg is used in this series for simplicity. Furthermore, the observation made in the ME series that certain global characteristics like $v_{rms}(t)$ and $Nu(t)$ vary little between the models whereas variables more closely tied to the melting domain, such as T, f, or Γ (figure V.22) are notably affected also applies to this series.

As known from previous model series, a kink appears in the geotherm where it crosses the solidus. The solidi of models SF1 and SF1a are parallel, but the latter lies 35 K lower, which causes the plume and ridge geotherms of SF1a to cross the respective solidus about a dozen kilometers deeper. However, this does in no way result in parallel geotherms within the melting zone: as can be seen particularly well in the plume center geotherm, both geotherms converge at shallower depths. The reason is that the function $f(\tilde{T})$, which also controls the consumption of latent heat and hence the temperature drop, is not linear, but, in models SF1 and SF1a, concave-down, *i.e.* its slope $\mathrm{d}f/\mathrm{d}\tilde{T}$ decreases with increasing $\tilde{T}$, and so does the extent of enthalpic cooling. In contrast, the geotherms of SF1 and SF3, which cross the same solidus at the same depth, diverge at greater depth, because $f(\tilde{T})$ is concave-up in model SF3, which results in only weak cooling at the initial stage. Hence, a greater depth extent of the melting zone does not necessarily imply that the mantle has a lower temperature at the base of the lithosphere: different positions and shapes of the solidus can result in similar temperatures at the top of the melting region for a given potential temperature, as for models SF1, SF1a, and SF2, and a mantle with lower $\mathcal{T}$ can have the same temperature there if the melting function f is different, as can be assessed from model SF3. The melting segments of the ridge geotherms lie mostly in the spinel and plagioclase stability fields, where they soon enter the thermal boundary layer; hence, their shape is relatively regular. In contrast, the plume geotherms show some more interesting peculiarities due to the larger depth and depletion range covered by the plume melting zone. For instance, the SF3 geotherm shows the cpx-out kink at 60 km already seen in the VV series (see figure V.9); the melting rates, which are related to the depth gradient of f, are strongly reduced above it. Kinks in the solidus, however, are also reflected in the shape of the geotherm, as can be seen from model SF2 and the corresponding KLB-1 solidus with its pronounced cusp at the sp–gt boundary at *ca.* 74 km depth. This is because in this case, the cusp strongly reduces the change with depth resp. pressure of the width of the solidus–liquidus interval on which the homologous temperature is defined; hence, $f(\tilde{T})$ varies much less above the cusp than below at adjacent pressure levels, and the resulting melting rate is diminished accordingly, although not strongly enough to be turned into freezing, as hypothesized by some authors (Asimow *et al.*, 1995; Kushiro, 2001, see sect. II.3.2). Below the cusp, melting and concomitant cooling are stronger, because the slope of the solidus is larger, leading to a stronger widening of the solidus–liquidus gap and accordingly higher f. It has been conjectured, however, that in nature, the cusp is not as sharp as in the parameterized solidus used here, so that $T(z)$, $f(z)$, and $\Gamma(z)$ in figure V.22 would show smooth bends rather than cusps.

Comparison of the melting rate profiles at the plume and at normal ridge (see figure V.22) highlights the substantially different character of melting in both environments. The $\Gamma(z)$ profile of normal MOR is relatively flat and uniform, and is not divided into fundamentally different melting regimes; the maximum melting rates are practically the same in all models, and even the shapes do not differ largely among the models with the same f parameterization: Γ is largest at the bottom, except for model SF3 due to the small slope of its $f(\tilde{T})$ near the solidus. In contrast, the plume profiles show much larger differences between the models, in particular regarding the peak values, which are also reached at great depth, but are preceded by a steeper increase and followed by a much larger decrease than observed at the ridge; melting rates tend to be even lower at levels above 50 km in the plume than at the ridge due to stronger depletion resp. cpx exhaustion in case SF3. It seems reasonable to explain the fundamental difference between the ridge profiles on the one hand and the plume profiles on the other hand by the fact that at ridges, passive, plate-controlled mantle flow, which is virtually the same in all models, rules the transport of material through the melting zone, whereas in plumes, where melting begins at greater depth and occurs over a

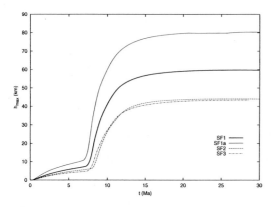

Figure V.23: Maximum crustal thickness as a function of model time, for model series SF.

larger depth range, active upwelling interferes in a complex way with the effect of different solidi and melting functions, resulting in a greater variability among the $\Gamma(z)$ profiles. The plume upwellings, which are also in principle similar, intersect the different solidi in different depths, but do so in a depth range with a steep vertical v_z gradient; hence, the higher mass flux at the deep solidus crossover of model SF1a results in a larger melting rate in this model than in model SF1, where the solidus is crossed at a shallower level with lower velocity. On the other hand, the peak values of models SF1 and SF2 are nearly the same, in spite of the different solidus depths; here, the significantly steeper solidus of SF2 at this depths, which results in a larger path derivative of f for a plume trajectory in $p, \tilde{T}$-space, seems to compensate the reduced mass flux. This is corroborated by the fact that $\Gamma(z)$ of SF2 becomes rapidly similar to that of SF1 above the sp–gt kink of the KLB-1 solidus, where the solidi have similar slopes and the plume has comparable supersolidus temperatures. In the case of model SF3, where the melting function f is concave-up, the peak is damped because of the initially low $\mathrm{d}f/\mathrm{d}\tilde{T}$; at shallower depth, the reduced upwelling in counteracted by the increasing productivity according to the f parameterization, so that the decrease of Γ with depth is not as strong as in the other models. – The small negative values at $z < 20\,\mathrm{km}$ in both the plume and the ridge profiles mark the crystallization of the retained melt fraction at the bottom of the thermal boundary layer and are also reflected by a slight re-enrichment in the $f(z)$ profiles. It should be noted that this crystallization zone would shift to deeper levels if the growing crust were fed back into the model as an influx through the top boundary.

Figure V.23 shows $h_{\max}(t)$ for the SF models. As expected, deeper onset of melting results in a thicker crust at a given potential temperature and melting parameterization, as can be seen by comparing the SF1 and SF1a curves. The $h_{\max}(t)$ plot has two major characteristics: first, models with different solidus and/or melting degree parameterizations produce substantially different amounts of crust both at normal MOR and plumes, even if all other parameters, including the potential temperature of the reference mantle, are the same; and second, certain combinations of T_{s} and f functions can nonetheless result in very similar crustal thickness values, as in the case of models SF2 and SF3. This highlights the need for a proper parameterization of the melting behaviour of the mantle source in the tectonic setting resp. at the location to be investigated in a model, which implies knowledge of its chemical composition. While evaluation of crustal thickness can possibly help to exclude certain parameterizations or combinations of them as unrealistic – in this case, those of model SF1a, which result in normal MOR crust of at least 9 km, and probably also those of model SF3, which yield only *ca.* 4.5 km –, it cannot be used to decide whether a petrological melting model is indeed realized in the mantle, because results can be ambiguous, as indicated by models SF2 and SF3; furthermore, a different choice for the potential temperature would also shift the size and position of the melting region and the amount of crust produced to other values. Comparison of model curves SF1 and SF3 shows that the concave-up *f* parameterization (SF3) leads to a thinner crust than the concave-down one (SF1); the reason is probably that in the former case, high productivity is only achieved at such shallow depth levels where the roughly triangular or rhombic melting region is already narrow and the upwelling velocity rather low. On the other hand, it should be recalled that in the original T_s parameterization of Katz *et al.* (2003), the 1 atm solidus was 35 K lower than in models SF1 and SF3, which is just the difference between the solidi of SF1 and SF1a. Thus, it can be estimated that the original anhydrous melting model of Katz *et al.* (2003) would give similar crustal thicknesses as the combination of the Hirschmann (2000) solidus and the McKenzie and Bickle (1988) melting degree parameterization.

The different depths of onset of melting also have geochemical implications: while the solidus intersection of the ridge geotherm of model SF2 lies well within the spinel stability field and those of models SF1 and SF3 lie at best at the edge of the spinel–garnet transition, melting clearly begins in the garnet stability field in model SF1a. Given the large range of estimates for the potential temperature of the mantle (see table II.1), for which a trade-off exists with the position of the solidus to a certain extent, a variation of 35 K is rather moderate, but can obviously have substantial impact on melt chemistry. Although inclusion of water in a melting model would certainly initiate melting below the sp–gt transition, this uncertainty must be kept in mind when interpreting geochemical signatures of MORB on the basis of experimental results.

V.1.8   Density and viscosity of the melt

From the melt momentum equation II.38a it can be seen that for the mobility of the melt, two material parameters are essential: the density and the viscosity of the melt. As discussed in sections II.1.2 and II.1.4, both of these parameters are p–T-dependent and can vary significantly over the depth range of melting, which should have some impact on the dynamics of the segregating melt, except in the unlikely case that both effects cancel out. This motivates the models series VM with one model where both the density and the viscosity of the melt are kept (essentially) constant (VM1), one model with a p–T-dependent melt density, but fixed viscosity (VM2), and one model with both parameters being functions of p and T (VM3).

For the melt, a reference density of $3000 \, \mathrm{kg/m^3}$ has been chosen, which corresponds practically to the value for PHN 1611 liquid by Courtial *et al.* (1997) (see appendix D.2.1, table D.8); the thermal expansivity was assumed constant, $\alpha = 6.3 \cdot 10^{-5} \, \mathrm{1/K}$, as given by the fitting formula for PHN 1611 liquid by Courtial *et al.* (1997) for $T = 1600\,°C$ (see appendix D.2.2, table D.9). Hence, melt densities show only slight T-dependent variations around *ca.* $2730 \, \mathrm{kg/m^3}$ in model VM1. In models VM2 and VM3, p dependence was included using the Vinet equation of state, eq. II.3c, with $K_{T0} = 22.3\,\mathrm{GPa}$ and $K_T' = 7$, which correspond to the basaltic di$_{64}$an$_{36}$ liquid of Rigden *et al.* (1989); although it might seem inconsistent not to use the PHN 1611 data given by Courtial *et al.* (1997) here as well, the values from Rigden *et al.* (1989) were preferred, because they seem to be more representative for the melts considered here, whereas the others are significantly higher and had been determined at much higher T (see appendix D.2.3, table D.10). The resulting $\varrho(z)$ profiles are shown in figure V.24 for the plume axis and at a normal ridge site. – For the viscosity, a

VM1	T-dependent ϱ_f, constant $\eta_f = 2\,\mathrm{Pa\,s}$
VM2	p–T-dependent ϱ_f, constant $\eta_f = 2\,\mathrm{Pa\,s}$
VM3	p–T-dependent ϱ_f and η_f

Table V.10: Model runs of series VM.

constant value of $2\,\mathrm{Pa\,s}$ has been used in models VM1 and VM2, which should be a reasonable guess for the average conditions of the melting region (see section II.1.4, figure II.2). In model VM3, p–T dependence has been assumed according to eq. II.7 (Bagdassarov, 1988), assuming a constant Saucier–Carron parameter of 150, which yields a scaling viscosity η'_{f0} of $4\,\mathrm{Pa\,s}$ as determined from that formula for $\mathcal{T} = 1410\,°\mathrm{C}$ at normal pressure; $\eta_f(z)$ profiles are shown in figure V.24 for the plume axis and at a normal ridge site.

The profiles in figure V.24 were taken at $t = 25\,\mathrm{Ma}$ and represent a stage in the evolution of the model, when the situation at the symmetry plane is practically stationary. The temperature differences between the plume and the background mantle of 75–$150\,\mathrm{K}$ within the melting region at a given depth are reflected by corresponding differences in density of about 16–$21\,^{\mathrm{kg}}/\mathrm{m}^3$ and in viscosity of up to $2\,\mathrm{Pa\,s}$ (a factor of 2). Given that the along-profile change in T is roughly as large as the contrast between the plume and normal mantle, it is clear that for the density, pressure is the more important variable in the mantle, as it leads to an almost $10\,\%$ increase over a depth range of *ca.* $100\,\mathrm{km}$; however, the plumes of models VM2 and VM3 are far from being molten down to the crossover depth postulated by Stolper *et al.* (1981) (see section II.1.2), because the density of the melt at the bottom of the plume melting region of $3061\,^{\mathrm{kg}}/\mathrm{m}^3$ is still much lower than the density of the solid at that depth of $3466\,^{\mathrm{kg}}/\mathrm{m}^3$. For the viscosity, both p and T are important: the combined effect results in a decrease of about one order of magnitude over the depth range of the

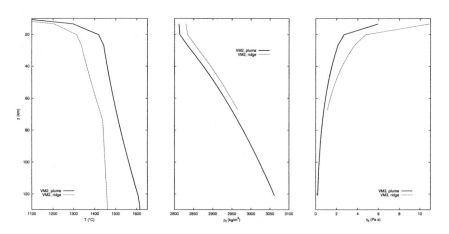

Figure V.24: Depth profiles of T, (left), ϱ_f (center), and η_f (right) in the melting region along the plume axis and at the spreading center of normal ridge, respectively. All profiles are taken at $t = 25\,\mathrm{Ma}$, and differences in the $T(z)$ profiles between the models are negligible, so that the VM2 profiles shown are representative. The ϱ_f and η_f curves are only plotted in the depth interval of the grid where melt is actually present (irrespective of how small φ is).

Figure V.25: Depth profiles of φ after extraction (left) and $v_{\mathrm{seg},z}$ (right) in the melting region along the plume axis and at the spreading center of normal ridge, respectively; all profiles are taken at $t = 25\,\mathrm{Ma}$.

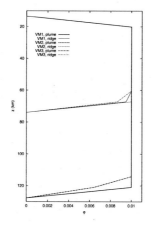

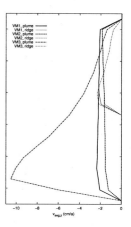

plume melting zone, as predicted by Kushiro (1986), and half an order of magnitude over the depth range of the melting zone beneath normal ridges. Hence, a significant increase of melt mobility with depth is expected, although it must be borne in mind that the possible influence of changes in melt chemistry have not been included here. Furthermore, if plume melts are indeed more water-rich, one would expect an even lower viscosity for them. – The strong viscosity increase at the top of the interval is due to the temperature drop in the thermal boundary layer; it was plotted here for completeness, but porosities are already $O(10^{-6})$ or less here, which suggests that the mobility of trace amounts of melts will be strongly reduced already at the very bottom of the lithosphere.

Figure V.25 shows depth profiles at the same sites and times as in figure V.24 of φ (after extraction) and of the vertical component of $\vec{v}_{\mathrm{seg}}$; the segregation velocity has been computed on the basis of a melt distribution limited by the extraction threshold, as outlined in section III.2. As usual in most models with a low threshold, the maximum retainable porosity is reached in a large part of the melting region, so that φ is homogeneous there after extraction. For the calculation of the segregation velocity, a grain size of $1\,\mathrm{mm}$, a geometry factor $b = 648$, and an exponent $n = 3$ have been used with the permeability–porosity relation eq. II.8; the dominating contribution to segregation is the buoyancy of the melt at these low porosities. This results in a relatively uniform $v_{\mathrm{seg},z}$ profile in model VM1, where the small downward increase is probably due to thermal expansion. In contrast, the compressibility effect is well visible in model VM2, although one would not expect stagnation of melt to occur above the transition zone, judging from the weakness of the decrease; extrapolation of the crucial parameter $\Delta\varrho_{\mathrm{f}}$ seems to confirm this rough estimate. However, model VM3 highlights the effect of the other parameter important to melt mobility, η_{f}: this model is distinguished by a strong increase of $v_{\mathrm{seg},z}$ with depth, which is clearly due to the viscosity drop; the effect is particularly strong in the plume, because melting begins at higher pressure there and occurs at higher temperatures throughout. Hence, this model series confirms the proposition of higher melt mobility at greater depth by Bagdassarov (1988) for this simplified system. However, it should be noted that the large segregation velocities at the bottom are also due to the availability of substantial amounts of melt, which in turn depend on the melting rate and the productivity. It is therefore possible that the use of a concave-up melting function as the one from Katz *et al.* (2003) might yield less melt at incipient melting, so that the resulting porosity, permeability, and eventually $v_{\mathrm{seg},z}$ would

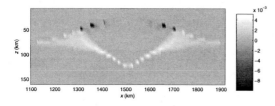

Figure V.26: $\varphi_{VM2} - \varphi_{VM3}$ after extraction in the ridge-perpendicular plane through the plume at $t = 25$ Ma. The large area with $\Delta\varphi = 0$ in the center is the volume where φ_c is reached or exceeded in both models.

also be lower; an impression of the importance of φ can be achieved by comparing the depth range $110\,\text{km} < z < 125\,\text{km}$ in the $\varphi(z)$ and $v_{\text{seg},z}(z)$ plots of figure V.25: a sudden drop of $v_{\text{seg},z}$ in model VM3 coincides with a strong decrease in φ, which itself is partly due to the fast ascent of the highly mobile melt.

The enhanced mobility of deep melts lets one expect a stronger drainage of melt from deeper and more remote parts of the melting region. Although in the current model setting, the porosity field does not look very different in all three models, the volume where the threshold is exceeded is larger in models VM1 and VM2, as can be seen by subtracting the porosity fields of VM2 and VM3 (figure V.26); the T field is virtually identical for these models at the time shown. Crustal thicknesses do not differ strongly: MOR crust is up to $450\,\text{m}$ ($\sim 8\,\%$) thicker in model VM3, whereas the plume crust is thinner by *ca.* $1.2\,\text{km}$ (*ca.* $3\,\%$).

As an aside, a speculation regarding the generation of the first plume melts can also be made. In nature, it is probable that the first melts which form in a plume head do not ascend to the surface as assumed here, because there is a certain timespan shortly before the plume reaches the MOR melting zone when the melting zone of the plume head has not yet merged with it, but is separated from it by an unmolten bar, which might be impermeable to melts; figure V.11 gives an idea how the constellation is shortly after this transitional two-zone melting stage. If the melts are very mobile and ascend fast within the deep plume head, it is possible that a high-φ lens of chemically distinct, enriched melts forms for a short time at the top of the plume head and forms a melt pulse when the two melting zones coalesce which travels upwards; although these pooled deep melts will probably diluted on their way, they might still form a chemically distinct precursor to flood basalts produced by later melting of the arriving plume.

V.2   APPLICATION TO THE ICELAND PLUME

This section is divided into three subsections. The first presents a set of models whose purpose is in the first place to probe a parameter space spanned by the threshold for melt extraction, φ_{ex}, and the excess temperature ΔT_P of a plume rising through the entire upper mantle, which is prescribed at the $660\,\text{km}$ discontinuity as a thermal anomaly and influx condition with a fixed radius and influx velocity. The second set of model series explores the effect of the volume flux of a plume in the upper $410\,\text{km}$, and the third considers changes caused by hydrous melting; a short supplement adds a tentative view on off-axis plumes. One central purpose of these models is to relate the crustal thickness to the processes in the mantle.

The initial condition for all models is a mantle with an adiabatic temperature–depth profile accounting for the depth dependence of α according to eq. V.1 and the typical $\sqrt{t}$ shape of cooling oceanic lithosphere symmetrical to the spreading ridge, which is here prescribed in the form of the analytical solution to the problem (*e.g.* Turcotte and Schubert,

Model	ΔT_P (K) at $z = z_m$	at $z = 200$ km	φ_{ex} (%)
DPRM1 (ICE1-135/1)	250	135	1
ICE1-220/1	350	220	1
ICE1-50/1	150	52	1
ICE1-135/0.1	250	136	0.1
ICE1-135/3	250	136	3
ICE1-135/100	250	136	100
ICE1-135/1c	250	136	1

Table V.11: Model runs of series ICE1. ICE1-135/1c is a variant of ICE1-135/1 with a background mantle potential temperature lower by 30 K.

1982, see eq. II.77). With this initial condition, the initial rifting and flood basalt phase of the opening of the ocean is missed, but this is acceptable, because the main interest lies on the later stages of joint plume–ridge evolution, with regard to observables in the area of the present Iceland plume. As in all previous models, the plume anomaly is located at the center of the model bottom and defined by a circular area with radius $2r_P$ and a certain temperature anomaly and vertical influx, both with a Gaussian shape.

V.2.1   Model set 1: influence of temperature and extraction threshold

In this subsection, a set of models is presented which examines the effect of temperature, especially of the temperature contrast of the plume, and of the melt extraction threshold on convection dynamics and melt and crust production (see also Ruedas *et al.*, 2004). The models of this series ICE1 cover the entire depth range of the upper mantle, because this is the largest volume accessible to regional geophysical studies (see section IV.2), and they include the α–β–γ transformations of olivine; for the convection–melting models considered here, the principal effect of the olivine transformations is an acceleration of the upwelling plume at intermediate depths. The anomaly was given a full radius of 250 km (*i.e.* $r_P = 125$ km), which lies in the main field of estimates from observations (*e.g.* Tryggvason *et al.*, 1983; Wolfe *et al.*, 1997; Foulger *et al.*, 2001), and a maximum influx velocity of 3 cm/a; according to eq. III.4, this gives a volume flux $\int_A v_z \, dA$ of 1.446 km³/a at the model bottom, which is close to the estimates of Schilling (1991) and Steinberger (2000) (see table IV.2). – From the technical point of view, it should be mentioned that model series ICE1 was computed with an older version of the program using the explicit, upwind-based time-stepping scheme (see section III.1.2) instead of the semi-Lagrangian technique used otherwise in this study. The exceedingly long runtimes were a major reason to limit the size of the model box, which is why the models of this series feature a numerical grid only 1500 km wide, in contrast to several of the other model series presented here. Furthermore, it is also noted that for the liquidus of forsterite, the older parameterization eq. C.5 was used instead of the now favoured eq. C.3; the difference results in a slightly broader solidus–liquidus interval at pressures between *ca.* 0.5 and 3 GPa and an interval larger by up to *ca.* 20 K at higher p (see figure C.2), but as a whole, the difference is not considered to be substantial.

As will have become obvious from foregoing sections, several field observables, *e.g.* crustal thickness, seismic velocities, and electrical conductivities, are particularly sensitive to variations in temperature and melt content, which itself is also T-dependent. Therefore it is of substantial interest to investigate the effect of different temperatures and melt contents on these observables. For this purpose a dynamic plume reference model, version 1 (DPRM1) is

defined and the plume excess temperature at the model bottom, ΔT_{P}, and the melt extraction threshold, φ_{ex}, are varied. With respect to the temperature estimates for the Iceland
plume listed in table IV.2, $\Delta T_{\mathrm{P}} = 250\,\mathrm{K}$ was chosen for the DPRM1 at the model bottom.
This yields a plume temperature well within the range of current upper-mantle temperature
estimates for this plume, because the plume cools during ascent to a temperature excess of
ca. 136 K at 200 km depth. Although it is commonly accepted that melting in the mantle
is nearly fractional, an adequate choice of the extraction threshold is even more difficult,
as is apparent from the discussions in sections II.1.5 and II.2.5. As a compromise between
the extremely low φ_{ex} of 10^{-3} or even less (*e.g.* McKenzie, 1985b) and higher values like
$\varphi_{\mathrm{ex}} \approx 0.02\ldots0.03$ for the onset of efficient melt separation (Faul, 2001), an extraction
threshold of $\varphi_{\mathrm{ex}} \approx 0.01$ was used for the DPRM1 (run ICE1-135/1). Altogether, the runs
explore the parameter space given by $\Delta T_{\mathrm{P}} = 150\,\mathrm{K}$, 250 K, and 350 K (corresponding to
52, 136, and 220 K at $z = 200\,\mathrm{km}$) and by $\varphi_{\mathrm{ex}} = 0.001$, 0.01, and 0.03. For better comparison with observations, model names are given according to the rounded 200 km-ΔT_{P},
i.e. runs ICE1-50/1, ICE1-135/1, ICE1-220/1, ICE1-135/0.1, ICE1-135/3, respectively (see
tables V.11 and V.13). Additionally, an unrealistic model ICE1-135/100 without melt segregation and extraction ($\varphi_{\mathrm{ex}} = 1$) was also considered in order to force a significant contribution
to upwelling from the retained melt; furthermore, the expected pronounced effect on seismic
velocities and electrical conductivities due to melt was of interest for the companion study
to the paper of Ruedas *et al.* (2004) by Kreutzmann *et al.* (2004). To assess the importance
of the background potential temperature, another run was made with parameters as for the
DPRM1, but with $\mathcal{T} = 1380\,^{\circ}\mathrm{C}$ (ICE1-135/1c). Some of the key variables discussed in the
following, *e.g.* the maximum crustal thickness, h_{max}, and the vertical upwelling velocity of
the plume, $v_{\mathrm{s},z}$, are estimated for a "steady state"; however, in most models, no real steady
state is reached. Therefore, the steady state is defined to be represented by a stage in the
evolution of the model, in which the plume has already spread significantly under the lithosphere. This situation is usually reached after 20–22 Ma model time, as is obvious from the
$h_{\mathrm{max}}(t)$ plot in figure V.28; most variables do not seem to vary appreciably at that stage
anymore. Further model parameters are listed in table V.12.

The DPRM1 plume starts at the model bottom with $\Delta T_{\mathrm{P}} = 250\,\mathrm{K}$ in its center and
crosses the solidus after *ca.* 9.2 Ma at a depth of about 110 km, having lost about half of its
excess temperature; the temperature distribution of this model and the temperature contrast
of the plume to the background mantle are shown in figure V.27a and b. During its ascent
through the upper mantle, the central part of the plume accelerates in the zone of the olivine
phase transitions, causing the conduit to thin to some extent as a consequence of mass
conservation, and then slows down again when approaching the lithosphere; the thinning
probably also contributes to the considerable extent of cooling. Apart from the decrease in
viscosity and the increase of thermal expansivity to shallower depth throughout the depth
range of the model, the reason for the increase in ascent velocity is the significant additional
buoyancy caused by the fact that the lower-density α-phase is formed in the hot plume at
a somewhat greater depth than in the normal mantle, whereby the density contrast and
concomitant buoyancy increase with plume temperature (figure V.29). Above the transition
zone, the effect disappears, as can be seen from the reduction in v_z and the localization of
the phase-related buoyancy in figure V.29. While the buoyant effect of the transition zone
certainly has an effect on ascent velocity in the deep upper mantle, and therefore also on
the shape and diameter of the plume, its immediate influence wanes towards the shallower
parts of the mantle, so that its effect on the melting processes is minor. The non-thermal
buoyancy components in the melting zone are mostly related to the depletion. The buoyancy
related to retained melt is only of some importance at those depths of the plume's melting

General		
a	grain size	$1\,\mathrm{mm}$
b	permeability scaling parameter	648
g	gravity acceleration	$9.9\,\mathrm{m/s^2}$

Model		
$x_\mathrm{m}, y_\mathrm{m}, z_\mathrm{m}$	size of convection model	$1500\,\mathrm{km} \times 1000\,\mathrm{km} \times 660\,\mathrm{km}$
$x_\mathrm{f}, y_\mathrm{f}, z_\mathrm{f}$	size of melt model	$556\,\mathrm{km} \times 1000\,\mathrm{km} \times (113,127,153)\,\mathrm{km}$
$\Delta T_\mathrm{P}(z_\mathrm{m})$	plume excess temperatures at z_m	$150\,\mathrm{K}, 250\,K, 350\,\mathrm{K}$
φ_ex	extraction threshold	$0.001, 0.01, 0.03, (1)$
$r_\mathrm{P}(z_\mathrm{m})$	plume radius at model bottom	$125\,\mathrm{km}$
$v_\mathrm{z,P}(z_\mathrm{m})$	max. vertical influx velocity of plume	$3\,\mathrm{cm/a}$
v_r	half spreading velocity	$1\,\mathrm{cm/a}$

Mantle		
α	thermal expansivity	$4.27\ldots 2.39 \cdot 10^{-5}\,\mathrm{1/K}$
η_0	reference shear viscosity	$5 \cdot 10^{22}\,\mathrm{Pa\,s}$
η_b	bulk viscosity	$5 \cdot 10^{21}\,\mathrm{Pa\,s}$
κ	thermal diffusivity	$6.072 \cdot 10^{-7}\,\mathrm{m^2/s}$
ϱ_0	reference density	$3660\,\mathrm{kg/m^3}$
c_p	isobaric specific heat	$1350\,\mathrm{J/kg\,K}$
ΔS_X	entropy change of phase transformation	$53.8\,\mathrm{J/kg\,K}$
$\mathcal{T}$	background mantle potential temperature	$1380\,^\circ\mathrm{C}, 1410\,^\circ C$
X_ol	fraction of olivine	0.65

Melt		
α_f	thermal expansivity	$6.3 \cdot 10^{-5}\,\mathrm{1/K}$
η_f	viscosity	$2\,\mathrm{Pa\,s}$
$\varrho_\mathrm{0,f}$	reference density	$2663\,\mathrm{kg/m^3}$
K_0, K_0'	isothermal bulk modulus and p derivative	$22.3\,\mathrm{GPa}, 7$
L_m	latent heat of melting	$550\,\mathrm{kJ/kg}$

Table V.12: Variables used for model series ICE1, and model parameters; values in italics are characteristic of the DPRM1. For better computational efficiency, z_f was varied for different ΔT_P, because the respective plumes cross the solidus at different depths; therefore a hotter plume needs a melt model grid which reaches deeper into the mantle.

zone where normal mantle is still solid and causes a deep local maximum of the buoyancy at *ca.* 100 km depth, whereas above that level, the porosity is virtually the same in the plume head and beneath the normal ridge due to the imposed extraction threshold (figure V.29).

A further temperature drop by *ca.* 60 K is due to melting enthalpy and leads to a slight viscosity increase in the whole melting region of up to *ca.* 0.2 log units. The temperature drop makes the plume head almost invisible in the temperature field in figure V.27a, but by subtracting the temperature at the model margin, $T(x,0,z)$, from $T(x,y,z)$ (figure V.27b), one can distinguish some details of the internal temperature structure of the thermal anomaly of the plume head. The onset of melting is marked by a decrease of the temperature anomaly above *ca.* 110 km depth, because melting is restricted to the plume there, and the consumption of latent heat causes its temperature to approach that of the normal mantle. At about mid-depth of the melting model, at 75 km, the reference mantle beneath normal MOR also begins to melt and cool, so that the temperature contrast increases again to smaller depths,

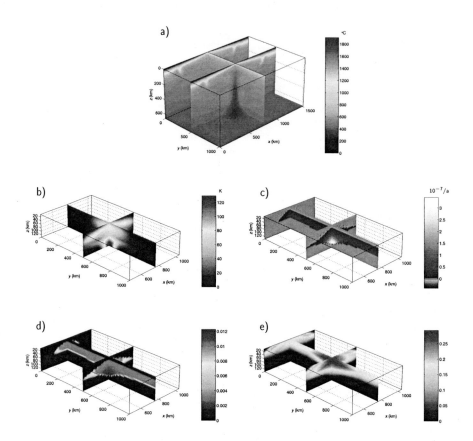

Figure V.27: The DPRM1 at $t = 21$ Ma. a) Temperature field in convection grid (in °C); b) temperature anomaly $T(x, y, z) - T(x, 0, z)$ in melting grid (in K); c) melt production rate in melting grid (in 10^{-7}/a); d) melt fraction before extraction in melting grid; e) melting degree (depletion) field in melting grid. (From Ruedas *et al.*, 2004, © 2004 Blackwell, reproduced with permission.) *See plate section for colour version.*

because the melt productivity of the strongly depleted plume head in the shallowest parts of the melting region is lower than that of the less depleted sub-MOR mantle at a given z due to the concave-down shape of $f(\tilde{T})$.

As in most models of this series, the vertical velocity of the plume peaks shortly below the top of the transition zone. It decreases to a somewhat constant lower value for the next 200–250 km, because the melting-related buoyancy increase at shallow depths counteracts the slowdown of the rising mantle and because the decrease of viscosity from 5–$6.5 \cdot 10^{19}$ Pa s at the plume axis in the transition zone to *ca.* $3.5 \cdot 10^{19}$ Pa s just below the solidus depth and the reduction of thermal buoyancy due to cooling of the plume compensate to some extent. v_z then drops to values of a few centimeters per year in the uppermost 100 km of the mantle (figure V.29), where the plume head gets slightly stiffer, spreads laterally and

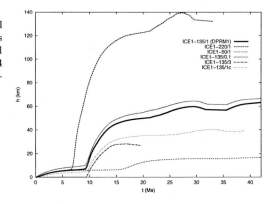

Figure V.28: Maximum crustal thickness in the models of series ICE1 as a function of time. (Adapted from Ruedas *et al.*, 2004, © 2004 Blackwell, reproduced with permission.)

develops a lobe ring at first. At a depth of 100–150 km, it already forms a small head, which is then slightly thicker than the stem; this is a feature which has also been observed by other workers (*e.g.* Feighner *et al.*, 1995), and is probably related to both the slowdown of the plume when approaching the cold boundary layer and the influence of the flow field of the spreading plates. In the plume center, T does not change significantly after 16 or 17 Ma, *i.e.* some 7 Ma after the plume head has begun to produce melt. In the first phase of plume–ridge interaction, there is a visible, though not strong elongation of the upper part ($z \lesssim 45$ km) of the plume head along the ridge, but after some million years drag by the plates takes over control of the head's shape and extends the thermal anomaly perpendicularly to the ridge. Inspection of the thermal anomaly field suggests that a waist of some 400 km forms. For channeling of plume material into the ridge larger viscosity contrasts would be necessary (Albers and Christensen, 2001), but cannot be reached with the algorithm used (see section III.1). Thermal erosion of the lithosphere by the spreading plume head is not strong.

   Figure V.27c and d give an overview of the spatial distribution of melting. At the top of the model, normal MOR melting takes place continuously in the roughly triangular area beneath the ridge between about 20 and 75 km depth. It is about 250 km wide at its base under normal ridge but broadens to more than 450 km during the spreading of the plume head and narrows only slightly at the later stages of plume evolution. The normal melting produces a layer of lighter depleted material beneath the lithosphere which moves to the sides, replacing the undepleted mantle (figure V.27e). Melting degrees in this layer reach a roughly constant level of around 18 % at intermediate depths. The plume starts to melt at *ca.* 9.2 Ma model time at a maximum depth of *ca.* 110 km and achieves a stronger depletion of 25–29 % in its central parts; the additional thermal and depletional density reduction in the plume enhances the upwelling in the melting zone. As one would expect from the melting parameterization of McKenzie and Bickle (1988) used in this series, melt production is largest in the initial stage of melting, thus most melts come from greater depths, especially those of the plume; higher temperatures and flow velocities cause the maximum melt production rate to be 5–7 times higher in the plume head than beneath normal ridge, but the stronger depletion and accordingly decreasing $f(\tilde{T})$ lead to a slightly lower melting rate in the top of the plume head. Vertical segregation velocities of the retained melt are around 2–2.5 cm/a as a maximum in both the plume head and beneath normal ridge; this is due to the prescribed threshold value. Between the melting zone and the thermal lithosphere, a thin freezing layer exists where the mantle is being re-enriched a bit. The extracted melts form a crust of 5.6 km

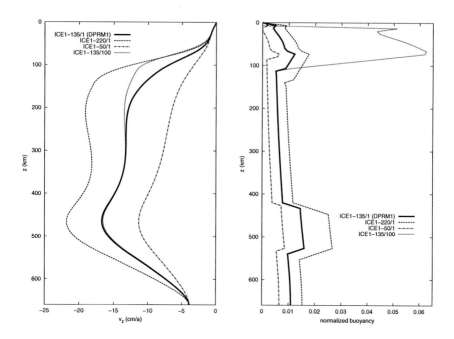

Figure V.29: Selected $v_z(z)$ (left) and buoyancy (right) profiles through the plume center at model times where crustal production has reached a quasi-steady state (30 Ma for ICE1-50/1, 21 Ma for the other models; see figure V.28). The $v_z(z)$ profiles of models ICE1-135/0.1, ICE1-135/3, and ICE1-135/1c do not differ appreciably from the DPRM1 profile and have therefore been omitted. Note the pronounced deviation of ICE1-135/100 from the DPRM1, which is due to the strong contribution of retained melt to buoyancy ($\varphi_{max} \approx 0.3$). The buoyancy has been computed relative to a mantle profile at normal ridge, and has been normalized with $-g\varrho_0$. (Adapted from Ruedas *et al.*, 2004, © 2004 Blackwell, reproduced with permission.)

at normal ridge, which is only slightly below the range of observed values (White *et al.*, 1992; Schubert *et al.*, 2001) and had therefore served for calibrating the model. As can be seen from plots of the maximum crustal thickness in the model at a given time (figure V.28), $h_{max}(t)$, crustal growth rates are largest at the beginning of melting – both at the beginning of the model run and at the beginning of plume melting – and approach a steady-state value after some million years. For the DPRM1, this value is 49 km, *i.e.* a bit larger than the maximum estimates of 39–46 km for the crust to the northwest of Vatnajökull in central Iceland as derived from the data from the ICEMELT and HOTSPOT projects (Darbyshire *et al.*, 1998, 2000b; Du and Foulger, 2001; Allen *et al.*, 2002b, see sect. IV.1.2, IV.1.3). The late-stage variability in crustal thickness and some other model variables are probably due to boundary effects of the model and should not be considered further.

All runs are summarized in table V.13. In the runs ICE1-50/1 and ICE1-220/1, the melt extraction threshold has been kept at 1 %, but the bottom excess temperature of the plume has been changed to 150 K and 350 K, respectively. The cold plume ICE1-50/1 ascends quite sluggishly and loses about two thirds of its original excess temperature until it reaches the solidus at 85 km depth after a bit more than 15 Ma; above 450 km, where the peak

	$h_{\max}$ (km)	z_{sol} (km)	$\Delta T_{\mathrm{P}}(z_{\mathrm{sol}})$ (K)	Δz_X (km)	$v_{\mathrm{s},z}\ \left(\frac{\mathrm{cm}}{\mathrm{a}}\right)$	$v_{\mathrm{seg},z}\ \left(\frac{\mathrm{cm}}{\mathrm{a}}\right)$
plume center						
DPRM1	49	110	115	15	17	~ 2.6
ICE1-220/1	119	133	~ 180	20	22	3.3
ICE1-50/1	15	85	~ 40	~ 5	11	2.4
ICE1-135/0.1	54	110	115		~ 16	~ 0.23
ICE1-135/3	27	110	115	15	16	18
ICE1-135/100	n.d.	110	115	15	~ 16	$O(\mathrm{km/d})$
ICE1-135/1c	34	~ 100	115	15	17	~ 2.5
normal ridge						
DPRM1	5.6	75	–	–	~ 1	~ 2.5
ICE1-135/0.1	7.7	75	–	–	~ 1	~ 0.08
ICE1-135/3	0	75	–	–	~ 1	9.5
ICE1-135/1c	4	65	–	–	~ 1	~ 2.5

Table V.13: Model results of series ICE1. $\Delta T_{\mathrm{P}}(z_{\mathrm{sol}})$ is taken at the centerline of the plume. The values of $v_{\mathrm{s},z}$ and $v_{\mathrm{seg},z}$ are maximum values; the entries for $v_{\mathrm{s},z}$ are estimated averages of the steady state. $v_{\mathrm{seg},z}$ for ICE1-135/100 is a rough estimate based on Darcy's law.

velocity of 10–11 $^{\mathrm{cm}}/_{\mathrm{a}}$ is reached, upwelling velocities decrease continuously to less than 2 $^{\mathrm{cm}}/_{\mathrm{a}}$ in the upper 100 km, thus yielding a $v_z(z)$ profile which is somewhat different from that of hotter plumes in that it lacks the constant interval in the upper half (see figure V.29). The maximum melting degree is only 23 %, and the final plume crust is not more than 15 km thick. As a whole, it is a rather unspectacular structure. In contrast, the hot plume ICE1-220/1 rises considerably faster than the others, especially in the uppermost 150 km, due to its larger thermal buoyancy and lower viscosity and crosses the solidus already after about 6.5 Ma at 133 km depth, with still two thirds of its bottom temperature excess in the center. However, the more extensive melting leads to stronger enthalpic cooling and lets the temperature contrast to the background mantle drop to only *ca.* 100 K at $z < 75$ km. The melting rates in the deepest parts of the plume melting region locally exceed those beneath normal MOR by up to one order of magnitude and eventually make possible a maximum melting degree of more than 30 % when the plume starts to spread beneath the lithosphere, which is probably unrealistic because the expected exhaustion of clinopyroxene would reduce the productivity more strongly than modelled here. In general, the hot plume qualitatively follows an evolution similar to that of the DPRM1; it takes it somewhat longer to approach its final crustal thickness of about 145 km, though, and the thermal erosion of the lithospheric lid is more pronounced.

The same threshold has been used for model ICE1-135/1c (cooler background $\mathfrak{T}$). In this run, a cooler background temperature also means that the absolute temperature of the plume is lower than that of the DPRM1 by the corresponding amount. Generally speaking, one would expect this model to be a lower-performance clone of the DPRM in terms of melting, because the downward shift of the mantle geotherm shrinks the melting regions of both normal ridges and the plume and moves the solidus depth to a shallower level. Indeed, all geotherm–solidus intersections lie roughly 10 km higher, plume melting starts slightly later, the melting degrees are 2–3 % lower, and the transition path of the material through the melting region is shorter, all of which leads to a thinner crust, in this case to *ca.* 4 km at the normal ridge and up to 34 km above the plume. Note that this effect is in a certain sense complementary to an upward shift of the solidus as investigated in model series SF (section V.1.7).

The other variable parameter of this study, the threshold porosity φ_{ex}, was changed to 0.1, 3, and 100 % (no extraction), respectively, in models ICE1-135/0.1, ICE1-135/3 and ICE1-135/100; ΔT_P was kept at 250 K. Dynamically, ICE1-135/0.1 and the DPRM1 ICE1-135/1 are very similar, because in both cases, the retained amount of melt is quite low and therefore has only a minute effect on the buoyancy, visible near the solidus intersection. The main difference is the crustal thickness: the final MOR crust thickness is 7.7 km in ICE1-135/0.1, and the final plume crust thickness 54 km. The reason obviously is that in ICE1-135/0.1 almost all melt is extracted from the whole melting zone, whereas in the DPRM1 case a certain fraction in the outer parts of the zone hardly segregates and never makes it to the top, but is rather transported outward and refreezes; moreover, as nearly everything is extracted in ICE1-135/0.1, there is hardly melt left which would recrystallize later. In contrast, in run ICE1-135/3, all melt is retained under normal MOR, so that only the plume produces a crust; a plume with an excess temperature of 3–4 K at $z = 200$ km is the weakest feature that would produce a crust with the other parameters unchanged (figure V.30). Freezing rates at the melting zone margins are accordingly higher, and the freezing layer is thicker at the plume. The dynamical buoyancy effect of the melt relative to the DPRM1 is still small in this model. In the unrealistic last model ICE1-135/100, where all melt is retained, the melt finally has a large effect on buoyancy: in agreement with the depletion, porosities of more than 25 % are reached and accelerate the plume by 20–50 % (*ca.* 1–3 cm/a) in the depth range from about 200 to 50 km; the velocity would be even greater if the effect of melt on viscosity had been included, which would be necessary at such high porosities (*e.g.* Kohlstedt *et al.*, 2000). For numerical reasons, segregation was not considered in this run.

An important general result of the models is shown in figure V.30, where the quasi-stationary crustal thicknesses of all models (except ICE1-135/100 and ICE1-135/1c) are plotted as a function[2] of ΔT_P for the different φ_{ex}. The principal feature of this plot is the strong non-linearity of $h(\Delta T_P)$, which is a result of the combined effect of the supersolidus temperature, active upwelling of the plume and the geometry of the melting region.

Preliminary discussion — The general discussion of the results of this section is deferred to section VI.2, but the design of an extension of series ICE1 in the following subsections requires a preliminary evaluation of its outcome.

One important result of series ICE1 is that, judging from MOR crustal thickness, potential temperatures around 1400 °C and rather low thresholds for rapid melt extraction are an adequate choice for models of sub-oceanic mantle melting, for they reproduce fairly well measurements from reality. The comparison with MOR is important for the calibration of plume models, because MOR melting is more tightly constrained by independent data, and it depends on less parameters. Inspection of maximum crustal thicknesses above the plume (figures V.28 and V.30) allows to further constrain the set of models promising to be reasonable approximations to the situation below Iceland: it turns out that the plume is likely to have a maximum temperature excess of about 135 K at a depth of 200 km, *i.e.* the DPRM1 as well as models ICE1-135/0.1 and ICE1-135/1c look like the best choices in terms of ΔT_P and φ_{ex} in this series, although none of them reproduces the observed plume values

---

[2]The data points of all runs from table V.13 except ICE1-135/100 and ICE1-135/1c have been used for fitting. The formula was constructed by: 1. finding the cubic polynomial through the points with $\varphi_{ex} = 1$ %; 2. finding the quadratic polynomial through the points with $\Delta T_P = 136.5$ K; 3. deriving the "stretchable" ΔT_P–φ_{ex}-dependent correction factor to fit the h_{max} values of the DPRM1 and ICE1-135/0.1 at normal MOR ($\Delta T_P = 0$ K).

Figure V.30: Maximum steady-state crustal thickness as a function of $\Delta T_{\mathrm{P}}(z = 200\,\mathrm{km})$ and φ_{ex}, for a background mantle $\mathcal{T}$ of $1410\,°\mathrm{C}$. Within the given data range, the formula $h_{\mathrm{max}} = (4.24 \cdot 10^{-6}\Delta T_{\mathrm{P}}^3 + 8.25 \cdot 10^{-4}\Delta T_{\mathrm{P}}^2 + 0.126\Delta T_{\mathrm{P}} + 5.6)(-0.038\varphi_{\mathrm{ex}}^2 - 0.07\varphi_{\mathrm{ex}} + 1.11) + 1.698(1 - \Delta T_{\mathrm{P}}/136.5)(1 - \varphi_{\mathrm{ex}})$ (in km) with ΔT_{P} in K and φ_{ex} in % can be used to approximate h_{max}. This formula should not be used for extrapolation much beyond the data range, because limiting effects such as cpx exhaustion are expected at very high ΔT_{P}. (From Ruedas *et al.*, 2004, © 2004 Blackwell, reproduced with permission.)

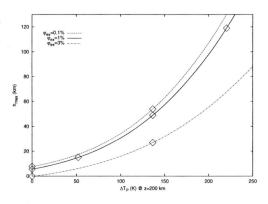

exactly; however, some of the simplifications in these models concerning the fast migration of melt and the transport of the crust could result in some redistribution of crustal material which would lead to a lower h_{max} (see section VI.2).

It is though a too narrow view of the problem to judge the models only from maximum crustal thickness. An adequate model would have to reproduce the overall shape of Iceland reasonably well. Figure V.31 shows a map of crustal thickness in the central part of the model. As expected when having in mind that in this model the plume crust is only at most *ca.* 12 Ma old, it shows a width and general structure of the crust – the greatest h values form a strip perpendicular to the ridge at the center of the plume plateau – comparable to the Moho map by Darbyshire *et al.* (2000b, see fig. IV.4), but the along-ridge dimension of the island is too small compared with Iceland. This is a problem which might rather be related with plume geometry resp. plume volume flux than with the variables investigated in this model series. Although the source region of the plume had been designed to match geometrical and flux constraints from observations in the hope to impose a sufficiently large dimension of the melting zone, plume dynamics are unruly and always lead to significant thinning of the stem upon ascent. In conclusion, in a forthcoming model set the variables controlling the plume's volume flux should be addressed.

As a whole, the DPRM1 is the best preliminary plume model from this series, considering crustal thicknesses from both plume and MOR as well as the results of Kreutzmann *et al.* (2004). Although model ICE1-135/0.1 is almost as good, the DPRM1 will be used as a starting point for the next series, because with regard to the observed velocity reduction and attenuation of seismic waves in the melting regions in nature (see sections II.2.2 and IV.2.1), the model with the lower extraction threshold seems to be the less probable alternative.

V.2.2   Model set 2: influence of plume volume flux

To improve and extend the preliminary reference model DPRM1 introduced and investigated in the previous section, it is necessary to probe other dimensions of the general parameter space, *e.g.* the volume flux. For this reason, a new set of models is devised in this section on the basis of the DPRM1, and a new version of the reference plume model is established. The design of this DPRM2 is based on the DPRM1 in many fundamental aspects, but makes

180   *Results of numerical modelling*

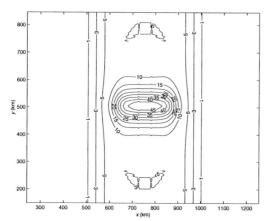

Figure V.31: Calculated crustal thickness (in km) of the DPRM1 at 21 Ma
model time. The two structures at
$y \approx 220\,\mathrm{km}$ and $y \approx 780\,\mathrm{km}$ are
weak artificial depressions. The ridge
axis is located at $x = 750\,\mathrm{km}$, and
is 60 km wide; plate drift has not yet
transported crust further than about
210 km from it. (From Ruedas *et al.*,
2004, © 2004 Blackwell, reproduced
with permission.)

use of some of the results from model series ICE1. The changes are discussed first before
proceeding with the models themselves.

The newer Iceland model series shown hereafter were all computed with the newer version
of the modelling program featuring semi-Lagrangian advection, which allows for somewhat
larger time increments. The possibility to perform the computations in a significantly shorter
time allowed it to use a broader convection model box to ensure that the central parts of
the model are not affected by the model boundaries in the more advanced stages of plume
evolution (see sections III.1.1 and V.2.1); therefore, x_m has been chosen twice as large as
in series ICE1. On the other hand, it had turned out that the inclusion of the transition
zone in the first model set, which demonstrably affects the dynamics of the deep plume
(also see section V.1.2), does not seem to have a direct influence on melting dynamics in the
shallow mantle, but introduced certain problems. For instance, the enhanced acceleration of
the plume in the transition zone makes it even more difficult to control the diameter of the
plume in the upper parts of the model, because the gain in speed induces a decrease in stem
thickness; as mentioned before, this has shrunk the actual size of the plume in the depth
range covered by observations to a value significantly below the one deemed representative
for the real structure and prescribed at the model bottom. As the transition zone is currently
poorly covered by observations and as the flow and temperature fields of series ICE1 were
not entirely satisfactory at that depth level, it was decided to omit the transition zone in
the upcoming models; this measure would also compensate in part the increase in runtime
and memory requirements imposed by the doubling of x_m. Therefore, the following model
series all have $z_\mathrm{m} = 410\,\mathrm{km}$.

The two quantities which can be expected to characterize the melting style of the plume
quite well, judging from the observations made in series ICE1, are the plume centerline temperature and upwelling velocity. Therefore, these two variables have been used to establish
the link between the DPRM1 and the newly defined DPRM2. Taking the steady-state excess
temperature of the DPRM1 plume center at a depth of 410 km (165 K) as the maximum bottom temperature anomaly of the DPRM2, the new reference model has a stationary plume
center geotherm which lies within a few degrees of that of the DPRM1 and can thus be
regarded as a very good reproduction. For the upwelling velocity depth profile, the part
close to and within the melting region is most important, whereas it is hardly possible to
reproduce v_z over the whole depth interval. On the other hand, the fact that it is apparently

x_m, y_m, z_m	size of convection model	$3000\,km \times 1000\,km \times 410\,km$
x_f, y_f, z_f	size of melt model	$806\,km \times 1000\,km \times 155\,km$
$\Delta T_P(z_m)$	plume excess temperatures at model bottom	$165\,K$
φ_{ex}	extraction threshold	0.01
$r_P(z_m)$	plume radius at model bottom	$75\,km,\ 100\,km,\ 125\,km$
$v_{z,P}(z_m)$	max. vertical bottom influx velocity of plume	$3\,^{cm}/_a,\ 5\,^{cm}/_a$
$\mathcal{T}$	background mantle potential temperature	$1410\,°C$
$\varrho_{0,f}$	reference density of melt	$3000\,^{kg}/_{m^3}$

Table V.14: Model parameters used for model series ICE2; values in italics are characteristic of the DPRM2. Parameters not listed here are the same as in table V.12.

chiefly thermal buoyancy which controls the general character of upwelling of the plume reduces the influence of the influx velocity at the model bottom on the shallower parts of the plume, and it is possible to tune $v_{z,P}(z_m)$ so that $v_z(z)$ converges to the velocity profile of the DPRM1 in the upper $150\ldots 200\,km$, where it is of greatest importance to melting; the test models TC2a and TC2c discussed in section III.1.3 also suggest that at shallow levels, the upwelling velocities of models with different basal velocities should converge (see figure III.7). A basal influx velocity of $5\,^{cm}/_a$ was found to result in a velocity–depth profile coming reasonably close to the DPRM1 profile, although it is noted that the upwelling velocities are a bit lower in this case; this is not only regarded as acceptable, though, but in fact desired in view of the crustal thickness of the DPRM1, which had been found to be a bit larger than the observed values (see figure V.28). To ensure a volume flux comparable to the one imposed in the DPRM1 and to the estimates of Sleep (1990), Schilling (1991), and Steinberger (2000), the radius of the influx area was decreased to $R_P = 200\,km$, yielding a volume flux of $1.54\,^{km^3}/_a$ for the DPRM2[3]; it was hoped that the smaller vertical variations in v_z would result in less thinning of the plume stem. – The other parameter important to crust production in series ICE1, the extraction threshold φ_{ex}, was kept constant at $1\,\%$, as in the DPRM1.

One further change is that in the forthcoming models, the newer parameterization of the forsterite melting curve, eq. C.3, which had also been used throughout the general models of section V.1, was applied again. For completeness it is also mentioned that the rheology law used in the newer models includes the dependence on melt fraction, and that a more appropriate choice of melt density than in the ICE1 models has been made. However, both changes are of only minor importance for the results. The new parameters in the models of series ICE2 are summarized in table V.14.

As can be seen from eq. III.4, there are two ways to modify the volume flux of a plume, namely to change the upwelling velocity and to change the cross-section area, *i.e.* the radius of the plume. In the models of series ICE2, both possibilities have been explored, but it should be noticed that changes of the radius are probably easier to compare with independent data, because other methods, especially seismology, also provide direct geometrical information, whereas a direct observation of mantle velocities is difficult. There are four

---

[3]The volume flux is preferred over the buoyancy flux (see appendix B.3) used by some authors, because its use seems to be more common (see table IV.2) and because it does not make any implicit assumptions on the cause of the flux, whereas the traditional buoyancy flux is ascribed to thermal expansion by definition, although there clearly is a compositional contribution to buoyancy at least at shallow levels; the strong depth dependence of this factor makes an extension of the standard definition including a chemical contribution inconvenient in this case. On the other hand, attributing all of the buoyancy to a thermal effect would result in an overestimate of the temperature contrast.

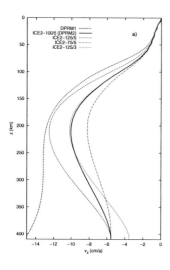

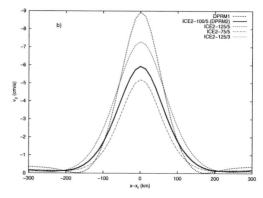

Figure V.32: Profiles of v_z through the quasi-stationary plumes of the models of series ICE2 and the DPRM1, vertically along the axes (left, a)) and horizontally in x direction at $z = 120$ km near the central axis (above, b)). The x coordinate is shifted so that the origin lies at the spreading/plume center for better comparison.

models in this series: in three of them, $v_{z,P}(z_m)$ is set to $5\,^{cm}/_a$, and the (full) radius R_P is varied by 50 km around the DPRM2 value of 200 km, *i.e.* the $^1/_e$-folding distance r_P is 100, 125, and 75 km, respectively (models ICE2-100/5, ICE2-125/5, ICE2-75/5); furthermore, a model with the radius of the DPRM1, $r_P = 125$ km, but a maximum influx velocity of only $3\,^{cm}/_a$ was considered to establish a closer link between this model series and series ICE1 (model ICE2-125/3). In principle, one could also consider a model where the basal influx velocity is set to the value observed in the $v_z(z)$ profile of the DPRM1, which is *ca.* $15\,^{cm}/_a$; to ensure that the flux is the same, the $^1/_e$-folding radius would have to be decreased to about 55 km then, whereas $r_P = 125$ km would lead to a volume flux of about $7\,^{km^3}/_a$, which is even more than the values of Ito *et al.* (1999) and Ito (2001). For this reason, and because it is regarded as physically more consistent to let the flow field evolve from the model itself in the relevant upper part, such a model was not attempted. – Table V.15 gives an overview of this model series.

Although the influx velocities in the models of series ICE2 are the same in three out of five cases, the differences in the volume fluxes and plume radii cause the axial v_z depth profiles to be noticeably different even for models ICE2-100/5 (DPRM2), ICE2-125/5, and ICE2-75/5, as can be seen in figure V.32a: above *ca.* 200 km depth, the broad plume ICE2-125/5 comes

Model	r_P (km)	R_P	$v_{z,P}$ $(^{cm}/_a)$	q $(^{km^3}/_a)$
DPRM2 (ICE2-100/5)	100	200	5	1.542
ICE2-125/5	125	250	5	2.409
ICE2-75/5	75	150	5	0.867
ICE2-125/3	125	250	3	1.446

Table V.15: Model runs of series ICE2. All values are for the model bottom. Note that the DPRM1 plume has the same basal volume influx design as model ICE2-125/3.

closest to the DPRM1, although it is significantly broader (figures V.32b and V.33), while the $v_z(z)$ profile of the DPRM2 is very similar to that of the initially broader, slower plume of ICE2-125/3; these plumes also produce crust of very similar thickness (figure V.34). The profiles of ICE2-125/5 and ICE2-125/3 are essentially parallel between 410 and *ca.* 250 km depth and offset by the difference in the influx velocity; at mid-model depths, the faster plume accelerates more, but at supersolidus depths, the velocities converge, and approximately the same v_z is reached at levels differing by 10 km or less in these two cases and the DPRM2. The clear differences between models with the same $v_{z,P}(z_m)$, but different radii must therefore be related to the larger volume flux resp. heat flux in broader plumes; inspection of simple analytical models such as the Stokes flow problem or the flow in plume heads and conduits with T-dependent viscosity (Schubert *et al.*, 2001, pp. 521ff., 531f.) confirms that the ascent speed increases with the radius of the Stokes sphere or with the heat flux of the plume, respectively, which itself is also radius-dependent at the model bottom (see appendix B.3). Nonetheless, the speedup thins the stems of all plumes, and the horizontal T profiles at 200 and 400 km in figure V.33 show that the differences between the radii shrink at shallower depths; the plumes of series ICE2 are less strongly focussed than the DPRM1. – Moreover, the horizontal T profile through the DPRM1 plume at 200 km depth shows a ring of slightly reduced mantle temperature around the hot plume stem which also weakens the upwelling a bit. This ring is absent in this form in the ICE2 plumes, because it is related to the ascent of the plume through the phase transition below 410 km; it coincides roughly with a hull of mantle material which is initially at ambient mantle temperature, but rises a bit faster than passively upwelling mantle, apparently due to drag from the plume. To explain this effect, one has to recall that in the ICE1 series, the vertical velocity of the background mantle became very low in the phase transition interval, so that the temperature shift was flattened out by conduction, leading to a decrease of temperature in that depth range and therefore retarding the phase transformation. In the surroundings of the rising plume, this cooled material is dragged along and undergoes the temperature decrease due to the transition, so that it develops into a weak low-T zone while passing the transition zone, which is then advected to shallower depth. Altogether, the phenomenon does though not seem to affect the general dynamics of the plume and of melt generation.

The radius variations do not only cause mantle rock to be transported through the melting zone more rapidly, but also broaden the volume where supersolidus temperatures are reached (figure V.33), so that the melt and crust production in a broader plume is expected to increase superlinearly as a function of plume radius, as suggested by the $v_{z,P}(z_m) = 5\,\mathrm{cm/a}$ curves in figure V.34. However, the depletion on the central axis, which is mostly a function of the homologous temperature, is essentially the same over the largest part of the melting depth range in all models (figure V.35), because the models were tuned to have similar centerline temperatures, and a common solidus and f parameterization had been used; the small differences in f near the lithosphere can be ascribed to minor factors such as the small temperature differences or the shallower final melting depth in series ICE2, respectively the deeper onset of recrystallization.

Radius and velocity variations lead to variations in volume flux q, and as all models have already demonstrated the importance of the flux through the melting zone for the amount of crust produced, it is of interest to consider crustal thickness and volume as an explicit function of q in this series, where emphasis lies on changes on the imposed based influx.

Figure V.36a shows a plot of $h_{\max}(q)$ for a time, where each plume has reached a quasi-stationary maximum crustal thickness according to the $h_{\max}(t)$ plot in figure V.34. The plots of figure V.36 highlight the trade-off between radius and influx velocity in terms of crust production: the two ICE2 data points near $q = 1.5\,\mathrm{km^3/a}$, which correspond to the

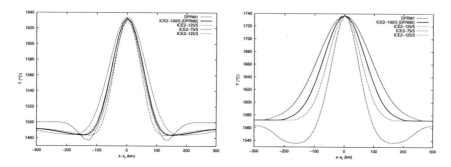

Figure V.33: $T(x)$ profiles through the quasi-stationary plumes of the models of series ICE2, at depths of 200 (left) and 400 km (right). Corresponding profiles through the DPRM1 plume are also shown; the x coordinate is shifted so that the origin lies at the spreading/plume center for better comparison. The ICE2 curves in the $z = 400$ km plot correspond practically to the basal T distribution, because that depth is only 10 km above their source, but 260 km above the source of the DPRM1.

DPRM2 and to model ICE2-125/3 (see table V.16), show that models with different r_P and $v_z(z_m)$ can produce very similar crusts; this is confirmed by the $h(x, y)$ contour plots in figure V.37, although it cannot be excluded that the patterns would differ a bit more if all the extracted melt were explicitly tracked to an eruption site at the ridge on the basis of some model of an oriented-channel network as outlined in section II.4.3. As expected, the DPRM1 value also lies close to these data points. For MOR, other authors (*e.g.* Scott and Stevenson, 1989) have already stated that, the higher the volume flux is, the more material is transported through the melting zone, and the thicker is the crust produced, whereby the dependence was found to be linear; in view of the pattern of the four data points of figure V.36a it seems therefore appropriate to fit $h_{max}(q)$ with a linear function. Notably, the $q = 0$ intercept of this line does not come near the crustal thicknesses for MOR around 6.2 km in these models, but lies at about 21.7 km on the h_{max} axis; the difference gives an estimate of the direct temperature effect of the plume on crust production, because the T anomaly – *ca.* 150 K at the solidus depth of the plumes – is what still remains of the plume

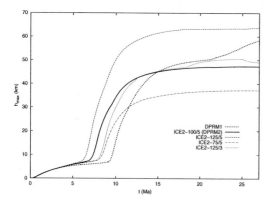

Figure V.34: Maximum crustal thickness in the models of series ICE2 as a function of time; the DPRM1 curve up to 27 Ma model time is also plotted for comparison.

Model	q $(^{\mathrm{km^3}}/_{\mathrm{a}})$	$h_{\max}$ (km)		y_{cP} (km)	y_{P0} (km)	V_{cP} $(10^6\,\mathrm{km^3})$	
		plume	MOR			at 21 Ma	at 27 Ma
DPRM2	1.542	45.6	6.2	271	530	1.56	2.22
ICE2-125/5	2.409	61	6	333	690	2.59	3.62
ICE2-75/5	0.867	35.9	6.3	218	420	0.95	1.38
ICE2-125/3	1.446	46	6.2	276	564	1.58	2.27
DPRM1	1.446	49	5.6	220	500	1.15	1.7[a]

[a]extrapolated

Table V.16: Model results of series ICE2. The along-ridge width of the plume crust, y_{cP}, was defined to be the distance between the $h = 10\,\mathrm{km}$ isolines at the spreading center (see figure V.37). The excess crust produced by the plume was calculated by subtracting an estimate of the total MOR crust from the total crust in the model; the estimate was calculated by multiplying the cross-section area of a slice near the margin of the model with the model length y_{m} and does hence not include the small along-axis variations of $h(x, y)$.

after reducing the excess volume flux to zero, *i.e.* for passive upwelling of anomalously hot material.

It has repeatedly been mentioned that plume material flows along the ridge (see sections I.3.2 and IV.2.3), and the along-axis flow is also expressed by enhanced crust production, as obvious *e.g.* from the V-shaped ridges in the case of Iceland (see sections IV.1.1 and IV.2.3). Ribe *et al.* (1995) had derived a relation of the type $y_{\mathrm{P0}} \sim q^c$ with $c = 0.553$ for the waist width of the plume head[4] (see section I.3.2). For this reason, a similar function model has been used for fitting the plume waist width y_{P0} – here defined as the greatest along-axis distance between the 10 K isolines of the temperature anomaly at the spreading center – and the along-axis width of the anomalous crust y_{cP}; the latter was chosen to be the distance between the intersections of the $h = 10\,\mathrm{km}$ isolines with the spreading axis, anticipating that the $h = 7\,\mathrm{km}$ isoline has a less simple shape in the hydrous models of

---

[4]Ribe *et al.* (1995) actually do not express y_{P0} as a function of q only, but rather in the form $y_{\mathrm{P0}} \sim \Pi_{\mathrm{b}}^{c'} \sqrt{q/v_r}$, whereby their "buoyancy number" Π_{b} is also a function of q, and they seem to fit their model data with y_{P0} as a function of Π_{b}, thereby neglecting the correlation between Π_{b} and other factors.

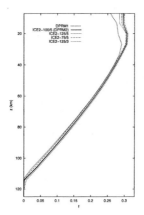 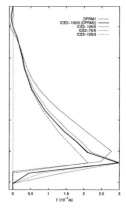

Figure V.35: Vertical profiles of f (left) and Γ (right) through the quasi-stationary plumes of the models of series ICE2 and the DPRM1.

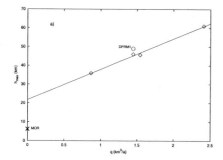
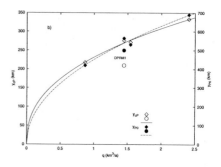

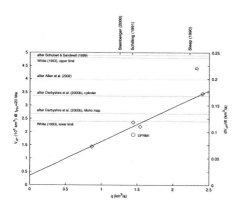

Figure V.36: Maximum crustal thicknesses (a), plume and crust waist widths (b), and excess crust volumes after 20 Ma of plume melting (c) as a function of volume flux for models of series ICE2 and the DPRM1. For fitting, the data points represented as diamonds have been used in all cases; the DPRM1 values are also given for comparison (circles). The derived scaling laws are: $h_{\max} = 16.236q + 21.7$ (in km, q in km^3/a); $y_{cP} = 231.77q^{0.411}$ (in km) and $y_{P0} = 451.3q^{0.48}$ (in km); $V_{cP} = 1.285q + 0.35$ (in $10^6\,km^3$). In figure c), a few estimates from literature data for the plume crust volume in excess of 7 km-thick normal oceanic crust are added (thin horizontal lines); see the text for details. The volume growth rate on the right ordinate is the 20 Ma-average as determined from the volume.

series ICE3 in section V.2.3 and having in mind that this is about the maximum crustal thickness of the ridges to the north and south of Iceland (see section IV.1.1), so that this isoline should give a useful impression of the along-axis extent of the Iceland shelf within the restrictions of a model without north-south asymmetry. The data and fitting curves are shown in figure V.36b. The exponent of $y_{P0}(q)$ is 0.48 and thus a bit smaller than the one of Ribe *et al.* (1995). This might partly be due to the fact that y_{P0} is expressed as a function of q instead of the buoyancy number used by these authors and that they define y_{P0} in a different way, but the fact that the temperature field in the models of this series included the effect of latent heat in the melting zone, which was not the case in the other study, could have reduced the spreading of the plume along the axis as a consequence of increasing viscosity; it cannot be excluded either, that in the later stages of waist evolution, the plume head is affected by the model boundaries at $y = 0$ and $y = y_m$. Furthermore, it must be noted that four data points are not too much data to construct a tight fit, compared with the 19 data points in the other study. – The "crustal waist width" y_{cP} has been fitted with a function of the same type as y_{P0}, because it is a direct consequence of the along-axis flow of the plume head; it was therefore expected that the exponents of both fitting formulae would not differ too much (see caption of figure V.36b), while the discrepancy can be explained by

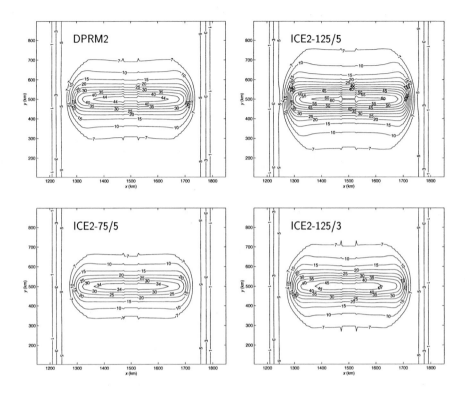

Figure V.37: Calculated crustal thickness (in km) of the models of series ICE2 at 27 Ma model time. The ridge axis is located at $x = 1500$ km; the crust has been transported by plate drift up to 270 km from the margins of the volcanic zone at $x = 1500 \pm 30$ km.

the arbitrariness of the isoline chosen for measuring y_{cP}.

Contrary to crustal thickness and waist width, the volume of the crust does not reach a stationary value, but approaches an approximately linear growth when $h_{\max}$ and $y_{\text{P0,cP}}$ cease to change, because continued melting and plate spreading result in a constant growth in x direction. The total amount of crust obviously depends on the size of the model, so that it is more reasonable to try to separate out the excess crust. This is done by integrating the total amount of crust in the model at a given time and subtracting the cross-section area of a normal MOR segment of the model times the model length y_{m}; at the later stages of model evolution, the resulting growth of excess crust is indeed linear to a good approximation. The oldest parts of Iceland, the western fjords, are a bit more than 20 Ma old (*e.g.* Sæmundsson, 1979, see sect. IV.1.1), therefore the amount of crust after 20 Ma of plume melting will be considered here. This volume was determined by taking the calculated thicknesses at 21 and 27 Ma model time, subtracting the timespan before the onset of plume melting, which can be assumed to be pinpointed fairly well by the kink in the $h_{\max}(t)$ curves in figure V.34, and interpolating (or extrapolating) linearly between the two calculated values. The resulting excess volumes are plotted in figure V.36c as a function of q. For comparison with observations, some estimates from results by Schubert and Sandwell

(1989), White (1993), Darbyshire *et al.* (2000b), and Allen *et al.* (2002b) are also included[5]; there is considerable scatter in the V_{cP} values, which is partly due to the different size of the regions taken into consideration, but with regard to the total productivity of the plume, which in reality affects areas beyond the Iceland shelf, it seems more likely that the higher values reflect the productivity potential of the plume better. The figure reveals that most of the models are rather at the lower end of the range of estimates or even below; in particular, the low-flux model ICE2-75/5 with a thin plume produces too little crust, indicating that a thin plume would need a much higher upwelling velocity *and* an efficient mechanism to distribute the material along the ridge in a way that prevents the generation of a too thick crust above the plume center, because the maximum thickness of ICE2-75/5 is not too far off the observed value of *ca.* 40 km.

Chemistry and petrology — At this opportunity, the DPRM2 melting degrees are used for a short illustration of applying parameterizations of chemical and petrological quantities of the type described in section II.3.3 to the data delivered by convection models. This application has to be rudimentary and is restricted to normal ridges, where anhydrous melting can be assumed in this case and will be confined to the spinel stability field, because for melting in the garnet stability, parameterizations are even less reliable, and the compositional change introduced by the phase transition would complicate the matter further.

Figure V.38a shows on-ridge depth profiles of f and Cr# of spinel from abyssal peridotites according to eq. II.34 (Hellebrand *et al.*, 2001); note that, although f does not exceed 0.2, the range of calibration of eq. II.34 (up to Cr# = 0.6) is slightly exceeded. Most of the samples used for constructing eq. II.34 were calculated near fracture zones, but Hellebrand *et al.* (2001) estimated a melting degree of at least 15 % for the samples least influenced by it, using samples from the Kane Fracture Zone area spanning an f range as large as 0.1. It is reasonable to regard $f = 0.15$ as a lower bound for the melting degree at MOR, because it cannot be excluded that the fracture zone also had an effect on the most depleted samples found there; hence, it seems that the melting degrees computed by the numerical model are reasonable and do not exceed observed values intolerably. This gives additional confidence in the applicability of the melting calculation method and allows for the prediction of compositional characteristics of mantle residuals from the numerical model.

The mineral mode fractions calculated from f after eq. II.33b using the melting proportions of Kostopoulos (1991) are plotted in figure V.38b. In agreement with observations, the fraction of olivine increases significantly, but on the other hand, clinopyroxene is not exhausted, but remains in the residue; this confirms the observation of Dick *et al.* (1984) that

---

[5]The estimate after Schubert and Sandwell (1989) was derived by subtracting the normal oceanic crust from their total volume of $14.151 \cdot 10^6$ km^3 and multiplying the result with 20/35 to account for the fact that the area considered by them includes the Greenland–Iceland–Faeroe Ridge with crustal ages up to ~ 35 Ma; it is expected that this crude calculation will underestimate the true volume to some extent. The estimates after White (1993) use his melt production rates in excess of those creating a 7 km-thick oceanic crust, which range from 0.12 to 0.24 km^3/a, multiplied by the proposed age of 20 Ma. The "Moho map" estimate after Darbyshire *et al.* (2000b) makes use of the Moho map in figure IV.4. As the dataset itself was not available, an attempt was made to extract the quantitative information from the digital form of the picture by dividing the Moho depth range into several data bins corresponding to a certain colour range on the colour bar, counting the number of coloured pixels in each bin, and multiplying by the average depth of the bin and the approximate area represented by a pixel on the map; however, noisy colour values and coverage gaps are certain to be sources of error in this estimate, which must be regarded as a lower bound anyway, because a part of the Iceland shelf and much of the adjacent anomalous ridges are missing from the map. The "cylinder" estimate after Darbyshire *et al.* (2000b) uses the same data, but approximates the Iceland shelf by a cylinder with a radius of 240 km and a depth equal to the average depth from the Moho map, which is *ca.* 25.5 km. The estimate after Allen *et al.* (2002b) works the same way, but uses their estimate of an average thickness of 29 km instead.

Figure V.38: Vertical profiles of f and spinel Cr#
(left) and fractions of olivine,
orthopyroxene, clinopyroxene,
and spinel (right) beneath normal ridge in the DPRM2
model.

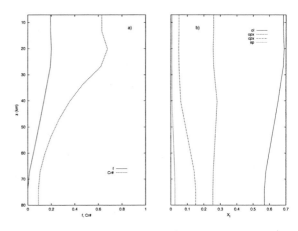

four-phase lherzolite exists beneath MOR to shallow levels (see section II.3.2). However, the
fact that with the independent melting parameterization by Katz *et al.* (2003) it is possible
to exceed cpx-out, albeit only slightly, as shown in section V.1.4, it must be asked if as much
as 5 % cpx really remain in the mantle, as calculated here. Altogether, more experimental
work needs to be done, but in principle, this procedure would also provide a possibility to
predict physical properties of the mantle from the melting model on the basis of calculated
modal composition.

Preliminary discussion — As with model series ICE1, the results of this section are shortly
evaluated here to prepare the models of the following series.

The basic result of series ICE2 is the dilemma that the models with an intermediate
volume flux of the order as suggested by Steinberger (2000) or Schilling (1991) produce crust
of a maximum thickness roughly compatible with observations, and even the low-flux model
does not fall short of these values too much, whereas the high-flux model ICE2-125/5, which
exceeds slightly the estimate by Sleep (1990), produces too thick crust. On the other hand,
it is rather this model, whose crustal volume agrees best with independent estimates, and
even larger volumes would probably be preferable, but to achieve them by increasing the
volume flux, one would have to leave the range of estimates based somehow on observations
(see table IV.2). However, there is not much freedom for doing this by broadening the plume,
because all observations which can be expected to have a reasonable spatial resolution point
to values of 150 km or less for r_P; this radius would result in a volume flux of $3.47\,^{km^3}/_a$
for a basal influx velocity of $5\,^{cm}/_a$, and in $h_{max} = 78\,km$, $y_{P0} = 820\,km$, $y_{cP} = 386\,km$ and
$V_{cP} = 4.81 \cdot 10^6\,km^3$, using the formulae in the caption of figure V.36 for extrapolation. While
this would match well the volume estimates and even yield more realistic waist widths than
all of the Iceland models presented so far – y_{P0} and y_{cP} are notoriously too small compared
with the values from table IV.2 –, the maximum crustal thickness is unacceptably large. To
achieve a similar flux in a plume with $r_P = 100\,km$, an influx velocity of more than
$11\,^{cm}/_a$ would be necessary, but given that the calculated upwelling rates in the melting zone
(figure V.32) are partly near the upper bound of independent estimates of this quantity,
which range from 2 to $10\,^{cm}/_a$ (*e.g.* Allen *et al.*, 2002b; Maclennan *et al.*, 2001a, also see the
discussion in sect. VI.2), it is suspected that such a high $v_{z,P}(z_m)$ would enforce even higher
values within this model setup. This would not solve the problem of the crustal thickness,
though, and it is questionable if such an upwelling velocity makes sense in connection with

a temperature anomaly of the magnitude of the present models.

The need to achieve a broader waist together with a crustal thickness no larger than the one of the DPRM2 with volume fluxes higher by a factor 1.5 or 2 rather suggests that the problem lies in the horizontal distribution of the melting material, *i.e.* the plume must flow along the ridge to a larger extent, as has been suggested by many authors to explain the V-shaped ridges. However, if the upwelling is too strongly concentrated at the central axis of the plume, one would still expect a thick crust there, even if the material flows fast along the ridge after that; furthermore, that material would already be depleted and therefore less productive. Hence, it seems to be more promising to assume that the flow of the material is already spread over a broader area when entering the melting zone, which would be the case if hydrous melting is considered, as already demonstrated by Ito *et al.* (1999); the models of set VV4 (see section V.1.4) confirm this idea, and in particular the $h(x_r, y)$ profile in figure V.14 shows that a larger waist width would come out of such a model. On the other hand, the reduced upwelling in the melting zone will probably result in a lower total melt productivity, so that *e.g.* a hotter plume is needed to reproduce the observed total volumes. This problem will be tackled in the next section.

V.2.3 *Model set 3: hydrous melting*

So far, the influence of several physical quantities on convection and melting dynamics of the Iceland plume has been considered, but the composition of the mantle has been regarded as uniform. However, a great many of geochemical investigations indicates that the mantle in general does not consist entirely of peridotite, but also contains a minor, pyroxene-rich component (see section II.2.5), and that in particular the Iceland plume is chemically quite heterogeneous and also features sources of basaltic, crustal origin (see section IV.2.2). Furthermore, in the models of series ICE1 and ICE2, a dry mantle has been assumed as an approximation, while the real sub-oceanic mantle contains small amounts of water and plumes are apparently often even enriched in water (see section II.3.2). While an eclogitic component will influence the melting style and the chemistry of the melts produced notably, but probably not change the large-scale dynamics of the plume substantially, the inclusion of water has a strong impact on both convection and melting style, as shown in the VV4 model subset in section V.1.4. For this reason, some models have been made in order to assess the influence of water-undersaturated melting on the Iceland plume: the first model, ICE3-MP1, is identical to the DPRM2, with the exception that the anhydrous melting parameterization by Katz *et al.* (2003) was used and that the solidus was lowered by 35 K to account for the possibly neglected low-f tail, *i.e.* it combines the solidus shift of model SF1a with the parameterization of model SF3 in section V.1.7; in the second model, ICE3-MP2, the effect of water is then included, whereby water concentrations of 142 and 500 wt ppm have been assumed for the normal mantle and for the plume, respectively, as in model VV4b, in order to account for the result from geochemical investigations that the plume source of Iceland has an elevated water content (Jamtveit *et al.*, 2001; Nichols *et al.*, 2002, see sect. IV.2.2). As there is some uncertainty on how strong the dependence of the viscosity on water content actually is, a third model, ICE3-MP3, has been run with $b_C = 0.5$ instead of $b_C = 1$ in eq. III.16, as in model ICE3-MP2, so that the viscosity variations with dehydration are weaker. From the $h_{max}(t)$ curves of models SF1, SF1a, and SF3 (figure V.23), it can be anticipated that $h_{max}(t)$ of model ICE3-MP1 should be very similar to that of the DPRM2, because the alternative $f(\tilde{T})$ parameterization and the lower solidus will approximately cancel out in terms of crust production. Table V.17 gives an overview of the results of this series.

Model	C_{H_2O} (ppm) plume MOR	b_C	h_{max} (km) plume MOR	y_{cP} (km)	y_{P0} (km)	ΔV_{cP} (10^6 km^3) at 21 Ma at 27 Ma
ICE3-MP1	0 0	–	45.5 5.5	253	540	1.56 2.22
ICE3-MP2	500 142	1	21.2 6.7	391	526	0.99 1.42
ICE3-MP3	500 142	0.5	21.2 6.7	414	514	1.26 1.48

Table V.17: Model parameters and results of series ICE3. y_{cP}, y_{P0} and ΔV_{cP} were determined in the same way as in series ICE2.

To establish the connection with the previous models series, one must first compare the DPRM2 with the anhydrous model ICE3-MP1. The depth profiles in figure V.39 show that both models are very similar indeed; the small local differences are due to the different solidus depths – the lower solidus of model ICE3-MP1 leads to an onset of melting about 6 km deeper than in the DPRM2, similar as in models SF1 and SF1a (section V.1.7) – and due to the cpx exhaustion criterion in the models of this series, which comes into effect at around $z = 60$ km in the plume and reduces enthalpic cooling and the concomitant stiffening at shallow depths, similar to model SF3[6]. With respect to model set VV3 (effect of φ on η, section V.1.4), the $\eta(z)$ profiles of the anhydrous models confirm the earlier conclusion that the effect of retained melt on mantle viscosity at the low porosities prevailing here is small. The effect becomes manifest in the small steplike viscosity drop around 125 or 80 km in the plume and beneath the ridge, respectively, *i.e.* at the onset of melting, in these two profiles; the $\eta(z)$ profiles in figures V.8 and V.9 do not display it, because the feature was disabled in model set VV2, while it is strongly masked by the dehydration effect in model set VV4, which is also the case in the hydrous models of this series.

In the details of melting, however, there are some notable differences between the two anhydrous models, as can be seen from figure V.40. As mentioned, melting begins at greater depth in model ICE3-MP1, but the productivity at low melting degree is significantly higher in the DPRM2 due to the concave-down shape of the McKenzie and Bickle (1988) parameterization. Furthermore, the DPRM2 is not subjected to an explicit cpx-out criterion as are the models of series ICE3, so that the maximum melting degrees reached in model ICE3-MP1 are *ca.* 3–5 % lower than in the DPRM2; especially in the plume, the lower f are likely to be more realistic, because there cpx-out is exceeded significantly. Nonetheless, the lower productivity is approximately counterbalanced by the larger volume of the melting zone, so that the maximum crustal thicknesses and the overall $h(x,y)$ distributions are very similar for both models, which becomes obvious from figure V.41 and by comparing the respective maps in figures V.37 and V.42; this is an unexpectedly good confirmation of the earlier prediction that the combination of the McKenzie and Bickle (1988) melting parameterization and the original Hirschmann (2000) solidus would lead to a similar crust as the anhydrous Katz *et al.* (2003) melting parameterization together with their 35 K downward shift of the solidus (see p. 167).

Having stated the relatively close similarity between the two anhydrous models DPRM2 and ICE3-MP1, it is now possible to consider the hydrous derivates of the latter, models ICE3-MP2 and ICE3-MP3. As known from the basic discussion in section II.3.3 and model set VV4 in section V.1.4, hydrous melting begins at greater depth, because water lowers the solidus; hence, consumption of latent heat in the depth interval between the hydrous and the dry solidus causes the geotherms of the hydrous models to differ by up to about

---

[6]The $\eta(z)$ profiles of the DPRM2 and of model ICE3-MP1 in figure V.39 have been reconstructed *a posteriori* from the $T(z)$ and $\varphi(z)$ data using eq. III.16.

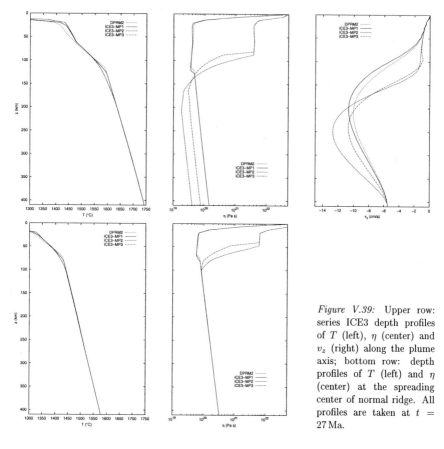

Figure V.39: Upper row: series ICE3 depth profiles of T (left), η (center) and v_z (right) along the plume axis; bottom row: depth profiles of T (left) and η (center) at the spreading center of normal ridge. All profiles are taken at $t = 27\,$Ma.

a dozen kelvins from those of the dry ones. In the normal mantle, melting therefore begins at *ca.* 100 km depth instead of 90 km as in model ICE3-MP1, and in the plume, the deepest melts form at more than 200 km instead of 128 km as in model ICE3-MP1 due to the higher temperature and the higher water content; nonetheless, the melting degrees hardly differ in most of the melting region. For better comparability, the reference viscosities and hence the viscosities of unmolten normal mantle are defined to be the same in both the water-free and the hydrous models; therefore, the subsolidus parts of the $\eta(z)$ profiles are identical beneath the ridge, whereas the unmolten plume stem has a lower viscosity in the hydrous models than in the dry ones, the higher b_C, the lower η (figure V.39). It is therefore not astonishing that the upwelling velocity in the deeper parts of the hydrous plume stems are correspondingly higher than in the anhydrous models, but on the other hand, dehydration stiffening lets v_z drop to about the passive upwelling rate rapidly above the solidus depth, whereas in the dry models, a relatively high upwelling rate is preserved through a large part of the melting depth range. – As a detail illustrating the different factors which influence the viscosity, it is worthwhile to consider the depth range between *ca.* 100 and 200 km in the $\eta(z)$ profiles of the plume. While the anhydrous models show the abovementioned rather abrupt, yet small decrease at the bottom of the melting zone due to the effect of retained melt, no such step is

Figure V.40: Depth profiles of T (left) and f (right) for model series ICE3.   Upper row: plume axis; bottom row: at the spreading center of normal ridge.   All profiles are taken at $t = 27$ Ma.

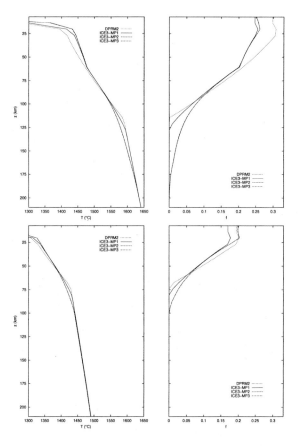

visible in the hydrous profiles, because dehydration stiffening overcompensates (ICE3-MP2) or at least cancels out (ICE3-MP3) the φ effect. However, as the porosity ramps up to the threshold over a much larger depth interval in the hydrous models – *ca.* 20 km below the ridge and *ca.* 40 km in the plume – than in the dry ones, both the φ and the water effect are only small at great depth, as in ICE3-MP2, or cancellation as in ICE3-MP3 pretends that melting begins at a shallower level, if one would only judge from the viscosity profiles. It is only after having reached the extraction threshold that the dehydration effect dominates the shape of $\eta(z)$.

In view of the strong influence of v_z on melt productivity, it is clear that the strongly reduced crustal thickness in the hydrous models (figure V.41) is largely due to the weakened mantle flow in their melting zones; in this context it is important to note that only active upwelling is affected, whereas the greater volume of the melting zone in models ICE3-MP2 and ICE3-MP3 results in a thicker crust at normal MOR, compared to model ICE3-MP1 (see table V.17), because the upwelling rate is the same there in all models. The shutdown of active upwelling in most of the melting zone in the hydrous models turns these plumes into "plumes with zero excess volume flux", *i.e.* features which are essentially characterized by their thermal anomaly, in the sense that much of the plume flux does not reach the main

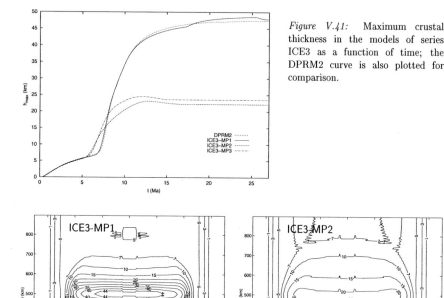

Figure V.41: Maximum crustal thickness in the models of series ICE3 as a function of time; the DPRM2 curve is also plotted for comparison.

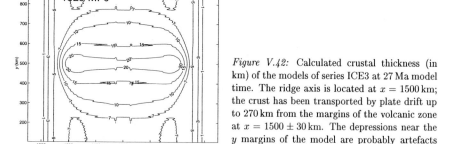

Figure V.42: Calculated crustal thickness (in km) of the models of series ICE3 at 27 Ma model time. The ridge axis is located at $x = 1500$ km; the crust has been transported by plate drift up to 270 km from the margins of the volcanic zone at $x = 1500 \pm 30$ km. The depressions near the y margins of the model are probably artefacts related to the boundaries.

melting region, but is diverted at greater depth, so that the material remains unmolten or at least does not melt to such degrees as in the anhydrous models. As the plumes of series ICE3 have the same basal volume flux as the DPRM2 plume, one can compare the observed maximum plume crustal thicknesses of this model with the $h_{\max}(q)$ plot in figure V.36a. While the two anhydrous models plot close to each other, the two hydrous models should fall close to the intercept of the fitting line with the ordinate at 21.7 km;

the value of 21.2 km from table V.17 shows that this is indeed the case, give or take a few percent due to the weak active upwelling component in the deep low-productivity region on the one hand and different melting parameterizations on the other hand. – However, the h maps for ICE3-MP2 and ICE3-MP3 in figure V.42 reveal another important property of the hydrous models, namely the ability of hydrous melting models to create a plume plateau with a greater along-ridge extent than their anhydrous counterparts; this was already demonstrated in model set VV4 (see figure V.14) and seems to offer a solution for the problem of the narrow plateau mentioned in the preliminary discussion of section V.2.2. Nonetheless, comparison of the values in table V.17 and the observational data in tables IV.1 and IV.2 and in figure V.36c show that the maximum thicknesses, waist widths and crustal excess volumes are too small, which indicates that models with higher volume flux, and possibly also higher excess temperature, are required.

For completeness, it is noted that melt segregation velocities are generally lower in the models of series ICE2 and ICE3, because the melt had been given a higher reference density, but it is still of the same order of magnitude as in series ICE1, and it also is still buoyancy-dominated in all models. In the models with the concave-up $f(\tilde{T})$ parameterization, v_{seg} is very small in the deepest parts of the melting region, in agreement with its porosity dependence via eq. II.8 and eq. II.38a, and it ramps up in a similar way as φ. Hence, the hydrous melts would hardly separate from their source and be nearly immobile, but the results of series VM (see section V.1.8) indicate that this would change if the pressure dependence of the melt viscosity would be taken into account. While the calculated segregation velocities in the lower parts of the melting region are well resolved, they become increasingly noisy at shallower depth in the hydrous models; the reason is likely that the discretization of the model is too coarse too capture well the large velocity and/or stress gradients in the strongly sheared mantle beneath the spreading center. Therefore, the melt segregation results will not be considered further here; unfortunately, the RAM requirements needed for a substantial increase in resolution are very probably beyond the reach of the computers available for this study, and computation times would also rise significantly.

Stress field and dike orientation — The outstanding differences between the dynamics, melt generation, and crust productions of models ICE3-MP1 and ICE3-MP2 are expected to result in pronounced differences in the associated stress field in the mantle. The stress field, however, has been shown to have significant influence on the orientation of dikes and on the formation of high-porosity channels (see section II.4.2), both of which have been implicitly taken as existing in the mantle in many models of this study for reasons discussed previously; different stress fields will therefore result in different patterns of fast melt migration. In section II.4.2, the importance of the orientation of the channels has been considered mainly in the context of melt focussing towards a normal mid-oceanic spreading center, and the corresponding appearance of anisotropic permeability has been discussed with respect to this setting in section II.4.3, partly following the work of Phipps Morgan (1987). Here, the formalism outlined in section III.2.2 is applied to the matrix velocity fields and melt distributions at $t = 27$ Ma of models ICE3-MP1 and ICE3-MP2 assuming that channels/dikes form everywhere in the partially molten region, but not outside of it, and that melt flow in these conduits is entirely buoyancy-driven. Note that this calculation was done *after termination* of the model runs themselves, *i.e.* the computed crusts of these runs shown in figures V.41 and V.42 are not a result of these considerations, but were derived under the default assumption of melt transport in planes perpendicular to the ridge; as will be shown below, they would likely look a bit different if the melt had been transported through oriented conduits, whereas the mantle convection pattern would not be affected, because neither the details of the rapid extraction mechanism nor the crust feed back into the convection/melting model. Hence, the following examination can help to judge the validity of this often-made simplification.

Figure V.43 shows the pattern of mantle flow and the viscosity for the two models. Comparison with the streamlines in figure II.10 indicates that the mantle flow under normal ridge (figure V.43c,d) is quite similar to the analytical corner flow solution, in spite of the presence of a thermal and compositional lithosphere and the rheological differences; it seems that the mantle flow is controlled by the boundary condition at the top in the first place. In contrast, active upwelling in the plume (figure V.43a,b,e,f) is visible as a zone with high matrix velocities, reaching a value of ten times the passive upwelling rate or more at *ca.* 200–250 km depth (cf. figure V.39); it decreases towards the passive upwelling rate at depths $\lesssim 100$ km in the dry model and $\lesssim 150$ km in the wet model; in the depleted region, dehydration stiffening has forced the flow velocity in model ICE3-MP2 to rates comparable to those at normal ridge, as already seen in the $v_z(z)$ profile in figure V.39. Comparison of the ridge-perpendicular and the along-ridge cross-sections shows that the flow system of the plume has roughly cylinder symmetry, at least within some 150 km of the axis; however, the shape and position of the low-viscosity zone of not or not fully dehydrated plume material suggest that plume material tends to be able to rise to shallower depth beneath the ridge than off-ridge, because perpendicular to the ridge, the background flow and the lithosphere hamper upward motion of the plume material. – The irregular pattern in the viscosity field of the plume head off-ridge around 100 km depth (figure V.43b) seems to be related to the irregular porosity in that region (see below).

Figure V.44 shows the pattern of channelized melt flow along with the depletion for both models. Under normal ridge (figure V.44c,d), the flow pattern also resembles that from the analytical corner flow solution, represented by the arrows in figure II.10, in that defocussing occurs for $x < z$ and focussing for $x > z$; the effect is not strong, though. Near the ridge, channels are oriented vertically, which is probably due to the extensional region in the 60 km-wide strip where the spreading velocity increases from zero to v_r; this is corroborated by the orientation of the extensional stress at shallow depths shown in figure V.45c,d and of the dikes (figure V.46). In the plume (figure V.45a,b,e–i), the situation is more complicated and differs between the dry and the hydrous plume in that the general pattern is shifted to greater depth in the latter, because the effective lithosphere is defined by composition, *i.e.* water content, instead of temperature, and therefore thicker. In the dry plume of model ICE3-MP1, the hot low-viscosity material flows relatively fast beneath the thermal lithosphere (see the black double-arrow in figure V.44a) and results in a stress distribution where the compressional stress (σ_1) is slightly tilted away from the plume axis (also see the red arrows in figure V.45a and the left picture in figure V.46). If melt ascent is purely buoyancy-driven, which was assumed here, it now depends on the orientation of the intermediate and the extensional principal stress (σ_2 and σ_3, respectively) whether it rises vertically in the tabular dike or parallel to the compressional stress. The x–z cross-section through the plume stress field in figure V.45a and the dike picture figure V.46 show that near the lithosphere, the greatest extension is in the plane of the page, hence a tensile dike extends perpendicular to this plane and is tilted away from the axis as indicated by the compressional stress and the melt flow vectors; at greater depth, however, σ_2 and σ_3 flip their directions, because the azimuthal extension caused by the radial spreading flow of the plume dominates, and the dikes lie in the plane of the page there, so that the melt can indeed rise vertically in them. At still greater depths of more than 140 km, the plume does not spread but essentially moves upward at relatively high velocity, as indicated by the white double-arrow in figure V.44a. As a result, dike orientation would flip again, and the dikes would become more horizontal

anhydrous   hydrous

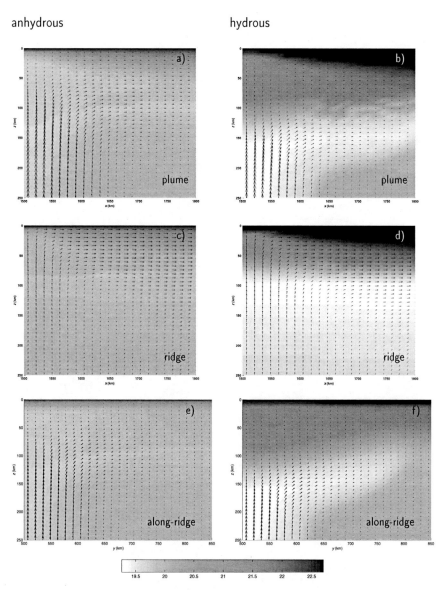

Figure V.43: Viscosity (gray-shaded, as $\log \eta$ with η in Pa s) and matrix velocity (arrows) in the upper 250 km of models ICE3-MP1 and ICE3-MP2. The spreading center is located at $x = 1500$ km, the plume center lies at $y = 500$ km.

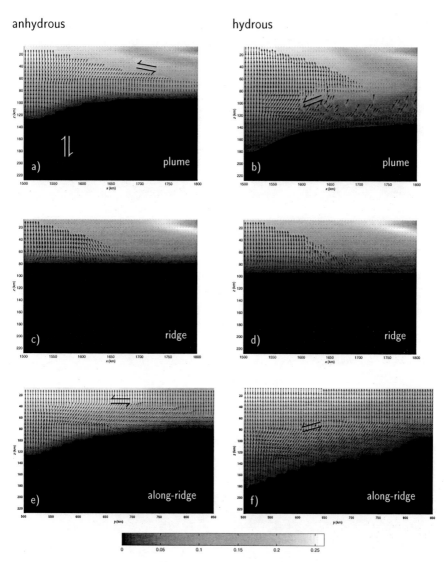

Figure V.44: Depletion (gray-shaded) and melt velocity in channels (arrows) in the depth range 6.7–228.5 km covered by the melt model grid of models ICE3-MP1 and ICE3-MP2. Velocity is only plotted where melt is actually present. The double-arrows in some of the pictures indicate the shear sense of matrix flow.

anhydrous   hydrous

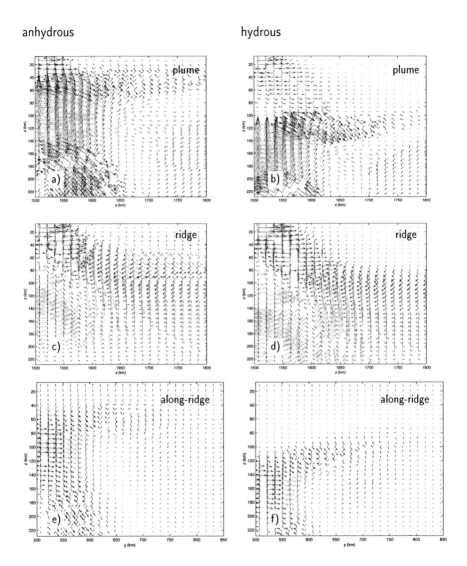

Figure V.45: Principal stresses in the depth range 6.7–228.5 km covered by the melt model grid of models ICE3-MP1 and ICE3-MP2 (red: maximum compression, green: intermediate, blue: maximum extension). The arrows are not to scale between colours and images. Note that their orientation is ambiguous by 180°. *See plate section for colour version.*

anhydrous   hydrous

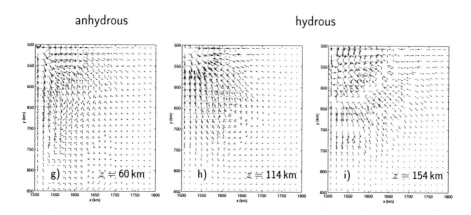

Figure V.45: continued.

if there were melt and dikes; this corresponds to the change in the melt migration pattern found by Ceuleneer *et al.* (1993, see sect. II.4.2, p. 57) for a plume under a stationary plate. In the hydrous plume, essentially the same stress patterns can be observed, but they lie some 30–40 km deeper, because dehydration by initial melting creates a compositional lithosphere which causes the plume head to partly spread outward at greater depth; the part of the plume head integrated into the hot, but dehydrated shallow mantle also moves outward more slowly than the deeper parts, because it is essentially mechanically coupled to the spreading plates, so that the plume head is not as broad as in the dry model in this case. The similarities between both models throughout the whole structure are further illustrated by the horizontal cross-sections in figure V.45g and h, which lie at analogous depth levels: near the plume, the horizontal component of σ_1 is directed radially outward and becomes more ridge-parallel

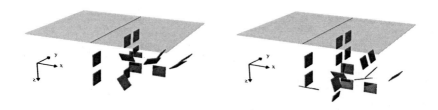

Figure V.46: Scheme of the orientation of tabular tensile dikes in a quarter of the model box for the dry model (left) and the hydrous model (right); the orientation of the dikes was derived by inspection of figures V.44 and V.45. The set of dikes in the foreground represents the pattern beneath normal MOR, the sets further in the background outline the plume head region. The spacing between the individual rectangles is *ca.* 100 km in the horizontal directions and *ca.* 40 km in the vertical direction, starting at $z = 20$ km depth; note that in the hydrous model, a deeper level of dikes is added, because the partially molten region reaches deeper. The spreading plates with the ridge are also shown for reference.

in the more remote realms of normal MOR spreading; the horizontal component of σ_3 is also approximately radial near the plume, so that the dikes lie on the interleaved surfaces of truncated vertical cones at that depth ($\lesssim 70\,\text{km}$ in ICE3-MP1, $\lesssim 110\,\text{km}$ in ICE3-MP2). Below this level, dike orientation is flipped, and the radial footprint of the plume expressed by σ_1 extends farther away (see figure V.45i for the hydrous case); the dikes form a system of spokes emerging from the plume axis, in which the melt can ascent vertically near the stem.

As a whole, the migration patterns suggest that the simplified instantaneous extraction used in this study is basically acceptable, because in several instances melt will ascend subvertically or be slightly focussed towards the ridge; however, for transport of melt from the outer parts of the melting domain to the eruption center, additional mechanisms have to be invoked, because there, the dikes are not tilted towards the spreading center strongly enough or even point away from it. Furthermore, defocussing is observed in the outer parts of the plume's melting domain, so that the crustal plateau built from it would actually be a bit flatter and more extended along the ridge. Interestingly, porous flow in a plume was predicted to be slightly defocussed in many cases as well (Ribe and Smooke, 1987; Li and Spohn, 1991, see sect. II.4.2, p. 55). This issue will be revisited in the discussion in section VI.2.

The irregular off-ridge patches of melt in the outer parts of the partially molten region in the hydrous model mentioned above (indirectly also visible from the η and f fields, see figures V.43b,d and V.44b,d) are not completely understood. The fact that they essentially appear at the outer, *i.e.* cooler, upper parts of the damp melting zone indicates that in principle, melting would cease there at least intermittently, because dehydration and low-degree melting have already led to slight depletion and to an increase of the solidus to values above the mantle temperature. On the other hand, minor amounts of low-degree melts are trapped below and form a low-porosity volume where melt migrates somehow by normal porous flow; this is confirmed by comparing porosity distributions of model ICE3-MP2 with those of the VV4 series, where segregation was not included and where these patches are absent. It is possible that the advection algorithm of the segregation routine does not always cope reliably with sharp transitions between partially molten and solid regions and possibly advects some melt into parts of the model where it should not; as an alternative, the streaks could also be the result of melt vein formation in the stress field of the mantle (Stevenson, 1989, cf. sect. II.4.2) in low-porosity mantle, which here of course is not properly resolved, because the grid is too coarse. Irrespective of whether the structures are caused by a physically consistent mechanism or by numerical shortcomings it can be stated that the comparatively small amount of hydrous melts produced in those marginal parts of the plume will certainly not have a substantial impact on the amount of crustal material produced from the model, and it is possible that they would be trapped in nature as well.

Seismic velocity anomaly, attenuation, and electrical conductivity — In sections II.2.2 and II.2.3, a short review of the effect of retained melt on seismic wave propagation and electrical conductivity had been given, and methods for the computation of the effects of temperature, depletion, and melt on the observables from the results of dynamical model runs had been described. In this paragraph, an attempt is made to apply these methods to two of the preceding models, namely ICE3-MP1 and ICE3-MP2.

Figure V.47 displays a number of depth profiles of Q_S^{-1}, $\delta v_P/v_P$, and $\delta v_S/v_S$ for the plume center and at a normal ridge site. The calculated Q_S^{-1} values are within the range of observations (cf. section II.2.2 and appendix D.1.8), although they could be lower in the deeper parts of the model; the importance of including the p term is highlighted in figure V.47a, where inclusion of the corresponding term has been omitted in the two additional $Q_S^{-1}(z)$ curves of model ICE3-MP1, resulting in a strong increase of absorption with depth. In figure V.47d, $Q_S^{-1}(z)$ has been plotted for model ICE3-MP2 with and without the effect of OH$^-$ ions included[7]. In the still unmolten mantle, the effect of water is very strong, especially in the plume, where the water content is more than 3.5 times as high as in normal mantle. The curves without the OH effect are very similar to those of the anhydrous model ICE3-MP1, except for minor differences in the interval below the cpx-out depth ($60\,\mathrm{km} \lesssim z \lesssim 150\,\mathrm{km}$) related to the different depths of incipient melting which are also visible in the $T(z)$ and $f(z)$ profiles in figure V.40. In the curves with the OH effect, the onset of melting provokes a change in Q_S^{-1}, which drops by a factor of about 5, until it merges eventually with the water-free $Q_S^{-1}(z)$, as expected. If the OH effect were misinterpreted as a temperature effect in Q_S observations, one would conclude that a very hot plume were present; however, the absence of the corresponding thermal anomaly at shallow depth should offer a means of distinguishing both alternatives.

The $\delta v_P / v_P$ and $\delta v_S / v_S$ profiles of model ICE-MP1 in figures V.47b and c have the same general shape; the main difference is that the v_S anomaly is stronger than the v_P anomaly, because P-waves are less sensitive to temperature and melt and experience a smaller attenuation. Both profiles have also been calculated with the melt effect excluded. The result is a weak decrease of the anomaly in the melting zone caused by enthalpic cooling of the mantle. This profile reveals that in the deepest part of the melting zone, the melt effect is dominant and causes a deep minimum in $\delta v_{P,S}/v_{P,S}$, whereas at shallower levels, the temperature effect is larger, at least partly because hot on-ridge mantle is compared with cool, older off-ridge lithosphere. The $\delta v_P / v_P$ and $\delta v_S / v_S$ profiles of the hydrous model ICE-MP2 in figures V.47e and f also share their essential features, again with S-wave velocities being affected more strongly. The ridge-perpendicular cross-sections of $\delta v_S / v_S$ in figure V.48 show that at shallow depth, the ridge anomaly does not appear much weaker than the plume head; the deep second velocity minimum, which is a distinguished feature of the plume, is rather narrow and does not extend laterally appreciably more than the stem. The remote parts of the plume head are still visible hundreds of kilometers away from the axis, but the anomaly is weak – barely $-3\,\%$ – and thin, and it may be difficult to detect it seismologically. The anomaly of the stem is not very strong either, which is in agreement with observations; this agreement is though at least partly coincidental, because low amplitudes can be caused by poor resolution and smearing in seismological imaging, whereas here, it is rather caused by the model assumptions. – Contrary to the $Q_S^{-1}(z)$ profiles, the velocity anomalies of ICE3-MP2 differ notably from those of ICE3-MP1 between *ca.* 80 and 200 km depth, because temperature and melt have a direct influence on velocity, and their pattern differs significantly between dry and wet melting. The effect of the OH$^-$ ions here is to blur the relatively strong decrease of $v_{P,S}$ visible in the OH-free curves, but to cause a strong weakening of the anomaly as the mantle is being dehydrated. This is related to the strong drops in $Q_{P,S}^{-1}$ and even leads to (very small) positive values in $\delta v_S / v_S$ at 80 km depth; however, this unexpected result might be due to the choice of the reference profile: it must be kept in mind that the normal mantle the plume profile is compared with mantle which is not fully dehydrated at that depth and would still have near-maximum Q_S^{-1} values, but had already been affected previously by damp melting while being in the melting region. Furthermore, it should be noted that,

---

[7]It should be noted that these profiles have been calculated without full knowledge of the full three-dimensional distribution of plume material by setting the original water content to the plume resp. the MOR value over the whole depth range of the profiles; this results in an overestimate of Q_S^{-1} in the uppermost 20–30 km of the model in the plume profile, because one would not expect undepleted plume material there. However, the attenuation is small there in any case, and the precise values are not of interest here.

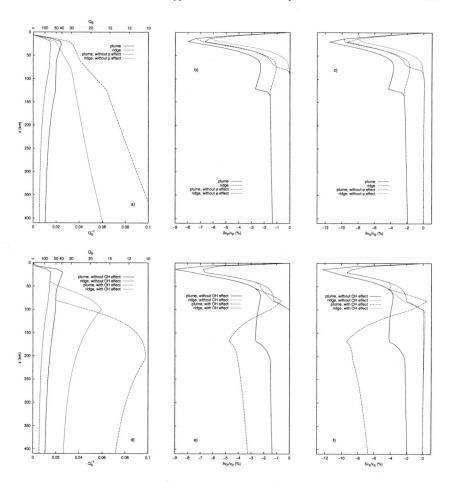

Figure V.47: Depth profiles of Q_S^{-1}, $\delta v_P/v_P$, and $\delta v_S/v_S$ at the plume center and at a normal ridge site, for models ICE3-MP1 (upper row) and ICE3-MP2 (lower row); Q_S^{-1} was calculated for a frequency of $2\,\mathrm{Hz}$, and the reference for the velocity anomaly is a profile through $20\,\mathrm{Ma}$-old, normal oceanic lithosphere. The plots include additional profiles which display the effect of omitting certain contributions in the observables.

especially in the hydrous model, a certain decrease of the anomaly with depth can be seen in the profiles, which might partly explain why plumes seem to fade out at greater depth in images of seismological data, apart from the obvious observational restrictions; it is to be expected that in the transition zone, the stronger drop in α would reinforce this effect.

These results can be compared to $\delta v_S/v_S$ profiles for the ICE1 series by Kreutzmann *et al.* (2004) shown in figure V.49, where the effect of water has also been estimated, although without explicit inclusion of water in the dynamical model (cf. section V.2.1). The curves of model ICE3-MP1 agree well with the anhydrous DPRM1 curve, and the differences can be ascribed to the differences in the model setup. The curves for ICE3-MP2 are though quite

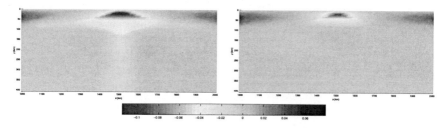

Figure V.48: x–z slices of $\delta v_S/v_S$ through the plume (left) and through a normal MOR segment (right) for model ICE3-MP1. *See plate section for colour version.*

different for depth greater than 50 km, and especially beneath *ca.* 120 km, which is probably due to different water contents; however, the seismic data suggest a weaker anomaly in the deeper parts of the model. This issue will be further discussed in section VI.2.

The calculation of the electrical conductivity follows the procedure of section II.2.3; the changes in mineralogy due to melting have though to be included in a very simplified manner. In the DPRM2, where the McKenzie and Bickle (1988) parameterization of f had been used, it would be the most straightforward option to apply eq. II.33b, but in the two models discussed here, where the Katz *et al.* (2003) parameterization was used, its explicit cpx-out criterion would clash with that formula. Inspection of $X(f)$ diagrams like figure II.8 for spinel lherzolite or *e.g.* the plot of Walter (1998, fig. 3) suggest though that for $f \lesssim 0.3$, X_{opx} is roughly constant, whereas X_{ol} and X_{cpx} are approximately linear, so that the modal composition for a three-phase peridotite can be described with linear $X_i(f)$ for $f < f_{\mathrm{cpx-out}}$; beyond that point, X_{ol} and X_{opx} are kept constant for simplicity.

Figure V.50 shows depth profiles of the electrical conductivity for models ICE3-MP1 and ICE3-MP2, including some additional profiles in order to test the importance of different contributions to σ. For the anhydrous model ICE3-MP1, the effects of Fe content, pressure, and mineral composition have been considered in figures V.50a and c. Under the assumptions made here, the effect of Mg# is only present for material which melts or has already traversed the melting zone; within the melting zone, it correlates with f by definition, and

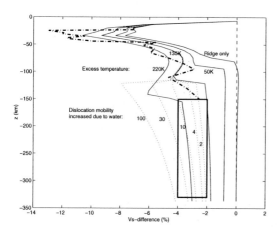

Figure V.49: $\delta v_S/v_S$ relative to the mantle beneath 20 Ma-old lithosphere at the plume axis of models ICE1-135/1 (DPRM1), ICE1-220/1, and ICE1-50/1, and at normal ridge; the dash-dotted line is a result from surface-wave inversions in central Iceland by Bjarnason and Sacks (2002), and the rectangle marks the approximate depth range covered by seismic tomography. The dotted curves are an estimate for the effect of water on the DPRM1 plume (Kreutzmann *et al.*, 2004, courtesy H. Schmeling.)

Figure V.50: Depth profiles
of σ resp. ρ for models ICE3-MP1 (left) and ICE3-
MP2 (right).   Upper row:
plume axis; bottom row: at
the spreading center of normal ridge. The plots include
additional profiles which display the effect of omitting certain contributions to the conductivity. Note the different
scales of the abscissas in the
left and the right plots.

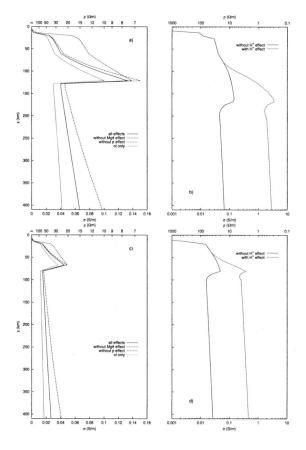

its omission would lead to an overestimate of σ by up to a factor 2 or more, as already
noted by Kreutzmann *et al.* (2004). As the effect is strongest at shallow levels, where f is
highest, it would favour penetration of electromagnetic waves into the earth for magnetotelluric observations, but its magnitude is probably not essential for real observations. The
pressure effect, which is only present in olivine, leads to a diminished increase of σ with
depth, but is not significant at shallow levels, to which most measurements are restricted. In
contrast, composition should probably always be accounted for, because the conductivity of
unmolten lherzolite is roughly a factor 1.5 higher than that of olivine, although the difference
decreases obviously as melting increases the fraction of olivine in the rock at smaller depths.
The most marked effect, however, is caused by the onset of melting, *i.e.* by the melt itself,
which was assumed to have a conductivity 400 times higher than that of the matrix. The
effect is largest in the deepest parts of the melting region, especially in the plume, whereas it
decreases towards shallower depth, possibly because it is counteracted by the consumption
of latent heat with melting.

In figures V.50b and d, similar $\sigma(z)$ have been calculated for the hydrous model ICE3-
MP2, with and without the effect of H$^+$ ions after eq. II.16. As expected, the solid curves
are quite similar to the "all effects" curve of figures a and c, with the obvious difference

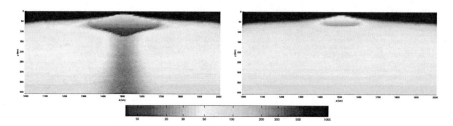

Figure V.51: x–z slices through the plume (left) and through a normal MOR segment (right) of $\log \rho$ (ρ in Ωm) for model ICE3-MP1. For the image, resistivities larger than $1000\,\Omega$m have been clipped. *See plate section for colour version.*

that melt affects a larger depth interval due to the deeper onset of hydrous melting, and that the increase is not as sharp, but rather ramping up smoothly, because the porosity is very low at the onset of melting. Nonetheless, dehydration already works in this region and counteracts the effect of melting, leading to a weak minimum in the dashed curves below the nose-shaped maximum of σ related to the melt itself; it must be kept in mind, though, that the hydrogen ions are likely to be still present in the melt there, so that in reality, there might be a stronger *increase* of σ there instead due to the higher conductivity of melt, if H or another hydrous species also acts as a charge carrier in the melt or enhances the mobility of whatever other charge carrier in it. The effect of water, resp. hydrogen, is a very strong increase of σ in the deep, undepleted upper mantle of more than one order of magnitude (note the logarithmic scaling of the abscissa); the effect is stronger in the plume, because it contains more water than the ambient mantle. At shallower depths, dehydration causes the hydrous profile to merge with the anhydrous, though; hence, it would need very long-periodic magnetotelluric measurements to detect the wet plume stem. The ridge-perpendicular cross-section in figure V.51 confirms that at the bottom of the melting zone of both the plume and normal ridges, resistivity reaches a minimum of around $10\,\Omega$m, which is likely to shield the deeper mantle from electromagnetic waves. In contrast, the thermal lithosphere should be relatively transparent to waves of reasonably short length, so that one should be able to map its thickness, but also the lateral extent of sublithospheric thermal anomalies such as a plume head with magnetotellurics, unless other factors, *e.g.* sea water, deteriorate the conditions.

Taking the seismic and the electrical models together, it can be said that the plume presents itself to both seismological and electromagnetic observations as a target with a moderate amplitude in the deeper, unmolten upper mantle and that the anomaly has a peak at the deeper levels of the melting region. At shallower depth, however, it fades, and especially in the water-bearing models it is so strongly reduced that the plume head might be hardly or not at all detectable in a certain depth range above *ca.* 100 km.

V.2.4   Concluding model: strong hydrous plume

In view of the shortcomings of the plume models in the previous series, a concluding attempt will be made in this section to design a model which comes close to the Iceland plume in terms of the observables considered here. To attain this goal, the results of series ICE1–3 are used to assess the parameters of such a model, and it is clear that a relatively strong plume in a water-bearing mantle would be needed. The most important model parameters of this final model, ICE-F, are listed in table V.18. In terms of temperature excess, the

x_m, y_m, z_m	size of convection model	$3000\,\text{km} \times 1200\,\text{km} \times 410\,\text{km}$
x_f, y_f, z_f	size of melt model	$861\,\text{km} \times 1200\,\text{km} \times 249\,\text{km}$
$\Delta T_P(z_m)$	plume excess temperatures at model bottom	$200\,\text{K}$
φ_{ex}	extraction threshold	0.01
$r_P(z_m)$	plume radius at model bottom	$150\,\text{km}$
$v_{z,P}(z_m)$	max. vertical bottom influx velocity of plume	$5\,^{\text{cm}}/_{\text{a}}$

Table V.18: Model parameters of model ICE-F.

model lies between the DPRM1 and model ICE1-350/1; this choice was made, because on the one hand, a higher temperature is needed to provide a certain crustal thickness in spite of the suppression of active upwelling by dehydration stiffening, but on the other hand, the results of series ICE3 suggest that a too large ΔT_P must be feared to leave too little freedom for the effect of water to be included in seismic attenuation, which must not be neglected in order to comply with geochemical results. The value of $200\,\text{K}$ at the model bottom was used, because it would still result in a $T(z)$ profile in reasonable agreement with the temperature anomalies derived from other studies (cf. table IV.2). A much higher value might also be difficult to bring the extent of thinning of the transition zone in agreement with a much hotter plume, even if the assessment of Shen *et al.* (2002, see sect. IV.2.1) is extended to non-olivine phase transitions, so that a higher temperature would result from their observations. The radius and volume flux have been chosen according to the constraints from observations (see table IV.2 and figure V.36c); the volume flux of $3.47\,^{\text{km}^3}/_{\text{a}}$ is significantly larger than that of the models in series ICE2 and represents something like a compromise between the lower values *e.g.* by Sleep (1990) or Schilling (1991) and the extraordinary high values by Ito *et al.* (1999) and Ito (2001). Note that the convection model grid has been enlarged in the y direction to provide more space for the plume flow, in the hope that the corresponding model boundaries will not affect the flow pattern of this bold structure substantially; from the computational effort perspective, this run is probably the most expensive of this study, needing about $700\,\text{MB}$ of RAM and taking *ca.* 15 days on a Pentium IV machine.

Table V.19 lists some key results of this model which can be compared with those of the previous series; the results in the table and the plots are taken at $24\,\text{Ma}$ model time, when the oldest plume-generated crust has reached an age of $20\,\text{Ma}$. Its temperature excess and maximum crustal thickness are very close to the results of Ito *et al.* (1999) (cf. table IV.2 and section IV.1.6), but fall still short of the observed maximum crustal thickness by a few kilometers; comparison with a similar model with a plume cooler by $10\,\text{K}$ (not shown here) suggests that, to produce a $40\,\text{km}$ thick crust thermally, the plume would have to have a $\Delta T_P(z_m)$ of about $250\,\text{K}$. The maximum upwelling velocity is reached near the hydrous solidus crossover depth, but drops to about the passive upwelling rate over the next 100–$150\,\text{km}$, with v_z lying at about $10\,^{\text{cm}}/_{\text{a}}$ around $185\,\text{km}$ depth, where the melting rate is highest. Judging from figure V.32, a larger influx velocity at the bottom might help to increase melt production and crustal thickness, but as it is a more or less fixed feature of hydrous models that practically passive upwelling prevails in much of the higher part of the melting region, this increase would have to be substantial; on the other hand, observations do not leave much room to increase the other flux-controlling variable, r_P, above the current value. If one equates the waist width of Schilling (1991) ($923\,\text{km}$, see table IV.2) with the calculated plume waist width[8], the value reached in this model is already considerably better than

---

[8]Comparison with the chemical field shows that the waist width of the chemical anomaly differs by no more than a few kilometers from the value in table V.19.

ΔT_P	maximum plume temperature anomaly (at $z = 200\,\mathrm{km}$)	$182.5\,\mathrm{K}$
z_sol	solidus depth (plume/MOR)	$228\,\mathrm{km}/100\,\mathrm{km}$
max. $v_{\mathrm{s},z}$	maximum upwelling velocity	$17.6\,\mathrm{cm}/\mathrm{a}$
q	volume flux	$3.47\,\mathrm{km}^3/\mathrm{a}$
h_max	maximum crustal thickness (plume/MOR)	$30.2\,\mathrm{km}/6.7\,\mathrm{km}$
y_cP, y_P0	waist width (crust/plume)	$640\,\mathrm{km}/800\,\mathrm{km}$
V_cP	excess crust volume	$2.89 \cdot 10^6\,\mathrm{km}^3$

Table V.19: Key model results of model ICE-F, taken at $t = 24\,\mathrm{Ma}$. Waist widths and excess crust volumes are defined as in section V.2.2.

those of the previous series, and the excess crust produced also lies well within the field of independent estimates plotted in figure V.36c; however, the along-ridge extent and size of the plume-generated crust are still too small, as can also be seen by comparing the Moho map figure IV.4 with the crustal thickness map in figure V.52a.

The crustal thickness map features an interesting detail, which is better visible in the along-ridge profile of h in figure V.52b: at about $100\,\mathrm{km}$ from the plume axis, where h is *ca.* $18\,\mathrm{km}$, the slope of the crustal thickness changes, being steeper near the plume and flatter farther away; in both subintervals, the slope can be approximated reasonably well by a straight line, *i.e.* it does not faithfully reflect the Gaussian-shaped anomalies prescribed in the deep mantle source, as one could have expected. This suggests that there is a relatively sharp structural change in the region where melt is produced and from where it is extracted, and it is reasonable to assume that this change is related to the shape of the thermal anomaly and to the strong gradients in the flow field caused by the viscosity pattern. The structural change would consist in the transition from a rather narrow partially molten region near the stem to a broader one, whereby a larger lateral (radial) component of mantle flow caused by the deflection of the upwelling at the base of the compositional lithosphere would distribute the melting mantle over a larger area; while this second domain would affect a larger region, it would sample a smaller depth interval and is less productive as a whole (figure V.52d), because upwelling, *i.e.* decompression, rates are reduced. The processes in the depth interval from *ca.* $180\,\mathrm{km}$ to $160\,\mathrm{km}$ depth seem to be crucial for this transition, as melt productivity reaches a first maximum there, and the increase in mantle viscosity begins at that depth as well (see the depth profiles in figure V.53). Looking at several quantities at that level, it is noted that the slope change in $h(y)$ coincides approximately with the width of that deep maximum of melt productivity, but also with the half-maximum positions of the temperature anomaly and of the mantle upwelling velocity, and with the maxima of the along-ridge component of the mantle current for $z \gtrsim 160\,\mathrm{km}$ (figures V.52b–d); furthermore, the second, shallower maximum of Γ has a similar width. This suggests that, in principle, a change in Moho slopes, which could be derived from Moho maps, could provide hints on the radius of the plume stem and the melting zone. The Moho maps of Darbyshire *et al.* (2000b) (figure IV.4) and Kaban *et al.* (2002) indicate that the slope of the Moho indeed flattens at greater distance from the plume center, but it is difficult to tell whether it changes abruptly at a certain distance or not, because the real Moho map is quite irregular and also still subject to errors due to sparse data coverage; furthermore, defocussing of melt flow as discussed in the previous section would transport melt away from the central axis and might blur the kink in $h(y)$.

For completeness, depth profiles of the usual variables are shown in figure V.53. Owing to the stronger temperature anomaly, the viscosity is slightly lower and the plume ascends

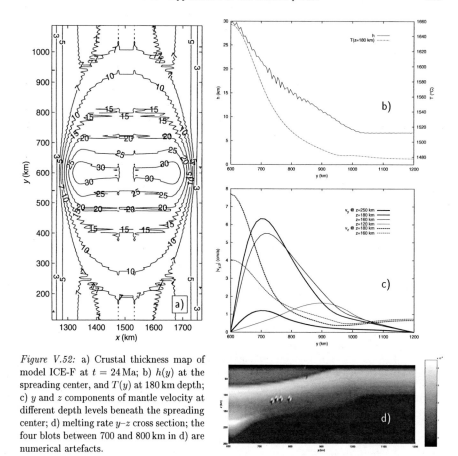

Figure V.52: a) Crustal thickness map of model ICE-F at $t = 24\,\mathrm{Ma}$; b) $h(y)$ at the spreading center, and $T(y)$ at 180 km depth; c) y and z components of mantle velocity at different depth levels beneath the spreading center; d) melting rate y–z cross section; the four blots between 700 and 800 km in d) are numerical artefacts.

faster; melting starts at greater depth, so that a slightly higher melting degree is reached, and dehydration is completed at greater depth as well. However, the general features are not substantially different from those of previous hydrous models, so that a more detailed discussion is omitted here.

Seismic velocity anomaly and electrical conductivity — The methods for the calculation of the seismic velocity anomaly and the electrical conductivity from sections II.2.2 and II.2.3 are also applied to model ICE-F, taking advantage of the availability of the chemical field, which serves as a measure of water content. The general picture is similar to the results of model ICE3-MP2, the differences result from the greater strength of the thermal anomaly.

For better comparison, the seismic anomalies have also been determined for a frequency of 2 Hz, and a mantle depth profile at a site 200 km away from the spreading center in an area unaffected by the plume has been used as reference; the results are shown in figure V.54. As in model ICE3-MP2 (figure V.47e and f, dash-dotted curves), a moderate decrease of $\delta v_{\mathrm{P,S}}/v_{\mathrm{P,S}}$ with depth in the unmolten stem and a very strong reduction of the anomaly in the melting region characterize the anomalies; again, positive values in the S-wave anomaly –

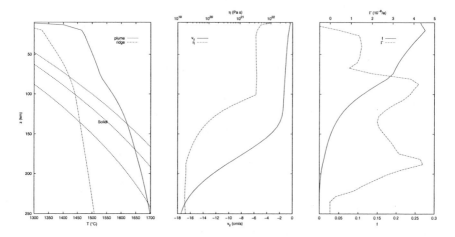

Figure V.53: Depth profiles for $250\,\text{km} \leq z \leq 10\,\text{km}$ of temperature (left), upwelling velocity, viscosity (center), and melting degree and rate (right) at the plume axis of model ICE-F, taken at $t = 24\,\text{Ma}$. In the T plot, a normal ridge geotherm as well as the solidi of dry melting and those of the background mantle and the plume (from shallowest to deepest) have also been plotted.

up to $1\,\%$ – are observed in a narrow depth interval between *ca.* 100 and 85 km. The P- and S-wave anomalies in the stem in this model are stronger by about $0.7\,\%$ and $1.5\,\%$, respectively, which is entirely due to the higher temperature of the plume; if the effect of water had not been included, the anomalies would match those from observations fairly well, as can be estimated from figures IV.5 and V.49, although the weakening of the anomaly at $z < 100\,\text{km}$ would not be as strong as observations suggest. Although the fading of the anomaly at shallow levels in seismic tomography has partly been explained by lack of resolution, additional seismic investigations, *e.g.* of surface waves such as the one by Bjarnason and Sacks (2002) shown in figure V.49, suggest that it is real, and the gaps in the anomaly within the plume heads of models ICE3-MP2 and ICE-F, which are caused by the lateral heterogeneity in water content, can help explaining it, especially with respect to its magnitude. A problem with the hotter plumes is that the deep velocity minimum shifts to greater depth the higher the temperature excess is; at a given temperature, this downward shift could be counteracted by moving the solidus depth to a shallower level by reducing the water content. – Although the minimum of $\delta v_S / v_S$ matches the observation of Bjarnason and Sacks (2002) (see figure V.49) remarkably well, the magnitude of the anomaly should be regarded with care, because it actually corresponds to inferred crustal depths, and it is probably a coincidence.

The electrical conductivity resp. resistivity images in figure V.54 correspond closely to the image gained from model ICE3-MP2. The plume axis profile is very similar to the ICE3-MP2 profiles with the H^+ effect (figure V.50b, dashed line), the difference essentially being a *ca.* 20 km downward shift of the "nose" of $\sigma(z)$ due to the deeper onset of melting; the directly temperature-related differences between corresponding parts of the profile do not exceed a few tenths of a siemens per meter, and are not significant. The cross-section emphasizes once more the effect of water, when compared with the section through the water-free model ICE3-MP1 in figure V.51: model ICE-F spans a resistivity range extending more than one and a half orders of magnitude further toward low values than model ICE3-MP1;

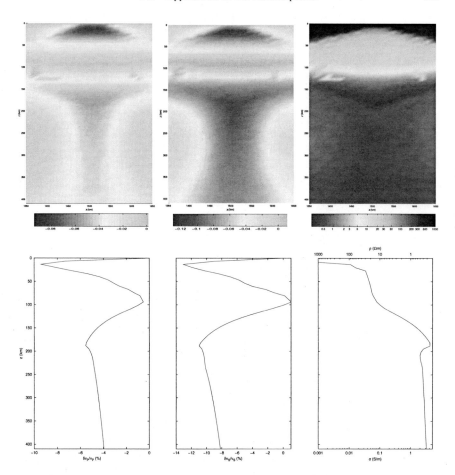

Figure V.54: Ridge-perpendicular cross sections through the plume (top) and depth profiles at the plume axis (bottom) of $\delta v_P/v_P$, $\delta v_S/v_S$, and σ resp. $\log \rho$ (ρ in Ω m) for model ICE-F. The seismic anomalies were calculated for a frequency of 2 Hz, and the reference for the velocity anomaly is a profile through 20 Ma-old, normal oceanic lithosphere, as in models ICE3-MP1 and ICE3-MP2; the resistivity in the cross section has been clipped at $1000\,\Omega$ m for plotting. The width of the cross sections corresponds approximately to what can be observed with an Iceland-based seismic array (cf. the ICEMELT tomographic image in figure IV.5). *See plate section for colour version of upper row.*

in particular, the background mantle also has a much lower resistivity in the hydrous model. At shallow depth, *i.e.* after dehydration, both models converge to quite similar values.

V.2.5 Supplementary model set: off-axis plumes

Although the focus of this study is on ridge-centered plumes, it is known that the Iceland plume has not always coincided with the ridge, but that initially, the plume was far off-ridge

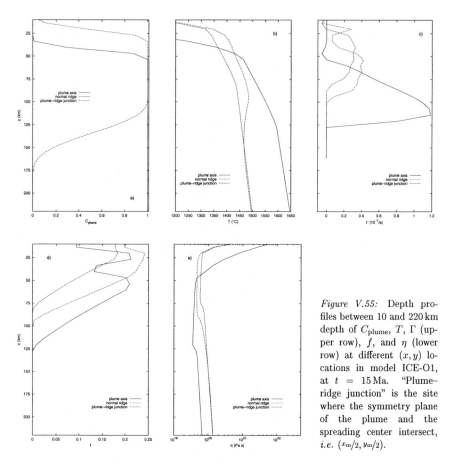

Figure V.55: Depth profiles between 10 and 220 km depth of C_{plume}, T, Γ (upper row), f, and η (lower row) at different (x, y) locations in model ICE-O1, at $t = 15\,\text{Ma}$. "Plume–ridge junction" is the site where the symmetry plane of the plume and the spreading center intersect, *i.e.* $(x_{\text{m}}/2, y_{\text{m}}/2)$.

and that the spreading center moved towards it over the last *ca.* 60 Ma (see section IV.1.1). While a full examination of the dynamics of ridge–plume convergence is beyond the scope of this study, the foregoing series shall here be supplemented by two models of a plume with a constant offset of 150 km from the spreading center. This offset is the only difference between the first model, ICE-O1, and the dry model from the previous series, ICE3-MP1; the second model, ICE-O2, is the water-bearing analogue of ICE-O1 and thus corresponds to model ICE3-MP2.

The wet plume rises a bit faster due to its lower viscosity in the unmolten state and begins to produce melt about 1.5 Ma earlier than the dry one; both plumes follow the shape of the lithospheric thermal boundary, but in the hydrous model, the plume does not rise to as shallow depths as in the dry model, at least not at that time (see also figures V.55a and V.56a). In both models, the offset is small enough for the plume head to rise partly beneath the melting zone of the ridge; hence, the partially molten plume head does not merge the ridge melting zone from the side, but from below, as in the ridge-centered models. Therefore, melt extraction will be performed with the usual method. This would be problematic if the ridge and the plume had two initially independent melting zones at the same depth level,

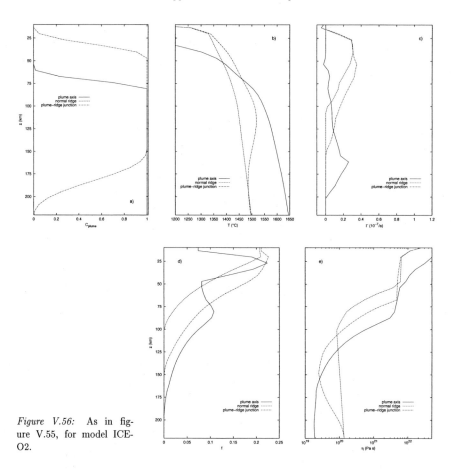

Figure V.56: As in figure V.55, for model ICE-O2.

because then, the plume melts would either have to be retained completely, possibly until establishment of a connection to the ridge melting zone, or they would have to be extracted at a separate site directly above the plume axis; in the present case, however, it can be assessed from the flow field of the plume directed upslope towards the ridge that the orientation of tensile dikes would actually favour melt transport to the ridge on the side of the plume head which flows toward the ridge, especially in the anhydrous model. It must be noted that the models discussed here have not reached a steady state; as they are only meant as a preliminary approximation to a plume rising under a moving lithosphere, the steady state is though not of much interest in this context anyway.

In agreement with the findings of several previous studies (*e.g.* Ribe, 1996; Sleep, 1996; Kincaid *et al.*, 1996, see sect. I.3.2), the plume establishes a connection to the ridge even if it is at a certain distance to it. With regard to the efficiency of lithospheric modulation of eruption activity, in particular for the generation of the V-shaped ridges (see section IV.1.1), it is of interest to investigate how strongly off-ridge dynamics and melting differs from the on-ridge processes. From the plume profiles in figure V.40 one would expect cpx-out to be reached at $f = 0.2$ at *ca.* 60 km depth in both models. The $f(z)$ profile along the plume axis in model ICE-O1 (figure V.55d) reveals that this point is barely exceeded[9], so that the plume is able to unfold almost its full productivity in spite of the thicker lithosphere; in contrast, the hydrous plume with its lower melting rates is far from reaching cpx-out at that site, probably because it does not rise as high as in the dry model, as can be seen from the marker field (figure V.56a), and temperatures and melting degrees in the thin layer of normal mantle material sandwiched between the plume head and the lithosphere are still a bit lower than in the plume head itself. Note again the substantial difference in viscosity structure (figures V.55e and V.56e): while in the dry model, viscosities remain low up to the bottom of the thermal boundary layer, whereby retained melt leads to a 70 km-high zone of even more reduced viscosity at the plume axis, η rises to lithospheric values at 100 km depth in the wet model, preventing the plume from rising much higher.

After impinging on the (kinematically defined) lithosphere, the plume flows upslope to-wards the spreading center and spreads along its strike. At the junction of the plume head and the ridge, clinopyroxene is in fact exhausted in both models between 40 and 50 km, but it is probably not pure plume material what melts above that depth in both cases, because again, plume material does not rise as high in the hydrous model as in the dry one (fig-ures V.55a and V.56a, dash-dotted curves), although it can be expected to heat up overlying mantle by conduction. In model ICE-O1, this might explain the presence of the shallow ($z \approx 20$ km) local maximum in $\Gamma(z)$ at the junction, which coincides with the upper margin of the plume head: the plume, which still has a significant temperature excess there, heats up adjacent non-plume material with a lower melting degree and triggers additional melt production in a region a few kilometers wide. The same probably happens in model ICE-O2 as well, but at greater depth, where melting of the normal mantle material is still occuring; therefore, a moderate increase of the background melting rate is observed there rather than a conspicuous peak, and the $\Gamma(z)$ profiles of the junction and of the normal mantle are virtu-ally the same above *ca.* 40 km depth (figure V.56c), where plume material is absent. – The influence of the plume on the opposite side of the spreading center, *i.e.* at $x > x_m/2$ is much smaller initially, but as the plume reaches the spreading center, the enhanced horizontal flow results in the transport of some of the plume material to the other side.

The remaining crucial question in the context of this short model series is: how does the thickness of the overlying lithosphere affect crust production? To assess the effect, one has to compare model ICE-O1 with ICE3-MP1 and model ICE-O2 with ICE3-MP2, respectively. In the former case, maximum crustal thickness and total volume in the off-ridge model are *ca.* 80–85 % of those of the on-ridge model at different times, with a weak tendency to increase as the model evolves. In the latter case, h_{max} differs by only a few percent in both models, but the excess volume in the off-ridge model is only about two thirds of that in the ridge-centered one. Altogether, these preliminary tests suggest that the thickness of the overriding lithosphere should indeed result in an observable modulation of crustal thickness and, at least in anhydrous models, also of the melting degree. However, it must be emphasized that the examples shown here do not represent a steady state, and to draw more detailed conclusions with regard to the specific situation of Iceland, a more extensive investigation including relative movement of ridge and plume is necessary.

---

[9]In the plume axis $f(z)$ profiles in figures V.55d and V.56d, a second maximum appears at 20–30 km depth. This maximum is the remnant of the initial melting event at model startup 15 Ma earlier, which happens to pass above the plume axis at this time, because it moves at approximately the half-spreading rate. In this context, it is to be regarded as an artefact of the initial condition, and is of no further interest.

Figure IV.4: Map of the Mohorovičić discontinuity of Iceland (from Darbyshire *et al.*, 2000b, © 2000 Elsevier, reproduced with permission).

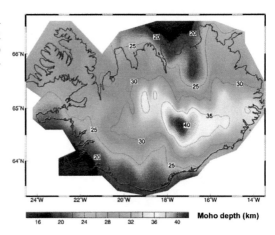

Moho depth (km)

Figure V.27. The DPRM1 at $t = 21$ Ma. a) Temperature field in convection grid (in °C); b) temperature anomaly $T(x, y, z) - T(x, 0, z)$ in melting grid (in K); c) melt production rate in melting grid (in 10^{-7}/a); d) melt fraction before extraction in melting grid; e) melting degree (depletion) field in melting grid. (From Ruedas *et al.*, 2004, © 2004 Blackwell, reproduced with permission.)

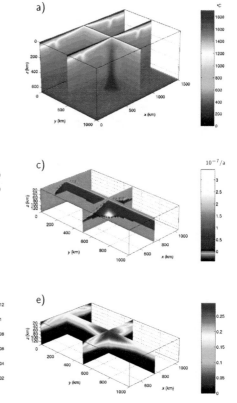

Plate 1

anhydrous   hydrous

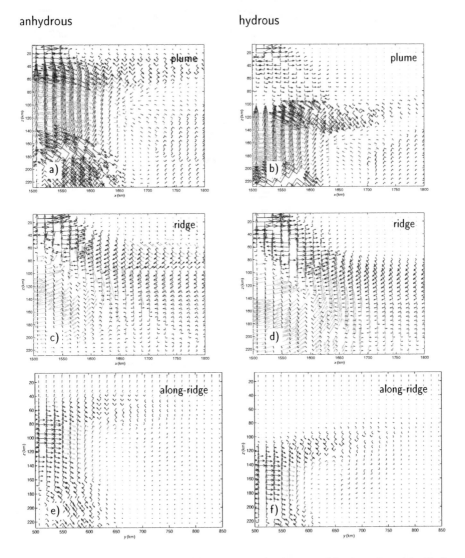

Figure V.45: Principal stresses in the depth range 6.7–228.5 km covered by the melt model grid of models ICE3-MP1 and ICE3-MP2 (red: maximum compression, green: intermediate, blue: maximum extension). The arrows are not to scale between colours and images. Note that their orientation is ambiguous by 180°.

Plate 2

anhydrous   hydrous

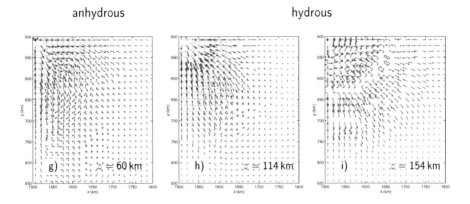

Figure V.45: continued.

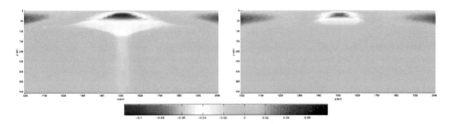

Figure V.48: x–z slices of $\delta v_{\mathrm{S}}/v_{\mathrm{S}}$ through the plume (left) and through a normal MOR segment (right) for model ICE3-MP1.

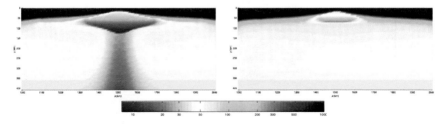

Figure V.51: x–z slices through the plume (left) and through a normal MOR segment (right) of $\log \rho$ (ρ in Ωm) for model ICE3-MP1. For the image, resistivities larger than $1000\,\Omega\,$m have been clipped.

Plate 3

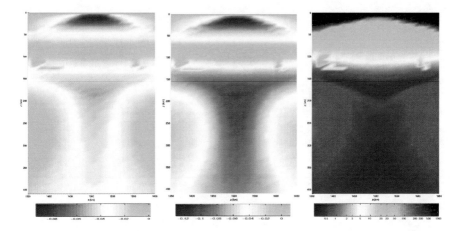

Figure V.54, upper row: Ridge-perpendicular cross sections through the plume of $\delta v_P/v_P$, $\delta v_S/v_S$, and σ resp. $\log \rho$ (ρ in Ω m) for model ICE-F. The seismic anomalies were calculated for a frequency of 2 Hz, and the reference for the velocity anomaly is a profile through 20 Ma-old, normal oceanic lithosphere, as in models ICE3-MP1 and ICE3-MP2; the resistivity has been clipped at $1000\,\Omega$ m for plotting. The width of the cross sections corresponds approximately to what can be observed with an Iceland-based seismic array (cf. the ICEMELT tomographic image in figure IV.5).

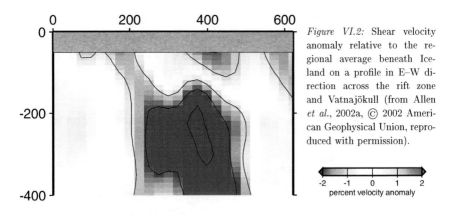

Figure VI.2: Shear velocity anomaly relative to the regional average beneath Iceland on a profile in E–W direction across the rift zone and Vatnajökull (from Allen *et al.*, 2002a, © 2002 American Geophysical Union, reproduced with permission).

Plate 4

CHAPTER VI

DISCUSSION AND CONCLUSIONS

What we know is not much. What
we do not know is immense.
—PIERRE-SIMON DE LAPLACE

VI.1 GENERAL ASSESSMENT

In section V.1, several model series designed to highlight effects of one particular quantity on convection and/or melt dynamics have been presented. Here, the results of these model series and of some more general outcomes of the Iceland-specific models from section V.2 are discussed with emphasis on possible relations between the results of different series.

One of the central issues in the mantle convection models of this study is the strength of active upwelling in the plume and its control by the direct and combined effects of temperature, composition, and viscosity; passive upwelling is essentially controlled by the spreading rate at the spreading center, which has been kept at a fixed value, so that all results discussed hereafter must be understood with the restriction that changes due to a different value of the passive upwelling component cannot be assessed from them. Model series SB, which tried to illustrate the importance of different contributions to buoyancy in simple, isoviscous models, demonstrates the domination of thermal buoyancy in general, but also reveals that chemical buoyancy can be similarly strong, especially if there is a chemically distinct volume with sharp boundaries, although in general, the effects of compositional buoyancy are smaller and more localized in the occasions considered here, *i.e.* melting and phase transition processes, because compositional heterogeneity exists only in certain depths in the model; this holds also for the buoyant effect of retained melt. While a considerable amount of melt would have to be retained in order to contribute significantly to the total buoyancy (see models SB3 and ICE1-135/100) and also to enhance the flow by reducing the viscosity (see models VV3), compositional differences of the solid material caused by depletion (model SB2) or phase transitions of major mineral constituents (model PT4) have been shown to be of some importance and should not be neglected in models striving for a close approximation of reality, even if it is admitted that the effect of depletional buoyancy in model SB2 is stronger than what one would normally expect in nature. On the other hand, the main importance of the phase transitions of Al-bearing phases seems to lie in the different chemical signatures they pass on to the melts generated in the different stability fields, while their dynamical effect is only small; however, the density changes due to depletion have been treated here in a still very simplified manner, and a more detailed inclusion of the changes in modal composition occurring when unmolten or molten peridotite undergoes *e.g.* the garnet–spinel transition might reveal an effect, but the general picture is not expected to change. The thermal effect of these transitions is also relatively small as far as melt production itself is concerned: in particular, models PM2 and PM2a demonstrate that the effect of the sp–gt transition does not affect anhydrous MORB production, and even in the case of plume melting, it results in a variability of less than 2 % (see figure V.21); the fact that in sufficiently hot plumes garnet is indeed exhausted before the phase change happens will certainly have consequences

SB – buoyancy effects
SB1   thermal buoyancy only
SB2   thermal and depletional buoyancy
SB3   thermal and porosity-related buoyancy
SB4   chemical buoyancy only (model with two sources)

PT – phase transitions, convection effects
PT1   no phase transitions
PT2   olivine–ringwoodite transition
PT3   olivine–ringwoodite and garnet–spinel transitions
PT4   olivine–ringwoodite, garnet–spinel, and spinel–plagioclase transitions

VA – variable thermal expansivity
VA1   constant α, linear adiabat
VA2   depth-dependent α, adiabat after eq. II.18
VA3 p–T-dependent α for fo–fa–oen–ofs mineralogy, adiabat after eq. II.18
VA4 p–T-dependent α for 18-endmember mineralogy of app. E, adiabat after eq. II.18

VV – variable viscosity, different model rheologies
VV1a   weakly T-dependent rheology, $b_T = 2.303$
VV1b   strongly T-dependent rheology, $b_T = 6.908$

VV2 p–T-dependent rheology, $b_T = 6.908$, $b_z = 2.303$; no enthalpic cooling during melting
VV2a   like VV2, but with a melting enthalpy of $550\,\mathrm{kJ/kg}$

VV3a   isoviscous
VV3b φ-dependent rheology, with $\leq 1\,\%$ retained melt; $s = 28$
VV3c   like VV3b, but with $\leq 3\,\%$ retained melt

VV4 p–T–φ–C_{H_2O}-dependent rheology, $b_T = 6.908$, $b_z = 2.303$, $s = 28$; no enthalpic cooling
VV4a   like VV4, but with a melting enthalpy of $550\,\mathrm{kJ/kg}$; chemically homogeneous
VV4b   like VV4a, but with a more hydrous plume

ME – melting entropy/enthalpy
ME1   constant melting entropy, $\Delta S_\mathrm{m} = 0.3\,\mathrm{J/g\,K}$
ME2   constant melting entropy, $\Delta S_\mathrm{m} = 0.3268\,\mathrm{J/g\,K}$
ME3   constant melting entropy, $\Delta S_\mathrm{m} = 0.4\,\mathrm{J/g\,K}$
ME2C   constant melting enthalpy, $L_\mathrm{m} = 550\,\mathrm{kJ/kg}$

PM – phase transitions (Al-bearing phases), effects on melting
PM1   no phase transitions; identical to model ME2
PM2   with garnet–spinel transition
PM2a   with garnet–spinel transition, garnet exhaustion during melting disabled
PM3   with garnet–spinel and spinel–plagioclase transition

SF – solidus and melting degree parameterizations
SF1 T_s of Hirschmann (2000), $f(\tilde{T})$ of McKenzie and Bickle (1988); identical to ME2C
SF1a   like SF1, with T_s lowered by $35\,\mathrm{K}$
SF2   KLB-1 solidus, $f(\tilde{T})$ of McKenzie and Bickle (1988)
SF3   solidus of Hirschmann (2000), $f(\tilde{T})$ of Katz *et al.* (2003)

VM – melt viscosity
VM1 T-dependent ϱ_f, constant $\eta_\mathrm{f} = 2\,\mathrm{Pa\,s}$
VM2 p–T-dependent ϱ_f, constant $\eta_\mathrm{f} = 2\,\mathrm{Pa\,s}$
VM3 p–T-dependent ϱ_f and η_f

Table VI.1: Summary of model runs of this study: general effects.

ICE1 – variations of excess temperature and extraction threshold	
ICE1-135/1 (DPRM1)	$\Delta T_P = 135$ K at 200 km, $\varphi_{ex} = 0.01$ (reference model 1)
ICE1-220/1	$\Delta T_P = 220$ K at 200 km, $\varphi_{ex} = 0.01$
ICE1-50/1	$\Delta T_P = 50$ K at 200 km, $\varphi_{ex} = 0.01$
ICE1-135/0.1	$\Delta T_P = 135$ K at 200 km, $\varphi_{ex} = 0.001$
ICE1-135/3	$\Delta T_P = 135$ K at 200 km, $\varphi_{ex} = 0.03$
ICE1-135/100	$\Delta T_P = 135$ K at 200 km, no extraction
ICE1-135/1c	as ICE1-135/1, but with a background $\mathcal{T}$ lower by 30 K
ICE2 – volume flux variations	
ICE2-100/5 (DPRM2)	$r_P = 100$ km, $v_z(z_m) = 5$ cm/a (reference model 2)
ICE2-125/5	$r_P = 125$ km, $v_z(z_m) = 5$ cm/a
ICE2-75/5	$r_P = 75$ km, $v_z(z_m) = 5$ cm/a
ICE2-125/3	$r_P = 125$ km, $v_z(z_m) = 3$ cm/a
ICE3 – hydrous melting	
ICE3-MP1	as ICE2-100/5, but with Katz *et al.* (2003) parameterization
ICE3-MP2	as ICE3-MP1, with hyd. melting; strong η–C_{H_2O} dependence ($b_C = 1$)
ICE3-MP3	as ICE3-MP1, with hyd. melting; weak η–C_{H_2O} dependence ($b_C = 0.5$)
ICE-F, ICE-O – concluding model and off-axis plumes	
ICE-F	concluding model
ICE-O1	dry off-axis model; derived from ICE3-MP1
ICE-O2	hydrous off-axis model; derived from ICE3-MP2

Table VI.2: Summary of model runs of this study: Iceland.

for the chemical signature of the melts and also increases the temperature contrast to the background mantle slightly (figures V.18 and V.19), but again, this latter effect is negligible. The larger spread in the h_{max} results of series PM is obviously rather a consequence of varying solidus positions due to different lithostatic pressure profiles, combined with different background mantle temperatures and corresponding shifts in the solidus depth, which have been shown to lead to considerable variation in crustal thickness in other model series as well, especially in models DPRM1 (ICE1-135/1) and ICE1-135/1c (see figure V.28 and section VI.2). – Furthermore, the effect of the olivine transitions in the deep upper mantle is probably overestimated, because the transition of pyroxene and garnet to garnet majorite, which occurs in the same depth range and whose Clapeyron slope has the opposite sign, is expected to counteract the impact of the olivine phase changes (see section I.2.2).

A substantial role in convection is also played by viscosity, and much depends on the rheological law chosen. The rheology of mantle rock is very complex and subjected to many uncertainties which have been discussed in sections I.2.4 and II.2.1, but clearly, pressure/depth and temperature dependence are substantial ingredients of a functional description of creep behaviour and should be included in advanced models. At low threshold porosities for melt extraction, the effect of retained melt is only minor, and the VV3 models suggest that it could be omitted, but this does not mean that the effect of melting on viscosity in general can be neglected. Apart from the also rather moderate stiffening caused by enthalpic cooling (cf. for instance the VV2 and VV2a profiles in figure V.8), which tends to compensate the softening by melt retention, a substantial change in convection style in the melting region is caused by dehydration. While experimental results show some scatter, it seems to be established that the rock is strengthened by at least one order of magnitude, if not two or more, as assumed in most of those of the models presented here, where dehydration stiffening is included. Inspection of the $v_z(z)$ profiles of model series ICE3 (figure V.39) suggests that for the melting region, it is of only minor importance whether the increase in η is of one or two orders of magnitude: upwelling velocities will drop to about the passive rate in both cases; however, for the unmolten plume stem, it does make a difference. Altogether, the models presented so far suggest that even if anhydrous melting might be regarded as a good approximation to melting beneath mid-ocean ridges and in plume heads, certain aspects of convection dynamics cannot be reproduced without taking into account the presence of water in the unmolten rock and its response to melting processes. As predicted by several authors, dehydration will result in a rheological boundary layer at the top of the mantle different from the one expected from the thermal structure, and this layer will be bounded relatively sharply. Internally, it features some heterogeneity nonetheless, because the thermal structure is still superimposed, and a further (essentially) vertical viscosity gradient would arise from rheological variations related to the depletion-controlled changes in the mineral composition of the composite rock (see sections I.2.4 and II.2.1), which have though not been included in this study; the internal rheological variability of this layer notwithstanding, the models shown here suggest that it would essentially behave as a uniform mechanical unit. On the other hand, the temperature dependence of the viscosity is still relatively moderate and restricted to Newtonian behaviour for technical reasons, and it is possible that the flow pattern within the compositional lithosphere would be more complex if the temperature dependence were stronger, at least in regions with active flow. – In summary, rheology is perhaps the most critical issue in mantle convection models, because it is difficult to constrain for the spatial and temporal scales relevant to convection on the basis of measurements on small, simplified rock assemblages in the laboratory or from postglacial rebound, but also because its large variations in three dimensions or properties such as nonlinearity make the numerical treatment of the equations of motion difficult and enforce several simplifications (see sections III.1 and III.4.1). The more complex models of this study, which after all include some essential characteristics of mantle rheology will provide a reasonable picture of the processes they are meant to represent, but much of the complexity of the real mantle is necessarily lost.

Another kind of simplifications pervasive in literature, partly related to the above, is the assumption that a number of thermoelastic properties of the mantle relevant to equations of momentum and energy conservation are constant. While it is clear that inclusion of p and T dependence of parameters such as α, c_p, κ, etc. makes the numerical implementation ever more difficult, it must be kept in mind that these quantities show sometimes significant variation even in the upper mantle (see figure E.2) and interact with convection dynamics differently when treated as variables than when kept constant, as was demonstrated exemplarily in section V.1.3 for the thermal expansivity; furthermore, the $\alpha(z)$ profiles in figure V.6c and the three-dimensional view in figure V.7 have substantial implications for the seismic observability of the plume in the transition zone, because α, which controls the anharmonic contribution to reductions in seismic velocities (see eq. D.17b), shows a marked decrease there. If the results of geodynamical models are to be compared with field observations, it is desirable to reproduce material properties of the real mantle which influenced the observations as closely as possible, but then it must be demanded for consistency that these properties be included in the dynamical model in the same way; in this sense, the prediction of material parameters and observables in the manner outlined in appendix E is of course not strictly self-consistent and only a stopgap, but it can be expected to provide a better approximation than a model with constant parameters, especially if more detailed data on the constituents are included. To be useful for parts of the mantle which have undergone partial melting, this model would have to be extended by melting proportions for the minerals and by taking into account the degree of depletion in each phase and major chemical component. If melt is present, the derivation of geophysical observables is even more complicated, but the use of experimental results and the application of theoretical models can be applied in order to compare model results with observations, as demonstrated in sections V.2.3 and V.2.4 for seismic velocity anomalies and electrical conductivity. These examples emphasize once more the importance of composition and compositional heterogeneity for physical properties and their geophysical observation; the conclusion to be drawn for instance with respect to seismological results and their interpretation is that it is probably too simple to explain all deep anomalies with temperature variation alone, but that differences in water content and/or even petrological heterogeneity should also be considered. Furthermore, the calculated seismic anomalies of those models suggest a physical explanation for the apparent fade-out of plumes at greater depth which adds to the influence of limited aperture of seismic arrays and might help to invalidate doubts on the deep origin of at least some mantle plumes (see the discussion in section IV.2.4); for a definitive statement, however, similar calculations would have to be performed for models comprising at least the transition zone and the uppermost lower mantle, if not the whole mantle, because several crucial material parameters, but also the water content, are known, or at least suspected, to change substantially there.

The results of model series ME and SF in sections V.1.5 and V.1.7 highlight once more the importance of constraining model setups with precise thermochemical information. For instance, different choices of melting entropy within the range of results regarded as realistic for mantle peridotite can result in temperature variations of up to some tens of kelvins and in depletion and crustal thickness variations of 10 % or more. Notable differences also occur if the consumption of latent heat is treated as constant or temperature-dependent. On the other hand, as long as there is still scatter in experimental data, and given the even larger uncertainty in the knowledge of the petrological and chemical composition of the mantle and the importance and abundance of a second member such as eclogite, the variability is still acceptable, and the best one can do is probably to stick to values in the middle of the range of proposed enthalpy or entropy values. Even more important is, in principle, the choice of the solidus and melting degree parameterization, which is though subjected to the same problems as the choice of melting entropy. Model series SF demonstrates that the spatial distribution of melt production as well as the total melt productivity of the melting zone are very sensitive to the choice of the T_s and $f(\tilde{T})$ functions, and it becomes clear that there is a trade-off between the position of the solidus in p–T space and the potential temperature. This matter is further complicated by the fine structure of the solidus function, exemplarily illustrated by the differences by the solidus of Hirschmann (2000) and the KLB-1 solidus, which still needs to be clarified more precisely for natural rocks, and by the substantially different $f(\tilde{T})$ functions proposed in the literature. On the other hand, as has been pointed out in section V.1.7 already, very similar results can be produced with certain combinations of T_s and $f(\tilde{T})$, as confirmed by models DPRM2 and ICE3-MP1 (see section V.2.3). The actual presence of water in the mantle and the fertile composition of many unmolten peridotites make it likely that the concave-up $f(\tilde{T})$ parameterization is a more realistic one for both dry and hydrous melting, so that the original Hirschmann (2000) solidus would have to be shifted to lower values and/or the potential temperature of the mantle would have to be raised to yield a given crustal thickness *e.g.* at MOR; as there does not seem to be very much support for still higher potential temperatures in a peridotitic mantle than those assumed in this study, a downward shift of the solidus seems to be the more appropriate choice. In this context, it is noteworthy that the maximum melting degree at normal ridges of the anhydrous model ICE3-MP1 agrees better with the estimate of Hellebrand *et al.* (2001) (cf. p. 188 and figure V.40), but the maximum f of the hydrous model ICE3-MP2 is almost identical with that of the DPRM2. Altogether, it is hoped that recent progress and ongoing efforts in experimental work on thermochemical properties and melting behaviour of mantle rocks improve the rock physical foundations on which numerical models are to be based in the future.

If it is assumed that at slow-spreading ridges the crustal thickness is below the worldwide average and that hydrous melting takes place, the results from table V.17 suggest that the combination of model parameters for the reference mantle used in model ICE3-MP2 – concave-up $f(\tilde{T})$, a potential temperature of 1410 °C, and a solidus lowered by 35 K relative to the standard solidus of this study – is close to reality, with actual $\mathcal{T}$ being maybe some 10 K lower; there is though still some trade-off with the extraction threshold. This issue has been treated in the framework of model series ICE1, where the DPRM1 was tuned to match observed MOR crustal thickness by adjusting the potential temperature and the threshold porosity, whereas the model with a lower background mantle potential temperature results in too low h values and run ICE1-135/0.1, with a lower extraction threshold, in turn produces slightly too much crust. Taking the results from that series together, the dataset suggests that $\mathcal{T}$ in the range from *ca.* 1380 °C to *ca.* 1400 °C, combined with extraction thresholds of a few tenths of a percent would give reasonable results in water-free models, in agreement with independent estimates (Lundstrom *et al.*, 2000; Faul, 2001); if the effect of water is added, the increase of melt production by damp melting in passive upwelling would allow for a slightly lower potential temperature not too high above the estimate of 1350 °C by White and McKenzie (1995). The models do not support the retention of melt fractions substantially larger than 1 %.

As far as the dynamics of melt migration is concerned, strong restrictions have been made in this study, the central one being the limitation of retained melt to small amounts in general, coupled with an admittedly not very sophisticated assumption about the fast extraction mode, which could not be modelled in a physically self-consistent way within the framework of the convection model due to the large difference in timescales (see the introduction to chapter III). Therefore, the variety of phenomena anticipated for unrestricted porous systems, in particular the formation of magmons, cannot be expected to develop here, and the most important mode of melt segregation modelled explicitly in the models of this study is slow buoyant ascent through an isotropic porous matrix, as predicted by Ribe (1985a, 1987, see sect. II.4.1) for most of the melting zone, which leads to a slow redistribution of melt in it. This redistribution is controlled by the physical properties of the melt, in particular by its density and viscosity. In model series VM, the importance of these factors has been shortly probed, and it turned out that it is worthwhile to take into account the compressibility and the viscosity changes of the melt in the depth interval of melting, because both, but especially the latter, have a pronounced influence on the mobility of the melt and hence on its distribution in the mantle. It is clear that for detailed models of melt chemistry, in particular those considering radionuclide equilibria, where residence and migration times are crucial, an appropriate representation of melt mobility is required, and more experimental constraints on the dependence especially of melt viscosity on p, T, and composition are urgently needed. Furthermore, the control of ϱ_f and η_f on melt distribution should also have an impact on geophysical observables, *e.g.* on seismic anomalies; as an example, one would expect the low-velocity zone under MOR to be smaller if η_f depends strongly on pressure than if it does not, judging from the differences in porosity visible in figure V.26. In systems without extraction thresholds, one would also expect a run-up of deep, fast low-viscosity melts at shallower levels and, as a result, a localized increase in porosity, at least unless other factors such as a low initial productivity and concomitant low porosities counteract this effect; it is left for future work to investigate if, and possibly under which conditions, this mechanism can support the development of magmon-like structures.

In section V.2.3, it was attempted to predict the pattern of channelized melt flow in a medium with anisotropic permeability on the basis of the method developed in section II.4.3, because such an approach could be extended to an algorithm which would at least allow to track melt flow physically consistently in an established channel network, although the formation of the instability creating the network could not be reproduced due to the low resolution. It should be noted that the orientation of tabular dikes along the principal axes of the stress, respectively strain rate, tensor is only one possibility of defining the orientation of the permeability tensor. The rock physical experiments *e.g.* by Zimmerman *et al.* (1999) or Bussod and Christie (1991) (see section II.1.5) rather showed a melt pocket orientation of *ca.* 30° to σ_1, and if this orientation is inherited to the channel network on larger lengthscales, a different flow pattern would probably result, because the influence of buoyancy would give one of the two possible orientations preference over the other in the general case where σ_1 is not vertical. In general, as long as the orientation of the channels can be related to the stress field prevailing in the mantle, the method outlined here could be applied and would only have to be extended by the use of a different permeability model or by an additional rotation of the tensor. – The application of the classical model to the flow field of models ICE3-MP1 and ICE3-MP2 shows that extracting melt in ridge-perpendicular planes is not too bad, because near the spreading center, melt rises almost vertically, and in the outer parts of the MOR melting zone, focussing is aided by dike orientation for $x > z$; however, the boundary between those two domains seems to depend on the width of the spreading zone defined by $\mathrm{d}v_r(x)/\mathrm{d}x \neq 0$, which might well be narrower under normal MOR in nature, and additional mechanisms, *e.g.* a melt layer at the bottom of the thermal lithosphere, or maybe viscous stresses in the high-viscosity part of the melting zone in the water-bearing model would probably also be operative. In the outer parts of the plume, a melt flow component parallel to the ridge is expected from figures V.44e and f and would redistribute the melt away from the plume axis; while the maximum thickness of the plume crust resulting from the convection models of this study can still be regarded as consistent with such a channel orientation, because it coincides with the (x, y) position of the plume axis where melt ascends vertically, one would expect the off-center crust along the ridge to have a smaller thickness close to the plume and a greater thickness farther away. Nonetheless, plume melts would not be erupted at much greater distance from the plume center than if they would not be channelized along the ridge, because melt ascent is still vertical in the upper part of the partially molten zone.

To summarize, the following main conclusions can be drawn from the general aspects addressed in this study:

- While thermal buoyancy is certainly the most important driving force for plumes, several compositional contributions to buoyancy – both those related to melting and those related to solid-state phase changes – are locally of some importance as well and should be included in models intended to reproduce real-world settings. The transitions of aluminous phases, however, are mainly important for the geotherm and with respect to the chemistry of melts, but not so much for buoyancy.

- In addition to pressure and temperature, water content and its change due to melting has a fundamental impact on the dynamics of plumes and on the style and extent of melt generation in general; while more experimental information is still needed to clarify the numerous effects of water, it can be considered a robust result that dehydration stiffening of the mantle and water-controlled shifts of the solidus should be included in melting models.

– This study confirms the importance of water for a proper interpretation of geophysical observations such as seismic or electromagnetic images of the mantle: the introduction of water can substantially change the shape and magnitude of anomalies, and it would be inconsistent to include water in a dynamical model, but to neglect it in models of seismic velocities or electrical conductivities derived from it.

– Melt and crust production depend strongly on the solidus and on the shape of the melting function; different parameterizations found in, or derived from, the literature produce a wide range of results, although they were all meant to characterize peridotite. Further uncertainty is introduced by the scatter in experimental data on melting entropy resp. enthalpy. With such problems arising already for a purely peridotitic mantle, it can be foreseen that the presence of a second, eclogitic component would increase the ambiguity unless there are good knowledge of source composition and tight constraints on its melting behaviour.

– With all the ambiguities and uncertainties in mind, the following set of parameters seems appropriate for melting in mantle with a small amount of water below slow-spreading MOR, assuming a concave-up melting function similar to the one proposed by Katz *et al.* (2003): a potential temperature around 1380 °C, a melt extraction threshold of a few tenths of a percent, with higher φ_{ex} corresponding to higher $\mathfrak{T}$, and a solidus a few dozen kelvins lower than the one originally derived by Hirschmann (2000). It should be recalled here that these values presume a melting function "stretched" between the solidus of lherzolite and the liquidus of forsterite, as an approximation to fractional melting. It should also be emphasized that the underlying petrology here is a purely peridotitic one, *i.e.* no eclogitic component is included.

– Segregation of melt leads to its redistribution in the partially molten mantle, which also can affect geophysical observables; to model the related phenomena realistically, melt density and viscosity should not be taken as constants, but rather as p–T–X-dependent quantities. Segregation velocities in the porous flow regime are on the order of some centimeters per year, at low viscosities possibly a few decimeters per year. With respect to the distribution of crust production, it is also desirable to take into account the redistribution by rapid flow in channels of some large-scale orientation which can be derived from the stress field, although instantaneous extraction in ridge-perpendicular planes serves as a first approximation.

VI.2 ICELAND PLUME MODELS

One of the most important quantities for the validation of the models is the crustal thickness h, because it is rather well known *e.g.* from seismic measurements. In particular, the thickness of normal oceanic crust is agreed to lie at 6–7 km, with a tendency to be thinner at half-spreading rates below *ca.* 1 cm/a (*e.g.* White *et al.*, 1992; Schubert *et al.*, 2001); hence, the spreading rates around Iceland are close to this limit, and without the influence of the plume, a crustal thickness slightly below the normal values would be possible. As can be seen from the summary in table V.13, the normal DPRM1 was constructed to match this constraint.

However, some caveats are necessary when considering these values. First, the phase transitions of the Al-bearing phase, in particular the spinel–garnet transition at about 80 km depth, have not been included for the sake of simplicity. Its importance will though depend on the depth of initial melting and the extent of garnet exhaustion, and is probably minor or nil especially in the plume, which certainly starts melting in the garnet stability field. Inspection of the results of models PM2 and PM2a (figures V.18, V.19, and V.21) suggests that the effect is indeed not important, but that the potential temperature would have to be lowered to compensate for the T shifts in the geotherm caused by the transitions, *i.e.* including both transitions would move $\mathfrak{T}$ to *ca.* 1395 °C, which is closer to the common estimates listed in table II.1. Besides, its effect on buoyancy is unclear: Scott and Stevenson (1989) did not find a significant influence of the sp–gt transition, whereas Niu and Batiza (1991) think that it does have a relevant effect; the results of series PM support the former view (cf. figure V.3). Second, the melting enthalpy, which is of substantial importance for the amount of melt produced, is not well constrained (Hess, 1992; Hirschmann *et al.*, 1999), but may introduce further ambiguity in addition to the $\mathfrak{T}$–φ_{ex} tradeoff; in model series ME it was shown that not only the enthalpy, but also its dependence from temperature have a visible effect and can result in crustal thickness differences of several kilometers above the plume, and of still 1 km at normal ridge. Third, $f(\tilde{T})$ was calculated with the widely used parameterization of McKenzie and Bickle (1988) in series ICE1 and ICE2, which is concave-down; however, theoretical investigations (Asimow *et al.*, 1997) and some experiments (Walter and Presnall, 1994; Schwab and Johnston, 2001) showed that concave-up $f(\tilde{T})$ are also possible; the parameterization by Katz *et al.* (2003), which takes into account these features and was used in several models here, leads to a different spatial depletion and porosity pattern and possibly also to a different crustal thickness, as shown in series SF (see figures V.22 and V.23), although adjusting other parameters such as the solidus temperature appropriately can help fixing the crustal thickness at a given value, as demonstrated for the DPRM2 and model ICE3-MP1 (see *e.g.* figure V.41). Finally, purely anhydrous melting and a homogeneous peridotitic mantle source were assumed in most of the models; both assumptions are though likely simplifications, and mainly the latter might cause melt productivity to be underestimated, especially if the plume contains some old recycled oceanic crust, as proposed for the Iceland plume in several geochemical studies (*e.g.* Hanan and Schilling, 1997; Breddam, 2002; Chauvel and Hémond, 2000; Scarrow *et al.*, 2000; Korenaga and Kelemen, 2000, see sect. IV.2.2). In contrast, while the volumetric contribution of possible hydrous initial melting leads to an increase in melt production (see table V.17) and could justify a further downward shift of the potential temperature, but is not exceedingly large, the effect of water on viscosity changes the dynamics of the plume in the melting region and reduces total melt production so radically that its omission is indeed substantial, as demonstrated by model set VV4 and, with regard to Iceland, by series ICE3, ICE-F, and ICE-O.

With the results for the normal oceanic crust in mind, the plume crust can now be considered. In the following, a crustal thickness of 35–41 km will be assumed to be realistic for the central part of Iceland, especially above the plume (Darbyshire *et al.*, 2000b). Looking at the anhydrous models first, the hottest plume, run ICE1-220/1, can then be discarded at once, because it produces far too much crust; by contrast, the coolest plume, run ICE1-50/1, produces too little, and is unlikely to generate substantially more at a lower extraction threshold (figure V.30). The other models though fall into a range of values which requires a closer look. At first sight, the DPRM1 and ICE1-135/0.1 also seem to be too productive, whereas ICE1-135/1c and ICE1-135/3 come closest to the observed $h_{\max}$ for the plume. However, accepting ICE1-135/3 would require to use a lower φ_{ex} for the normal crust, which would otherwise be absent, but there is no obvious reason to do that. Furthermore, the seismic anomalies calculated by Kreutzmann *et al.* (2004) and their comparison with the seismological observations by Bjarnason and Sacks (2002) do not support the presence of such a high porosity in any part of the plume head except, maybe, the very top (see figure V.49), but with regard to the extreme anomalies observed in less than 50 km depth, it must be borne in mind that it is probably crustal root material which is compared with lithospheric mantle, so that their magnitudes might be overestimated. By contrast, ICE1-135/0.1 yields a slightly too large thickness for normal oceanic crust due to the low extraction threshold and is not in good agreement with the seismological data either, making it a less than ideal candidate as well. On the other hand, all models had been based on a plume volume flux estimate which in series ICE2 turned out to be probably too low, although it is difficult to assess from the observational data how much too low. Increasing the flux would result in an overly thick crust (see figure V.36), requiring a lower excess temperature. This, however, is related to the dilemma already encountered by the early models of Ribe *et al.* (1995) and Ito *et al.* (1996), who had found a cool, broad, high-q plume to reproduce crustal thickness, gravity, and topography better than a hot, narrow one, which would be in better agreement with seismological and geochemical observations (see section IV.2.3); it is therefore also in conflict with the seismological constraints (see section V.2.3 and below).

The picture changes if the models of series ICE3, especially the hydrous ones, and model ICE-F are considered. If the rheological law includes the effect of water, active up-welling is reduced so strongly in most of the melting region, that crustal thickness drops to half the value of the DPRM1 and its anhydrous descendants, DPRM2 and ICE3-MP1. From this perspective, a somewhat hotter plume would thus be needed to achieve the observed maximum h. Moreover, the greater along-ridge extent of the plume-generated crust in the hydrous models ICE3-MP2 and ICE3-MP3 indicates that, by including water, the problem posed by the too narrow crustal waist widths of the DPRM1 and DPRM2 could be overcome. For this reason, it is necessary to get back to the hot plume of model ICE1-220/1 and recon-sider the reasons for this one as well as the DPRM1 and ICE1-135/0.1 producing too much crust. Apart from being water-free, they have in common that no explicit cpx-out criterion is implemented; inspection of figure V.40, where the DPRM2 as a model similar to those of series ICE1 is contrasted with the models of series ICE3, in which the McKenzie and Bickle (1988) parameterization was substituted by the Katz *et al.* (2003) parameterization, shows that even the plumes of intermediate temperatures go well beyond cpx-out, so that a hotter plume would maybe not generate as much more crust as expected from the $h_{max}(\Delta T_P, \varphi_{ex})$ fits in figure V.30. Another possibility not implemented in any of the models is that the thick crust above the plume center would thin by creep, as proposed by Buck (1996); this mechanism is expressed as the diffusion term in the crust transport equation II.65. A rough estimate by this author indicated that even a fairly cold, high-viscosity lower crust would thin by several kilometers over some million years, although there is some uncertainty in estimating the reference viscosity of the crust. This effect could lower the h_{max} values of the DPRM1 to those observed in real data as in the Moho map of Darbyshire *et al.* (2000b, see fig. IV.4), both above the plume center and away from the volcanic zones. It should also be noted here that the extracted melt forming the crust is not returned to the model by an influx through its top and does not form a crustal root which would reach into the melting zone and depress its upper boundary; in nature, this effect also reduces the produc-tivity of the melting region of the plume, although the effect is not very strong (Marquart and Schmeling, 2003), given that it would affect the region where less melt is produced be-cause of clinopyroxene exhaustion anyway. If dehydration is included, these remedies turn though into problems, because the otherwise crucial role of active upwelling mentioned in several occasions becomes largely inoperative, and the only way of enforcing a higher melt production is to rise the temperature, unless one assumes a much weaker effect of water on creep behaviour than used even in model ICE3-MP3, but there seems to be no experimental foundation to do this. The option of raising the plume's temperature was therefore the one chosen for model ICE-F.

As mentioned in the previous section, these arguments are all subjected to the numerical restrictions on the underlying rheological model, for if viscosity is more sensitive to temperature, a larger active upwelling rate would probably be maintained to shallower depths. In this case, the channeling of the melting plume head along the ridge axis modelled by Albers and Christensen (2001) would also become a viable explanation for the influence of the plume hundreds of kilometers along the ridge, which it is not in the setup presented here or in the work of Ito *et al.* (1999) and Ito (2001); one problem in particular with the anhydrous models is that the excess crust produced covers a too short interval along the ridge, which means that the melting rock would have to be distributed more widely along it in order to match the real extent of the crustal anomaly. Such a plume-channeling mechanism would probably decrease the crustal thickness. Another possibility for enhanced along-ridge transport of melt would be the flow of melt in dikes with a stress-dependent orientation, as calculated for models ICE3-MP1 and ICE3-MP2. The difference between reality and the model restricted by these factors can be illustrated by comparing the spatial distribution of the calculated crust plotted in figures V.31, V.42 and V.52a with the Moho map of Darbyshire *et al.* (2000b): while the width of Iceland is reproduced quite well, the along-ridge length *e.g.* of the calculated crustal body thicker than 25 km in the anhydrous models and in ICE-F is only about two thirds of the observed along-ridge extent of crust with a Moho deeper than 25 km, taken across Vatnajökull. – The problem is exacerbated by the fact that most of the models, except ICE-F and a few others, less favoured ones, do not or hardly match the total excess volume of crust from independent observations (see figure VI.1b), the hydrous models usually being even a bit less productive than the DPRM1 (cf. tables V.16 and V.17); even model ICE-F, which lies within the range of estimates, has still significantly less excess crust than the probably most appropriate values from Schubert and Sandwell (1989) and White (1993). Figures V.36a and VI.1 suggest that the well-known estimates for plume flux by Schilling (1991) and Steinberger (2000), and probably even the one by Sleep (1990), underestimate the volume flux of the Iceland plume, because no plume with a temperature within the accepted range and a flux of $1-2\,\mathrm{km^3/a}$ comes close to matching the constraints from both crustal thickness and excess volume if the effect of water is included, the plumes from hydrous models even less than those from dry models. This discrepancy has already been found in the models by Ito *et al.* (1999), who needed a plume with a much a higher volume flux to achieve their relatively close reproduction of several observables from a numerical model (see table IV.2 and section IV.2.3). As an increase of the volume flux would result in an unreasonably thick crust in anhydrous models, as can be seen from V.36a, dehydration has to be invoked to suppress the formation of a very thick crust in a small region and rather lead to the melt and crust production being distributed over a larger region, which would fit the constraints better.

As an additional constraint for the temperature of the plume, geochemical estimates for the pressure resp. depth of initial melting can be used. Shen and Forsyth (1995) found a value of about 100 km for the Iceland plume, which would indicate a ΔT_P of about 200 K at 660 km and of about 100 K at 200 km depth, judging from the results for the DPRM1 in table V.13 and several geotherms of other models; hydrous melting would of course begin much deeper, but for the dry solidus depth, this estimate holds anyway. The downward shift of the 410 km discontinuity relative to the undisturbed mantle, Δz_X, for a Clapeyron slope of about $2.9\,\mathrm{MPa/K}$ (Bina and Helffrich, 1994) is also listed in table V.13 and can be compared to the results of Shen *et al.* (2002), who find a reduction of transition zone thickness of $\sim 19\,\mathrm{km}$ for the same Clapeyron slope at 410 km, corresponding to a temperature excess of at least 140 K. The models of series ICE1 do not include the effect on the 660 km discontinuity, but the Δz_X provide an upper bound for ΔT_P, because the Δz_X from the models cannot be

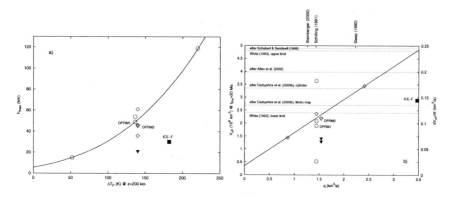

Figure VI.1: Selected Iceland models from series ICE1 (circles: DPRM1, ICE1-220/1, ICE1-50/1, ICE1-135/0.1), all models of series ICE2 (diamonds) and ICE3 (triangles), and model ICE-F (square); empty symbols: dry models, filled symbols: hydrous models. a) Maximum crustal thickness as a function of maximum plume excess temperature in 200 km depth, and fitting curve for the $\varphi_{ex} = 0.01$ models of series ICE1 (cf. figure V.30). b) Plume crust in excess of 7 km-thick oceanic crust as a function of bottom volume flux, and fitting curve for the models of series ICE2 (cf. figure V.36c).

larger than the total thinning of the transition zone, but should rather be about half that value. Therefore, a plume with a ΔT_P between *ca.* 100 and 135 K at 200 km depth would fit the observation of Shen *et al.* (2002) best. It should though be noted that their estimate did not include the effect of the (px+gt)–mj and mj–Al–pv transitions, nor the effect of water, which is expected to be abundant in the transition zone, on the phase transitions of olivine; both would raise the 410 km discontinuity and lower the 660 km boundary (Yusa *et al.*, 1993; Weidner and Wang, 2000; Hirose, 2002; Wood, 1995; Smyth and Frost, 2002; Higo *et al.*, 2001), thereby shifting the temperature required to match the observed thinning to higher values, which would be consistent with the conclusion to be drawn from models ICE3-MP2 and ICE-F.

While Kreutzmann *et al.* (2004) found magnetotelluric observations to be little promising for the clarification of the deep structure of the plume, the interpretation of seismological observations has been used for this purpose in many occasions, of which several of those dedicated to Iceland had been summarized in section IV.2.1. Kreutzmann *et al.* (2004) concluded from figure V.49 that an anhydrous plume with a temperature between that of the DPRM1 and model ICE1-220/1 would match observed S-wave velocity anomalies from seismic tomography and surface wave data fairly well, with the lower temperature to be preferred if a moderate increase of dislocation mobility due to water is included; it should be noted that they estimated water to be effective to considerably shallower depths than expected from the position of the hydrous solidus. Their anhydrous anomalies agree basically with the $\delta v_S/v_S$ calculated for model ICE3-MP1 (figure V.47c), but there are differences between their estimate for the effect of water and the hydrous anomalies calculated for model ICE3-MP2 (figure V.47f) and those for model ICE-F (figure V.54): the latter two, which are based on geochemical results for the water content of the Iceland plume, are much larger in the unmolten plume stem than the seismic data let expect, and they would not shift to much lower values if the lower bound on the water content of the plume source of 300 wt. ppm had been used; furthermore, the strong reduction of the negative anomaly between *ca.* 120 and 60 km is not observed in the data of Bjarnason and Sacks (2002), let alone a change of its sign. As mentioned previously, the latter might be due to the choice of the reference mantle, but it should be noted that those strongest parts of the anomaly which are in blatant conflict with observations affect a depth range of only about two dozen kilometers; a seismic resolution test would have to be conducted to clarify whether or not a narrow layer of relatively high velocity at that depth can be resolved by surface waves, but this is beyond the scope of this study. Judging the shallow parts of the anomaly from the regional tomography models of Wolfe *et al.* (1997) or Foulger *et al.* (2001) is somewhat difficult, because these models do not have sufficient resolution in all or part of the depth interval < 100 km, but they both seem to feature a decrease of the anomaly near the upper edge of the well-resolved depth range, which might reflect the existence of a gap in the plume head anomaly. More reliable information can possibly be derived by considering the lateral v_S variations relative to the average velocity–depth model for the Iceland region as determined by Allen *et al.* (2002a). The computation of this model is roughly comparable to the computation of $\delta v_S/v_S$ in sections V.2.3 and V.2.4 and also reveals a strong decrease of the negative anomaly and even a change in sign for $z < 100$ km (see figure VI.2). While those authors attributed this feature to higher degrees of partial melt, the laterally heterogenous water content and extent of dehydration might hence be an alternative explanation for it. – The conflict between model and observation in the deep mantle also needs to be resolved. If the geochemical estimate is reasonably accurate, the discrepancy between the calculated and the measured anomaly points towards an effect of the limited resolution of seismic imaging, which would smear a narrow, strong anomaly into a broader, weaker one. If this is the case, it is probably necessary to assume a lateral gradient in the water content of the plume stem, contrary to the very primitive assumption of a homogenous water content in the unmolten plume used here; a concentration of the chemical anomaly also seems to be in better agreement with geochemical profiles *e.g.* of ^{3}He/^{4}He along the ridge on Iceland (Breddam *et al.*, 2000), which suggest that the mantle source responsible for the peak is confined to a narrower region than a homogeneous plume like in ICE-F would imply (see figure IV.9). If the seismological image is blurred, the higher value for the temperature anomaly required from crustal thickness in water-bearing models could also be made plausible. It is suspected that this would also apply to the synthetic seismic inversion model by Ito *et al.* (1999) if they had included attenuation more comprehensively. Furthermore, recent tomographic investigations using methods which take into account the finite frequency resp. wavelength of seismic waves instead of the traditional ray-method (*i.e.* infinite-frequency) approach (Hung *et al.*, 2003) indicate that the older ray tomographies could have underestimated upper-mantle anomalies by a factor of two, which would strengthen the case of a wet plume.

Furthermore, the ascent velocities of the plumes can be compared with independent estimates for the ratio of active to passive upwelling. For this upwelling ratio, values of 2–4 have been deduced for the depth range of melting (Allen *et al.*, 2002b; Holbrook *et al.*, 2001), but at the base of the melting region it might be as high as 10 (Maclennan *et al.*, 2001a). These results are in good agreement with v_z at the plume axis in the DPRM1 and in models ICE1-135/0.1 and ICE1-135/1c, which corresponds to upwelling ratios of 7–8 at their respective solidus depths and drops to a ratio of 1 only near the top of the melting zone (see figure V.29). In contrast, the hot plume of ICE1-220/1 has a ratio of 17 at the solidus depth, whereas the cold plume of ICE1-50/1 does not reach a ratio of more than ~ 2. The pronounced effect of active upwelling is also reflected by the superlinear increase of h with ΔT_P (figure V.30), whose cause is that the amount of produced and extracted melt does not just depend straightforwardly on supersolidus temperature, but is also affected indirectly

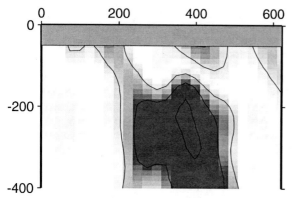

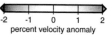

Figure VI.2: Shear velocity anomaly relative to the regional average beneath Iceland on a profile in E–W direction across the rift zone and Vatnajökull (from Allen *et al.*, 2002a, © 2002 American Geophysical Union, reproduced with permission). *See plate section for colour version.*

by consequences of a high ΔT_{P}: a hot plume also has a larger thermal (and depletion) buoyancy, and the supersolidus part of the thermal anomaly affects a larger volume in the mantle, because melting starts at greater depth. Again, these conclusions do not hold for most of the melting zone if dehydration stiffening comes into play (see figure V.39), but if the estimates are restricted to the bottom of the melting region, they are still roughly valid. In fact, the lower estimates support the notion of strongly weakened active upwelling in the melting zone, if regarded as an average of its whole depth range.

The segregation velocity $\vec{v}_{\mathrm{seg}}$ depends strongly on the porosity of the rock, as can be seen from eq. II.8. Therefore, variations in φ_{ex} have the strongest effect on $\vec{v}_{\mathrm{seg}}$: the maxima of $v_{\mathrm{seg},z}$ are on the order of $1\,^{\mathrm{mm}}/_{\mathrm{a}}$ for $\varphi_{\mathrm{ex}} = 10^{-3}$, a few centimetres per year for $\varphi_{\mathrm{ex}} = 0.01$, essentially independent from T, and 10–$20\,^{\mathrm{cm}}/_{\mathrm{a}}$ for $\varphi_{\mathrm{ex}} = 0.03$; for ICE1-135/100, where segregation has not been modelled, $\varphi = 0.25$ would result in velocities on the order of kilometers per day. Similar ratios apply to the horizontal components of $\vec{v}_{\mathrm{seg}}$, but they are much smaller, because melt flow is chiefly buoyancy-driven. They reach their highest values in particular beneath the freezing front of the lithosphere, where the ascending melt is dammed and would form a high-φ layer, were it not extracted instantaneously; this layer, which was also observed by Ghods and Arkani-Hamed (2000) in a plain MOR model, would drain the melt towards the ridge. Although these values are considered to be reasonable estimates of the order of magnitude of both absolute and relative segregation velocities, it should be kept in mind that there are several factors which were not included here, but will have some impact on the velocity distribution of segregating melt, *e.g.* the viscosity of the melt, as shown in model series VM (section V.1.8). It is therefore to be expected that the viscosity in the deeper parts of the melting zone is lowered strongly enough to overcompensate the reduction of $v_{\mathrm{seg},z}$ due to the buoyancy decrease of the compressed melt.

Considering all Iceland models together, the results of this study essentially confirm the conclusion of Ito *et al.* (1999), in whose study the focus had been a bit different and who put less emphasis on the details of melting processes; in particular, the role of water was also found to be crucial, and hydrous plume models are preferred over anhydrous ones, because there is good evidence for the presence of water and the importance of its effect on mantle rheology, and the hydrous models have shown a greater potential to approximate the shape of the Icelandic crust, as demonstrated in series ICE3 and model ICE-F. Assuming a melt extraction threshold of $1\,\%$, the hydrous Iceland models suggest that the plume has an excess temperature of some $200\,\mathrm{K}$ at $200\,\mathrm{km}$ depth and a significantly larger volume flux than assumed *e.g.* in model series ICE1; the flux is probably even larger than in model ICE-F $(3.47 \, \mathrm{km^3/a})$ and might well lie between 5 and $6 \, \mathrm{km^3/a}$. To yield the observed crustal thickness of *ca.* 30–40 km and match the constraints on the plume radius of 100–150 km, the upwelling velocity of the plume near the base of the melting zone would probably lie between 15 and $20 \, \mathrm{cm/a}$, but decay over a depth range of a few dozen kilometers to values near the passive upwelling rate, as can be estimated from model ICE-F (cf. figure V.53). – Calculated seismic anomalies for hydrous models suggest that $\delta v_S / v_S$ could reach 8–10 % in the deep upper mantle, which is about 2–3 times higher than reported values; in contrast, a gap in the anomaly could exist within the plume head. The electrical conductivities computed for these models increase with depth, and the plume head is not expected to be visible as a strong good conductor. The abovementioned conflict between the seismic anomalies measured and calculated for water-bearing models suggests that the plume is either thinner than assumed here, and/or that there is a chemical gradient in its stem, and that seismic observations yielded rather a lower bound for the anomaly up to now. Furthermore, the models of this study also support the notion that the Icelandic crust is thick.

Future work could attempt to clarify more the effect of the water content on plume dynamics and bring it into better agreement with the seismic anomaly. For this task, it would be sufficient to actually model a quarter plume, thereby either saving three quarters of the otherwise used RAM and having shorter run durations or reaching a four times finer resolution; from the numerical point of view, modelling an entire plume in a laterally closed box is a less than optimal approach, which had been taken here only for technical reasons and due to the original intention to put more emphasis *e.g.* on off-ridge plumes. The major task for forthcoming research should be to model the temporal evolution of the relative migration of the plume from beneath the Greenland craton to its present location and the possibility to explain the formation of V-shaped ridges by control from above related to thickness variations of the overlying lithosphere, as described in section IV.1.1; the preliminary results of series ICE-O in section V.2.5 suggest that this mechanism could result in significant variations in crustal thickness. Furthermore, a consistent integration of the crust into the convection–melt model is desirable in order to be able to estimate the role of the crustal root in limiting the size of the melting region from above; preliminary attempts made during the work on this study, which have though not reached a sufficient maturity to be included here, have tackled the problem via an influx mechanism through the model top, making use of the same flux balancing mechanism as the plume at the bottom, and this, combined with a chemical marker field to account for the different density, is probably the road to follow in further development.

APPENDIX A

NON-DIMENSIONALIZATION RULES

*I have no satisfaction in formulas
unless I feel their numerical
magnitude.*

—LORD KELVIN

In fluid dynamics, it is common practice to simplify the governing differential equations by non-dimensionalizing the variables and combining several scaling parameters into non-dimensional scaling numbers such as *e.g.* Ra. Here the non-dimensionalizations used in this study are listed; in ambiguous cases, dimensionized variables are primed.

$$\text{length:} \qquad x = \frac{x'}{z_\mathrm{m}},\ y = \frac{y'}{z_\mathrm{m}},\ z = \frac{z'}{z_\mathrm{m}} \qquad (A.1)$$

$$\text{time:} \qquad t = \frac{\kappa}{z_\mathrm{m}^2} t' \qquad (A.2)$$

$$\text{velocity:} \qquad v = \frac{z_\mathrm{m}}{\kappa} v' \qquad (A.3)$$

$$\text{density:} \qquad \varrho = \frac{\varrho'}{\varrho_0} \qquad (A.4)$$

$$\text{viscosity:} \qquad \eta = \frac{\eta'}{\eta_0},\ \eta_\mathrm{f} = \frac{\eta_\mathrm{f}'}{\eta_0},\ \eta_\mathrm{b} = \frac{\eta_\mathrm{b}'}{\eta_0} \qquad (A.5)$$

$$\text{temperature:} \qquad T = \frac{T'}{\Delta T} \qquad (A.6)$$

$$\text{pressure/stress:} \qquad p = \frac{z_\mathrm{m}^2}{\kappa\eta_0} p',\ \sigma = \frac{z_\mathrm{m}^2}{\kappa\eta_0} \sigma' \qquad (A.7)$$

$$\text{entropy/enthalpy:} \qquad S = \frac{S'}{c_p},\ L = \frac{L'}{c_p\Delta T} \qquad (A.8)$$

MASS FLUX BALANCE AT THE MODEL BOTTOM

In section III.1.1, several fluxes through model boundaries have been discussed in connection with boundary conditions for the convection model. In this appendix, an analytical description of the fluxes is derived, and some consideration is given to the question how the total flux into the model is balanced against the total flux out of it to achieve mass conservation.

If an influx is prescribed at a model boundary for physical reasons, mass conservation has to be preserved by defining a corresponding mass flux of the same magnitude out of the model box. The only boundary where this can be done in a reasonable way in this study is the bottom of the model. The volume flux through a surface A is defined as

$$q = \int_A \vec{v}(x, y, z) \cdot \vec{u}_N \, dA. \tag{B.1}$$

In the following it is assumed that the flux vector through the boundary is perpendicular to the bottom plane with normal unit vector $\vec{u}_N$ and parallel to the z axis, $\vec{e}_z || \vec{u}_N$, hence only the z component of $\vec{v}(x, y, z_m)$ is relevant and most conveniently expressed with a scaling velocity and a weighting function: $v_z(x, y, z_m) := v_0 w_i(x, y)$.

B.1 WEIGHTING FUNCTIONS AND THEIR WAVENUMBER SPECTRA

The spectral properties of the weighting function $w(x, y)$ are of certain interest, because they have influence on the resolution of the structure in the numerical grid; this is particularly important in the case of the plume influx function. As pointed out in section III.1.1, the preferred function is a Gaussian curve; in polar coordinates, which are appropriate for a plume, the Gaussian curve $w_0(r)$ and its amplitude spectrum $W_0(k_r)$ are given by

$$w_0(r) = e^{-\left(\frac{2r}{R_P}\right)^2} \qquad\qquad W_0(k_r) = \frac{\sqrt{\pi}R_P}{2} e^{-\left(\frac{k_r R_P}{4}\right)^2} \tag{B.2}$$

(Bronstein and Semendjajew, 1991). For a cosine or squared cosine function tapered with a box at subsequent roots,

$$w_1(r) = b(r) \cdot \cos\left(\frac{r\pi}{2R_P}\right) \tag{B.3a}$$

$$w_2(r) = b(r) \cdot \cos^2\left(\frac{r\pi}{2R_P}\right) = b(r) \cdot \frac{1}{2}\left[1 + \cos\left(\frac{r\pi}{R_P}\right)\right] \tag{B.3b}$$

$$b(r) = \begin{cases} 1, & -R \le r \le R \\ 0, & \text{otherwise,} \end{cases} \tag{B.3c}$$

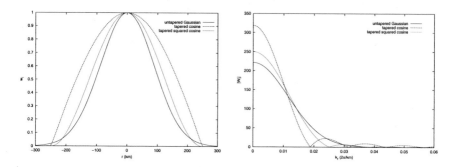

Figure B.1: Weighting functions eqs. B.2, B.3a, and B.3b (left) and absolute values of their spectra eqs. B.2, B.4a, and B.4b (right). $R = 250\,\mathrm{km}$ for all functions.

the spectra can be obtained by applying the convolution theorem of the Fourier transform (see Buttkus, 1991):

$$W_1(k_r) = \frac{4\pi R_P \cos(k_r R_P)}{\pi^2 - 4k_r^2 R_P^2} \tag{B.4a}$$

$$W_2(k_r) = \frac{\pi^2 \sin(k_r R_P)}{k_r(\pi^2 - k_r^2 R_P^2)} \tag{B.4b}$$

$$B(k_r) = \frac{2\sin(k_r R_P)}{k_r}. \tag{B.4c}$$

Comparison of eqs. B.2, B.4a, and B.4b shows that the spectrum of the untapered Gaussian curve and of the squared cosine decay more rapidly at high wavenumbers than the spectrum of the cosine (see figure B.1), thus favouring a coarser spatial sampling, although tapering at $\pm R$ would introduce a minor high-frequency component into the spectrum of the Gaussian. In the spatial domain, it is obvious that the Gaussian and the squared cosine concentrate the energy more in the center and provide a smooth (or, as the Gaussian has to be tapered at a nonzero value, almost smooth) transition to the surroundings, while the tapered cosine has a cusp at its border; they are therefore also expected to be better behaved *e.g.* in the computation of finite difference operators.

B.2   Flux functions

B.2.1   Plume influx

For a plume a circular influx area is the most obvious choice. In cylindrical coordinates the volume flux through the plume base area A_P with radius R_P is

$$q_{\mathrm{in}} = \int_{A_P} v_{z_P}(r, \varphi) \, \mathrm{d}A = \int_0^{R_P} \int_0^{2\pi} v_{z_P}(r, \varphi) r \, \mathrm{d}\varphi \mathrm{d}r. \tag{B.5}$$

With a radial velocity profile given by eq. B.2, $v_{z_P}(r) := v_{z0_P} w_P(r) = v_{0_P} e^{-\left(\frac{2r}{R_P}\right)^2}$, the volume flux into the model box is

$$q_{\text{in}} = \int\limits_0^{R_P} \int\limits_0^{2\pi} v_{0_P} r e^{-\left(\frac{2r}{R_P}\right)^2} \, \mathrm{d}\varphi \, \mathrm{d}r = 2\pi v_{0_P} \int\limits_0^{R_P} r e^{-\left(\frac{2r}{R_P}\right)^2} \, \mathrm{d}r = \frac{\pi v_{0_P} R_P^2}{4} \int\limits_0^{R_P^2} e^{-\varrho} \, \mathrm{d}\varrho$$

$$= \left(1 - e^{-4}\right) \frac{\pi v_{0_P} R_P^2}{4}. \tag{B.6}$$

The Gaussian curve does not differ significantly from zero beyond R_P; it has decayed to 1.8 % of its maximum value there. Hence, this flux is almost equal to the flux through an unbounded Gaussian-weighted area.

B.2.2 Outflux through lateral sinks

The compensating outflux is provided by sinks at the margins of the bottom boundary; in principle, outflux through the side boundaries would be preferable, but could not be implemented here. In a plume–ridge setting, the overall flow pattern suggests prescribing a ridge-parallel sink of width x_S at either side of the model; again, a suitably scaled Gaussian curve is used as weighting function. If the ridge and the sinks are parallel to the y axis, the total outflux (both margins) is given by

$$q_{\text{out}} = 2 \int\limits_{A_S} v_{z_S}(x, y) \, \mathrm{d}A = 2 \int\limits_0^{x_S} \int\limits_0^{y_m} v_{z_S}(x, y) \, \mathrm{d}y \, \mathrm{d}x,$$

$$= 2 y_m v_{0_S} \int\limits_0^{x_S} e^{-\left(\frac{2x}{x_S}\right)^2} \, \mathrm{d}x \approx 2 y_m v_{0_S} \int\limits_0^{\infty} e^{-\left(\frac{2x}{x_S}\right)^2} \, \mathrm{d}x$$

$$= \frac{\sqrt{\pi} v_{0_S} x_S y_m}{2} = \frac{\sqrt{\pi} v_{0_S} A_S}{2} \tag{B.7}$$

(Bronstein and Semendjajew, 1991); the integral cannot be solved exactly in the interval $[0; x_S]$, but again, the contribution of the parts of the Gaussian beyond x_S to the flux are small. Setting $q_{\text{in}} = q_{\text{out}}$ for mass balance, the reference outflux velocity v_{0_S} can be tuned to fit the plume influx:

$$v_{0_S} = \left(1 - e^{-4}\right) \frac{\sqrt{\pi} R_P^2}{2 A_S} v_{0_P}. \tag{B.8}$$

For application in the numerical algorithm, the fluxes can be computed numerically directly from eq. B.1 and adjusted as necessary.

B.2.3 Couette flow

To calculate the plate-driven flow associated with the spreading ridge, the mantle can be idealized as a channel of height z_m, whose upper boundary moves relative to the lower one with a velocity v_0; if there is no horizontal pressure gradient, the shear stress is constant (Turcotte and Schubert, 1982), and a flow known as Couette flow results from this setting. To estimate the flux towards the lateral sinks in the model bottom, one can calculate a vertical profile of $v_x(x_S, y, z)$ in a reference frame attached to the model bottom, assuming that Couette flow is induced by the moving plates at the model top and that no-slip boundary conditions prevail at both the top and the bottom of the model. The flow profile will then depend on the viscosity of the model fluid, and if its only non-zero component is parallel to the x axis, it can be calculated by

$$\sigma_{xz} = \eta \frac{\mathrm{d}v_x}{\mathrm{d}z} \Rightarrow \int_{v_{x0}}^{v_x(z)} \mathrm{d}v_x = \int_0^z \frac{\sigma_{xz}}{\eta(z')} \, \mathrm{d}z' \tag{B.9a}$$

$$\Leftrightarrow v_x(z) = v_{x0} + \sigma_{xz} \int_0^z \frac{\mathrm{d}z'}{\eta(z')}, \qquad \sigma_{xz} = -\frac{v_{x0}}{\int_0^{z_m} \frac{\mathrm{d}z}{\eta(z)}} \tag{B.9b}$$

(Turcotte and Schubert, 1982); eq. B.9b could be solved numerically in the general case.

Constant viscosity — For an isoviscous model, the integrand in eq. B.9b is constant, and the velocity profile is linear:

$$v_x(z) = v_0 \left(1 - \frac{z}{z_m} \right) \tag{B.10a}$$

(Turcotte and Schubert, 1982); the flux through a vertical plane at x_S thus is

$$q_{\text{side}} = \int_0^{y_m} \int_0^{z_m} v_x(z) \, \mathrm{d}y \mathrm{d}z = y_m \int_0^{z_m} v_0 \left(1 - \frac{z}{z_m} \right) \mathrm{d}z = \frac{y_m z_m v_0}{2}. \tag{B.10b}$$

p–T-dependent viscosity — As a more realistic approach to the mantle, one can instead consider a fluid with p–T-dependent viscosity. Given that p is proportional to z, this dependence can be transformed into a simplified form of the general rheological law used in this study, eq. III.16, with $b_C = s = 0$. The thermal structure of the mantle is approximated by

$$T(z) = \begin{cases} T_0 + k_1 z, & z \leq z_1 \quad \text{(lithosphere)} \\ T_0 + k_1 z_1 + k_m(z - z_1), & z_1 < z \leq z_m \quad \text{(mantle)}, \end{cases} \tag{B.11}$$

where $T_0 = T(z = 0)$ and k_1 and k_m are constant temperature gradients for the lithospheric thermal boundary layer and the adiabatic mantle, respectively. The following calculation is similar to the one by Turcotte and Schubert (1982, p.318f.) for a single fluid layer with a viscosity with a T dependence of slightly different type; secondary effects such as shear heating are neglected.

For the lithosphere, insertion of the simplified rheological law $\eta(z) = \eta_0 \exp(b_z z - b_T T(z))$ into eq. B.9a yields

$$\int_{v_{x0}}^{v_x(z)} \mathrm{d}v_x = \int_0^z \frac{\sigma_{xz}}{\eta_0} e^{b_T(T_0 + k_1 z') - b_z z'} \, \mathrm{d}z'$$

$$\Rightarrow v_x(z) = v_{x0} + \frac{\sigma_{xz}}{\eta_0} e^{b_T T_0} \int_0^z e^{(b_T k_1 - b_z)z'} \, \mathrm{d}z' = v_{x0} + \frac{\sigma_{xz}}{\eta_0} \frac{e^{b_T T_0}}{b_T k_1 - b_z} \left(e^{(b_T k_1 - b_z)z} - 1 \right). \tag{B.12a}$$

At the transition from the lithosphere to the mantle, the velocity is continuous. Hence for the mantle,

$$\int\limits_{v_x(z_1)}^{v_x(z)} \mathrm{d}v_x = \int\limits_{z_1}^{z} \frac{\sigma_{xz}}{\eta_0} e^{b_T[T_0+k_1z_1+k_m(z'-z_1)]-b_z z'} \, \mathrm{d}z'$$

$$\Rightarrow v_x(z) = v_x(z_1) + \frac{\sigma_{xz}}{\eta_0} e^{b_T[T_0+(k_1-k_m)z_1]} \int\limits_{z_1}^{z} e^{(b_T k_m - b_z)z'} \, \mathrm{d}z'$$

$$= v_x(z_1) + \frac{\sigma_{xz}}{\eta_0} e^{b_T[T_0+(k_1-k_m)z_1]} \frac{e^{(b_T k_m - b_z)z} - e^{(b_T k_m - b_z)z_1}}{b_T k_m - b_z}$$

$$= v_{x0} + \frac{\sigma_{xz}}{\eta_0} e^{b_T T_0} \left[\frac{e^{(b_T k_1 - b_z)z_1} - 1}{b_T k_1 - b_z} + \frac{e^{b_T(k_1-k_m)z_1}}{b_T k_m - b_z} \left(e^{(b_T k_m - b_z)z} - e^{(b_T k_m - b_z)z_1} \right) \right].$$

$$\text{(B.12b)}$$

Inserting the boundary condition for the model bottom, $v_x(z_m) = 0$, into this result yields the shear stress

$$\sigma_{xz} = -v_{x0}\eta_0 e^{-b_T T_0} \left[\frac{e^{(b_T k_1 - b_z)z_1} - 1}{b_T k_1 - b_z} + \frac{e^{b_T(k_1-k_m)z_1}}{b_T k_m - b_z} \left(e^{(b_T k_m - b_z)z_m} - e^{(b_T k_m - b_z)z_1} \right) \right]^{-1}, \quad \text{(B.12c)}$$

which can then be inserted into eqs. B.12a and B.12b to calculate the velocity–depth profile and the flux similar to eq. B.10b. Note that $v_x(z)$ depends only on $\eta(z)/\eta_0$, because η_0 cancels out. Figure B.2 shows an example for isoviscous and p–T-dependent Couette flow in a 410 km deep channel; the p–T-dependent Couette flow can be computed numerically on the basis of eq. B.9b using the trapezoidal rule with a relative error to the analytical solution of less than half a percent with step of 10 km. This flux leaves the model through the lateral sinks as well, and returns into the model as a more or less weak background influx through the entire bottom plane between the sinks, as mentioned in section III.1.1 (see also figure III.1).

B.3 Buoyancy flux

In the plume literature, it is common to give flux estimates; to tune the numerical model for a given R_P accordingly, one would solve eq. B.6 for v_{0_P}:

$$v_{0_P} = \frac{4q_{in}}{(1 - e^{-4})\pi R_P^2}. \quad \text{(B.13)}$$

The volume flux q used can also be related to a dynamical quantity measuring the strength of buoyancy driving the mass flow. Assuming that buoyancy is wholly thermal, this buoyancy flux (*e.g.* Ribe and Christensen, 1994) can be defined as

$$\mathfrak{q} = \varrho_0 \alpha \Delta T_P q = \varrho_0 \iint \alpha(T - T_{ref}) v_z \, \mathrm{d}x \mathrm{d}y, \quad \text{(B.14)}$$

(after Albers and Christensen, 2001), where the possible p–T (*i.e.* position) dependence of α is accounted for. Obviously, $\mathfrak{q}$ depends on the depth level, which can make comparison between models difficult.

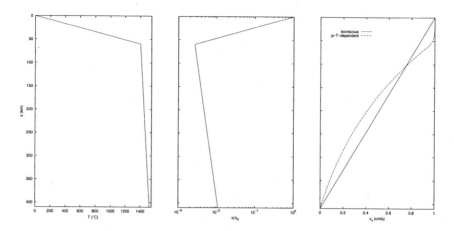

Figure B.2: Simplified geotherm (left) and depth profiles of the normalized viscosity for a p–T-dependent rheology (center) and of v_x for an isoviscous and a p–T-dependent Couette flow. The parameters used are $z_l = 60\,\text{km}$, $z_l = 410\,\text{km}$, $b_z = \ln(10)$, $b_T = \ln(1000)$, $v_{x0} = 1\,\text{cm/a}$, $k_l = 23.333\,\text{K/km}$, $k_m = 0.35\,\text{K/km}$.

It is justified to use indeed the maximum plume temperature anomaly as ΔT_P, because the flux has a narrower peak than $v_z(r)$ and $\Delta T_\text{P}(r)$; *e.g.* for a profile like eq. B.2 and constant α, it is given by

$$\mathfrak{q} = \varrho_0\alpha \int\limits_0^{R_\text{P}}\int\limits_0^{2\pi} \Delta T_\text{P,max} e^{-\left(\frac{2r}{R_\text{P}}\right)^2} \cdot v_{z,0} e^{-\left(\frac{2r}{R_\text{P}}\right)^2} r\,\mathrm{d}r\mathrm{d}\varphi = \frac{\pi}{8}\varrho_0\alpha R_\text{P}^2 v_{z,0}\Delta T_\text{P,max}\int\limits_0^8 e^{-\tilde{r}}\,\mathrm{d}\tilde{r}$$

$$= \frac{\pi}{8}\varrho_0\alpha R_\text{P}^2 v_{z,0}\Delta T_\text{P,max}\left(1 - e^{-8}\right) \approx \frac{\pi}{8}\varrho_0\alpha R_\text{P}^2 v_{z,0}\Delta T_\text{P,max}. \tag{B.15}$$

The approximation is very accurate. In the context of a numerical model, where the shape of the anomaly is not precisely defined as a function, it seems though more appropriate to evaluate the buoyancy flux numerically. The heat flux at the base of the plume follows from the buoyancy flux as $\mathfrak{Q} = \mathfrak{q}\,c_p/\alpha$.

APPENDIX C

PARAMETERIZATION OF THE MELTING OF PERIDOTITE

> *Errors using inadequate data are*
> *much less than those using no data*
> *at all.*
> —CHARLES BABBAGE

The onset of melting and the amount and chemical composition of melts is controlled by the form of the solidus T_s and the liquidus T_l of the rock, which are functions of several variables, mainly pressure resp. depth, chemical composition and volatile content. Too little is known about the precise form of the complex dependence of T_s, T_l and φ on these factors to summarize the melting behaviour in a generally applicable analytical expression for straightforward implementation and calculation, so it is necessary to put constraints on them by experimental data. In the following, a compilation of several datasets for typical mantle materials and parameterizations for T_s, T_l, φ and compositional features of the melting process are given.

The modal compositions of some of the natural and synthetic mineral assemblies used in several melting experiments are given in table D.1. Of all investigated multimodal systems, the melting behaviour of the spinel lherzolite KLB-1 (see section D.1.1) is the most thoroughly documented. As this rock has been characterized as a fairly representative fertile upper mantle rock with average contents of incompatible elements (Hirose and Kushiro, 1993; Hirschmann, 2000), most of the parameterization attempts will focus on it. Admittedly, using only one rock type bears the risk of lacking generality by introducing its individual signature; on the other hand, as Iwamori *et al.* (1995) also point out, the resulting parameterization may be expected to be more self-consistent, which justifies this approach.

C.1 DATA FITTING METHOD

For fitting, nonlinear and iterative linear least-squares (L_2) as well as a simplex method, which yields the best L_1 approximation of a dataset (*e.g.* Chvátal, 1983), have been tried out. The underlying assumption of least-squares fits is that the data are distributed according to a Gaussian curve, which implies that there are no large errors; for $n \to \infty$, L_n norm fits become increasingly sensitive to outliers and give them large weight, which might make the whole fitting function worthless (Menke, 1984). The L_1 method, which suppresses the influence of large deviations from the main trend (Huber, 1981), has been considered as an alternative, because inspection of the available melting experiment data lets one suspect that there is considerable scatter. To reduce further the influence of suspected outliers, only a hand-picked subset of data collected from several publications instead of all available published data have been used. The model function used for fitting was

$$T_{s,l} = c_0 + c_1 p + c_2 p^2 + c_3 p^3 + c_4 \ln p + c_5 \sqrt{p} \qquad (C.1)$$

with p in GPa and T in °C, or simplified descendants of it with some coefficients set to zero; inclusion of polynomials of higher degree than 3 gave obviously useless curve shapes, and simplified forms of eq. C.1 seem to be in common use for fitting melting data. The least-squares fits were calculated with the subroutine library STARPAC v.2.08 (Donaldson and Tryon, 1990) or with the built-in fitting function of the Gnuplot plotting software (Williams *et al.*, 1986). It turned out that the different methods yielded quite similar results in nearly all cases.

As mentioned, the used data are a selection from a larger dataset. The most important selection criteria were to preclude over-emphasis of single data points which had been measured multiply in several runs, and the purity of the experimental result materials; this means that results where the sample has been contaminated by metal or carbonate from the capsule or other obvious pollutants or where the natural composition of the initial material had been modified intentionally have not been used. However, there is still some scatter in the data which is due to methodical shortcomings: maybe the most obvious concerns differences in the capsule materials used; solidus temperature differences of several tens of degrees have been reported as a consequence of using different materials (C, Pt, Re), which cause different problems such as iron loss or reduction of iron (Harrison, 1981; Hess, 1992; Herzberg *et al.*, 2000).

C.2 ANHYDROUS SOLIDUS

The parent material of all melts considered here is lherzolite containing an Al-bearing phase, which will be either plagioclase, spinel, or garnet, depending on the depth resp. pressure and the temperature. There is a wealth of experimental data on the solidus of natural lherzolites and synthetic analogues which cover the whole pressure range from 1 atm up to 25 GPa and follow a rather clear trend up to about 10 GPa. However, at higher pressures the data show considerable scatter, even for samples from the single nodule considered here (KLB-1): the measurements by Takahashi (1986) and Takahashi and Scarfe (1985) give a solidus temperature which is more than 100 K lower than that from data by Herzberg and Zhang (1996), Zhang and Herzberg (1994) and Herzberg *et al.* (1990, 2000); it seems that the latter are more precise, although it apparently had been necessary to revise the values for $p = 5 \dots 9.7$ GPa of Zhang and Herzberg (1994) in Herzberg *et al.* (2000) after switching to a more sensitive observation technique, so they are preferred. Using them also permits to extend the assessment of the solidus consistently up to 20 GPa; in the experiments of Zhang and Herzberg (1994) at still higher pressures, all samples were partially molten, so that no lower bound could be determined, and the data were not used.

Published solidus curves display cusps at some points, which are related to phase transitions (see section II.3.1), although they are probably not as sharp in natural rocks as frequently drawn; thus, it seems to be reasonable to divide the whole pressure interval of interest into some smaller intervals according to the positions of some important phase transitions. It is somewhat difficult to assess the p–T values where the solidus intersects the phase boundary from the datasets reviewed here. Concerning the Al-bearing phases, it seems that no plagioclase is to be expected for pressures above 1 GPa and no spinel for pressures above 3 GPa; diagrams in Hirose and Kushiro (1993), Takahashi (1986), Takahashi *et al.* (1993) and Fei and Bertka (1999) for KLB-1 confirm that the intersections are at about (0.9 GPa, 1200 °C) for the plagioclase–spinel transition and at around (2.6 GPa, 1425 °C) for the spinel–garnet transition. Here, eq. D.14 has been used to constrain the position of the sp–gt cusp on the p axis. Similarly, it can be estimated that the cusps related with the olivine transitions in KLB-1 are at (15.5 GPa, 2050 °C) (α-ol/β-ol) (Herzberg and Zhang, 1996), (20.5 GPa, 2220 °C) (β-ol/γ-ol) and (22 GPa, 2250 °C) (γ-ol/pv+mw) (Zhang and Herzberg, 1994; Fei and Bertka, 1999).

stability field	c_0	c_1	c_2	c_3	c_4	c_5
plagioclase (0–0.89 GPa)	1119.069	139.22				
spinel (0.89–2.61 GPa)	1260.825		3.285		174.078	
garnet/α-ol (2.61–15.62 GPa)	1073.517	−329.927	18.651	−0.424	1165.902	
β-ol (15.62–20 GPa)	−11960.875	−753.979				6527.184

Table C.1: Coefficients for calculation of the solidus of KLB-1 with eq. C.1 (rounded to three decimals).

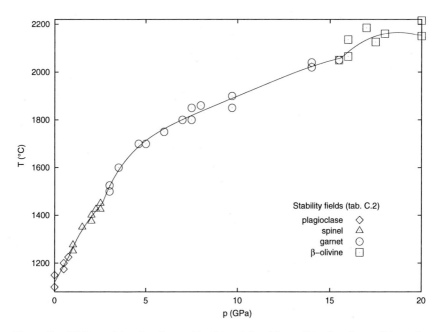

Figure C.1: KLB-1 melting data from table C.2 and fit with eq. C.1 using the coefficients from table C.1.

With these considerations, a selection of data on KLB-1 (table C.2) has been assembled and divided into subgroups according to the different stability fields. The data for each interval have been fitted as described in section C.1, yielding the curve shown in figure C.1; the coefficients are listed in table C.1. The kinks in the curve correspond to the locations of the phase boundaries and have been placed such that the whole curve is continuous.

C.3 LIQUIDUS

In contrast to the solidus, the liquidus to be considered depends on the melting mechanism assumed to be realized in the mantle. For batch melting, the liquidus of the lherzolite treated in section C.2 must be taken here; for fractional melting, however, the last solid phase to be present can be assumed to be forsterite (Mg_2SiO_4), which has a significantly higher liquidus (Iwamori *et al.*, 1995).

stability field	p (GPa)	T (°C)	state	Al phase	ref.[a]
plagioclase	0.0001	1100	solid	pl,sp	1,2[b]
	0.0001	1150	liquid present	pl,sp	1
	0.5	1175	solid	pl,sp	1
	0.5	1200	liquid present	sp	1
	0.75	1225	liquid present	sp	1
spinel	1	1250	solid	sp	1,3[b]
	1	1275	liquid present	sp	1
	1.5	1350	solid	sp	1,3[b]
	2	1375	solid	—	3
	2	1400	liquid present	—	1
	2.25	1425	liquid present	sp	1
	2.5	1425	solid	—	3
	2.5	1450	liquid present	—	3
garnet/α-olivine	3	1500	solid	—	1–3[b]
	3	1525	liquid present	—	3
	3.5	1600	solid	sp,gt	1
	4.6	1700	solid	gt	4
	5	1700	solid/liquid present	gt	5[bc]
	6	1750	solid	sp,gt	1
	6	1850	liquid present	—	1
	7	1800	liquid present	—	1
	7.5	1800	liquid present	—	1,2[b]
	7.5	1850	liquid present	gt	5
	8	1860	liquid present	gt	1
	9.7	1850	solid	gt	5,6[b]
	9.7	1900	liquid present	gt	5[bc]
	14	2020	solid		7
	14	2040	liquid present	gt	8
	15.5	2050	liquid present	gt	8[d]
β-olivine	15.5	2050	liquid present	gt	8[d]
	16	2065 ± 1	solid	gt	6
	16	2135 ± 1	liquid present	gt	6
	17	2185 ± 3	liquid present	gt	6
	17.5	2125 ± 1	solid	gt	6
	18	2160	liquid present	gt	8
	20	2150 ± 4	solid	gt	6
	20	2215 ± 2	liquid present	gt	6

[a]Data sources: 1: Takahashi (1986), 2: Takahashi and Scarfe (1985), 3: Hirose and Kushiro (1993), 4: Takahashi *et al.* (1993), 5: Herzberg *et al.* (2000), 6: Zhang and Herzberg (1994), 7: Herzberg *et al.* (1990), 8: Herzberg and Zhang (1996)

[b]data point only used once

[c]2 measurements: 1 solid, 1 with liquid

[d]2 measurements: 1 with α, 1 with β present

Table C.2: Data points for calculation of the solidus of KLB-1.

C.3.1 Lherzolite

In several melting experiments on natural peridotites and synthetic aggregates attempts have been made to bracket the liquidus up to 25 GPa. There is large scatter in the available data over the whole pressure range, even in data from aggregates of the same type. In particular, the two natural lherzolites KLB-1 and PHN 1611 (see section D.1.1), for which many data points are available, show significantly different trends, although they are both considered to be chemically quite similar to pyrolite; on the other hand, most of the data follow roughly the trend outlined by KLB-1. This makes it difficult to determine an "average" liquidus curve for a lherzolite composition. Thus, it seems particularly reasonable to restrict the evaluation to a single composition in this case, for which KLB-1 is the best choice due to the amount of available data and the large pressure interval covered by them.

The KLB-1 data points in table C.3 are a subset of the available data (Takahashi, 1986; Takahashi and Scarfe, 1985; Takahashi *et al.*, 1993; Herzberg *et al.*, 1990) and were best fitted by the function

$$T = 1806.82121 + 2.22842p^2 - 0.12564p^3 + 95.17188 \ln p. \tag{C.2}$$

Note that the intercept is somewhat higher than *e.g.* in figure 9 of Takahashi (1986) and figure 6 of Zhang and Herzberg (1994); however their extrapolation to exactly 1700 °C at normal pressure seems a bit arbitrary. Inclusion of the two additional points (10 GPa, 2200 °C) and (10.5 GPa, 2000 °C) did hardly change the fit, so they seem to be insignificant and were not taken into account; on the other hand, a data point at normal pressure from Takahashi (1986) could not be included either, because there was no corresponding upper bound, and no acceptable fit could be achieved using this point.

C.3.2 Forsterite

The pressure range from 0.55 to 16.7 GPa is well covered by the melting experiments of Davis and England (1964), Ohtani and Kumazawa (1981) and Presnall and Walter (1993) (table C.4 and figure C.2); two additional 1 atm data points by Navrotsky (1994) and Richet *et al.* (1993) (see also section D.2.4) have also been used to have a hook to surface conditions. Most of the data from Ohtani and Kumazawa (1981) have not been used, though, because the temperature data were unreasonably high and showed huge errors (up to 200 °C at higher pressures) in several cases due to technical problems; the data from Presnall and Walter (1993) seem to follow the expected trend better, as they indicate a flattening of the liquidus curve for $p \gtrsim 8$ GPa.

Considering the pressure interval up to 5 GPa as particularly important and the interval up to *ca.* 8 GPa as still significant,

$$T = 1895.5 + 60.9543p - 3.52246p^2 + 0.0801926p^3 \tag{C.3}$$

with p in GPa and T in °C was chosen as the best fit; the ambient pressure values of Navrotsky (1994) and Richet *et al.* (1993) are regarded as especially precise, so that the fit was forced to give their arithmetic mean at $p = 0$. This equation is used in most of this study unless stated otherwise. However, it should probably not be used for extrapolation of the liquidus beyond $p \gtrsim 14.6$ GPa, because it has a physically unjustified inflexion point there; with regard to the lower pressures, it seems to fit the data from Davis and England (1964) better than the Simon equation

$$T = 2171 \sqrt[11.4]{1 + \frac{p}{2.44}} - 273.16 \tag{C.4}$$

p (GPa)	T (°C)	state	Al phase	ref.[a]
1.5	1800	solid pres.	sp	1,2[b]
1.5	1850	liquid	sp	1
3	1900	solid pres.	sp	1,2[b]
3	1950	liquid	sp	1
5	2000	solid pres.	sp	1
5	2000	liquid	sp	2
7.5	2000	liquid	gt	1
8	2010	solid pres.	gt	1
8	2120	liquid	gt	1
12.5	2010	solid pres.	gt	1
12.5	2150	liquid	gt	1
14	2115	solid pres.	gt	3
14	2150	liquid	gt	1

[a]Data sources: 1: Takahashi (1986), 2: Takahashi and Scarfe (1985), 3: Herzberg *et al.* (1990)
[b]data point only used once

Table C.3: Data points for calculation of the liquidus of KLB-1.

p (GPa)	T (°C)	state	ref.[a]
0.0001	1890		1
0.0001	1901		2
0.55	1905	solid	3
0.55	1930	liquid	3
1.25	1950	solid	3
1.25	1975	liquid	3
1.8	1980	part.molten	3
2.5	2030	part.molten	3
3.05	2030	part.molten	3
3.95	2080	solid	3
3.95	2105	liquid	3
4.65	2105	solid	3
4.65	2130	liquid	3
4.7	2100 ± 10	liquid	4
6.2	2185 ± 10	liquid	4
7.7	2185 ± 30	solid	4
8.4	2230 ± 10	solid	4
9.7	2210[b]		5[c]
10	2245 ± 10	solid	4
10.6	2220[b]		5[c]
11.8	2240[b]		5[c]
12.8	2260[b]		5[c]
13.9	2310[b]		5[c]
13.9	2270[b]		5[c]
14.9	2280[b]		5[c]
15.6	2310[b]		5[c]
16.5	2280[b]		5[c]
16.7	2315[b]		5[c]

[a]Data sources: 1: Navrotsky (1994), 2: Richet *et al.* (1993), 3: Davis and England (1964), 4: Ohtani and Kumazawa (1981), 5: Presnall and Walter (1993)
[b]nominal T
[c]Error ranges are ± 0.2 GPa and ± 30 °C, respectively.

Table C.4: Data points for calculation of the liquidus of forsterite Mg_2SiO_4.

given by Presnall and Walter (1993). A Simon-type fit to the data in table C.4 did not yield a significant improvement over the fit of Presnall and Walter (1993); in particular, both tend to overestimate T for pressures between about 1.5 and 3 GPa.

The equation

$$T = 1718.15628 - 11.8091p + 187.91605\sqrt{p + 0.8} \qquad (C.5)$$

used in several early models was later discarded, because eq. C.3 gives a better fit for $p <$ 5 GPa.

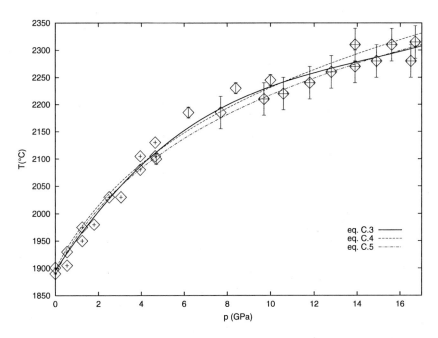

Figure C.2: Forsterite melting data of table C.4 and fits eqs. C.3–C.5. The error bars show the uncertainties in p and T measurements.

C.4   MODAL AND CHEMICAL COMPOSITION

As described in section II.3.3, complex changes in the petrological and chemical composition of the rock take place in the progress of melting. As it is too complicated to perform explicit modelling of these processes in the frame of convection/percolation modelling, an attempt is made to account for the compositional aspect by using parameterizations of the respective processes.

If the melting proportions $\mathcal{M}_i$ of the constituents are known, the modal composition of the residue from fractional melting can be calculated by solving eq. II.33a, which is derived as follows. The mass fraction X_{s_i} of the ith phase after a very small melting reaction increment δf is

$$X_{s_i}(f + \delta f) = \frac{X_{s_i}(f) - \mathcal{M}_i \delta f}{1 - c_\varphi \delta f} \tag{C.6a}$$

$$\Leftrightarrow \ \delta X_{s_i} - c_\varphi X_{s_i} \delta f - c_\varphi \delta X_{s_i} \delta f = -\mathcal{M}_i \delta f.$$

c_φ is the fraction of δf which actually becomes melt; this accounts for the fact that a part of X_{s_i} can recrystallize as another phase. The RHS in the first step is divided by $1 - c_\varphi \delta f$ to renormalize the solid mass fraction correctly according to $\sum X_{s_i} = 1$. For $\delta f \to 0$, cancellation of the mixed δ terms leads to the differential equation II.33a,

$$\frac{\mathrm{d}X_{s_i}}{\mathrm{d}f} = c_\varphi X_{s_i} - \mathcal{M}_i, \tag{C.6b}$$

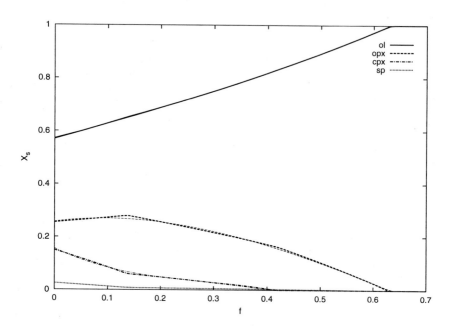

Figure C.3: Comparison of $X_s(f)$ parameterizations for spinel lherzolite with the analytical solution eq. II.33b (thick lines) and the polynomial fit of Kostopoulos and James (1992) (thin lines), for the starting modal composition of Kostopoulos (1991) listed in table D.1, the melting proportions of table C.6 and the fitting coefficients of table C.7. In this case, $c_\varphi = 1$.

which can be solved by separation of variables (eq. II.33b):

$$X_{s_i}(f) = \frac{(c_\varphi X_{s0_i} - \mathcal{M}_i)e^{c_\varphi(f-f_0)} + \mathcal{M}_i}{c_\varphi}. \qquad (C.6c)$$

In the course of melting, $\mathcal{M}_i$ changes at some points f_0, in particular when a phase is exhausted; at these points X_{s0_i} has to be reset, because $X_{s_i}(f)$ is continuous, but not dif-

	ol	opx	cpx	sp/gt
spinel facies				
sp lherzolite (cpx > 6 %)	0.5579	0.1744	−0.6137	−0.1186
cpx-bearing sp harzburgite	0.485166	−0.312409	−0.161078	−0.01168
cpx-free sp harzburgite	0.464233	−0.452554		−0.01168
(cpx+sp)-free harzburgite	0.462731	−0.462732		
garnet facies	0.0052p	0.0348p	−0.11p	0.07p
	+ 0.5724	− 0.0024	+ 0.107	− 0.677

Table C.5: X_{s0} for the parameterization of the modal composition in the spinel and the garnet stability field by eq. II.33b. The transitions from one interval to the next in the spinel field occur at $f = 0.1368459$, $f = 0.419145554$ and $f = 0.58664446$, respectively.

	ol	opx	cpx	sp/gt	ref.[a]
spinel facies					
sp lherzolite	0.0121[b]	0.0806[b]	0.7637[b]	0.1436[b]	1
	0.1	0.2	0.68	0.02	2
	−0.3[c]	0.4[c]	0.82[c]	0.08[c]	3
cpx-bearing sp harzburgite	0.0955	0.6388	0.2447	0.021	1
cpx-free sp harzburgite	0.1273	0.8517		0.021	1
(cpx+sp)-free harzburgite	0.13	0.87			1
garnet facies					
5 GPa	0.09	−0.57	0.71	0.2	4,5
4 GPa	0.0078	0.0522	0.47	0.47	1
3 GPa	0.013	0.087	0.36	0.54	1
	0.07	−0.16	0.68	0.25	4
	0.05	−0.49	1.31	0.13	6
	0.03	0.03	0.47	0.47	7
2.1 GPa	0.19	−0.5	0.68	0.63	3
garnet harzburgite	0.1	0.278	−	0.622	3

[a]Data sources: 1: Kostopoulos and James (1992), 2: Johnson *et al.* (1990), 3: Hauri and Hart (1994), 4: Walter (1998), 5: Presnall *et al.* (2002), 6: Salters (1996), 7: Schilling *et al.* (1999)
[b]cpx > 6 %
[c]at 1 GPa

Table C.6: Melting coefficients of peridotite for the spinel and the garnet facies. In the model by Kostopoulos and James (1992), the transition from spinel lherzolite to cpx-bearing spinel harzburgite occurs after 12.5 % melting, clinopyroxene is exhausted after further 23.5 % (Kostopoulos, 1991; Kostopoulos and James, 1992). The coefficients for the garnet facies at 2.1 GPa are given for completeness, although according to the parameterizations of the spinel and garnet stability fields, no garnet should be present at such low p. Walter (1998) gives $f = 0.1$ for gt-out, but this is not well constrained and probably too high for his low-garnet rock. The differences between the coefficients of Kostopoulos and James (1992) and Walter (1998) for garnet lherzolite at 3 GPa are probably mostly due to the fact that the former were constructed for a fractional melting model, whereas the latter stem from batch melting experiments.

ferentiable there. – Kostopoulos and James (1992) provide $\mathcal{M}_i$ values for fractional melting of spinel and garnet peridotite (table C.6) and give an overall fit with polynomials to the solution for the spinel stability field:

$$X_{\mathrm{s}_i} = \sum_{i=0}^{4} a_i f^i; \qquad (C.7)$$

the coefficients a_i are listed in table C.7, and their polynomials are compared with eq. II.33b in figure C.3; the agreement is very good. For the garnet stability field, the melting proportions are pressure-dependent and only valid until garnet is exhausted. A linear relation $\mathcal{M}_i(p) = a_{\mathcal{M}_i} p + b_{\mathcal{M}_i}$ has been assumed, for which the coefficients in table C.8 have been derived from the values in table C.6; this causes $X_{\mathrm{s}0_i}$ to be p-dependent as well.

	a_0	a_1	a_2	a_3	a_4
ol	0.57317	0.51939	0.24323		
opx	0.25742	0.27226	−1.5246	0.7228	
cpx	0.15587	−0.95957	2.7461	−3.2049	
sp	0.026582	−0.21345	0.82991	−1.4496	0.89734

Table C.7: Coefficients for the parameterization of the modal composition in the spinel stability field by eq. C.7 (Kostopoulos and James, 1992).

	$a_{\mathcal{M}_i}$ (^{1}/GPa)	$b_{\mathcal{M}_i}$
ol	−0.0052	0.0286
opx	−0.0348	0.1914
cpx	0.11	0.03
gt	−0.07	0.75

Table C.8: Coefficients for the p dependence of the $\mathcal{M}_i$ in the garnet stability field; after table C.6.

MATERIAL PROPERTIES

Many material properties relevant to mantle convection or melt flow are functions of physical variables such as pressure and temperature. Their chemical and petrological complexity and a number of physical phenomena as phase transitions make it impossible to describe these dependencies in a simple analytical form for the large p–T range covered in the earth's mantle. Therefore, petrophysical experiments are crucial in constraining the behaviour of geomaterials. This appendix compiles data from a number of experiments and gives parameterizations and approximations for several material parameters for use in the numerical models appearing in this study.

D.1 Minerals and solid rock

D.1.1 Upper mantle petrology

It is desirable to have a well-defined starting point for developping the specification of the petrological composition of the mantle in a numerical model. The most obvious approach is to use an average composition based on the analysis of a large number of representative rock samples or the composition of a synthetic model rock as pyrolite (Ringwood, 1975) derived from them. Table D.1 lists fractions of the major mineral phases of upper mantle peridotite as put forth in different publications; it is restricted to the upper mantle, because this study does not include models for the whole mantle, which would make it necessary to account for completely different high-pressure phases as well. Additionally, some real peridotites from different xenoliths are included.

Clearly, olivine is the most abundant mineral in almost all cases, as predicted by the pyrolite model; studies on the α–β phase transition of olivine and its depth, thickness, and reflectivity in the mantle as well as thermodynamic investigations also confirm that the mantle contains a pyrolite-like fraction of olivine of about 55 %, whereas piclogite, an alternative, garnet-rich model petrology proposed by Duffy and Anderson (1989), fits such models much less well in general (Ita and Stixrude, 1992; Gaherty et al., 1999; Shearer and Flanagan, 1999). However, at least orthopyroxene is also important enough that it should not be neglected when trying to construct averages of physical properties for spinel lherzolite from the properties of the individual components; it is of particular importance in connection with melt topology and the permeability of partially molten, but not yet strongly depleted peridotite, as discussed in section II.1.5. On the other hand, Walter (1998) reports that orthopyroxene is not present at the solidus in a large part of the garnet stability field. On melting, one phase after another is exhausted, in particular the Al-phase and diopside, although it can also happen that a phase actually crystallizes during melting, as is the case with opx in the garnet stability field (Longhi, 1995; Walter, 1998); experimental data on these phase-out lines in p–T space are not very abundant and show considerable scatter (Hess, 1992). Note that the mineral phases are solid solutions; they also undergo a variety of phase changes even in the upper mantle.

	ol (%)	opx (%)	cpx (%)	Al phase (%)	ref.[a]
plagioclase lherzolite	63.6	26.3	1.2	8.9	1
	61.9	22	11.2	4.9	after 2,3
spinel lherzolite	66.7	23.74	7.83	1.73	4
	65.3	21.8	11.3	1.5	5
	62[b]	24	12	2	6
	57.8	27	11.9	3.3	1
	57	25.5	15	2.5	3
	56[c]	22	19	3	6
pyrolite, spinel facies	55	25	18	2	7
	55.2	24.7	17.8	2.3	8
abyssal peridotite	75	21 (en)	3.5 (di)	0.5 (sp)	9
garnet lherzolite	67	12	11	10	5
	66[b]	17	12	6	6
	63	30	2	5	4
	59.8	21.1	7.6	11.5	1
	57	17	12	14	16
	57[c]	16	14	13	6
	60.1	18.9	13.7	7.3	10
	53	4	38	5	11
pyrolite, garnet facies	57 (fo_{88})	17 (en_{88})	12 (di)	14 (py_{70})	12
	55	20	15	10	7
	55.3	19.3	9.5	15.9	8
KLB-1	58	25	15	2 (sp)	13
KR-4003 (at solidus)	54.9	22.7	22.2	0.1 (sp)	14
PHN 1611	58.8	11.1	19.7	10.4 (gt)	15
HK-66	27	57	14	2 (sp)	13

[a]Data sources: 1: McKenzie and O'Nions (1991), 2: Green and Falloon (1998), 3: Kostopoulos (1991), 4: Maaløe and Aoki (1977), 5: Harris *et al.* (1967), 6: McDonough and Rudnick (1998), 7: Johnson *et al.* (1990), 8: Hauri and Hart (1994), 9: Dick *et al.* (1984), 10: Kostopoulos and James (1992), 11: Salters (1996), 12: Oxburgh and Parmentier (1977); Ringwood (1975), 13: Hirose and Kushiro (1993), 14: Luth (2002), 15: Jordan (1979), after Nixon and Boyd (1973), 16: Schilling *et al.* (1999)

[b]massifs

[c]primitive mantle

Table D.1: Petrological (modal) composition of average resp. model upper mantle rock and some lherzolites from xenoliths; the Al-bearing phase is plagioclase, spinel or garnet, respectively. The "abyssal peridotite" is an average of samples dredged from different slow-spreading on-ridge sites, including some near a hotspot; about a third of them additionally contains < 1 % plagioclase on average. The second entry for plagioclase lherzolite is an estimate from the CMAS reaction eq. D.12, using the spinel lherzolite proportions from Kostopoulos (1991) in this table; in particular, Na is neglected here. KLB-1 (Kilbourne Hole, New Mexico) is a relatively fertile spinel lherzolite, enriched in Al_2O_3 and CaO with Mg# = 89.5 and low in Cr_2O_3. KR-4003 (West Kettle River, British Columbia) is a spinel lherzolite with a composition also considered as a good approximation to primitive mantle (*e.g.* Walter, 1998). PHN 1611 (Thaba Putsoa, Lesotho) is a garnet lherzolite relatively close to the pyrolite composition (Mg# = 86.8), but with considerably less TiO_2 and Cr_2O_3 (Nixon and Boyd, 1973). HK-66 is a spinel lherzolite with depletion characteristics similar to those of KLB-1 (Mg# = 85.3), but with an opx-dominated modal composition.

D.1.2 Density

The mineral density of a solid solution formed by n end-members is calculated by

$$V_{\text{ss}} = \sum_{i=1}^{n} \underline{X}_i V_i \Rightarrow \varrho_{\text{ss}} = \frac{m_{\text{ss}}}{\sum\limits_{i=1}^{n} \frac{\underline{X}_i m_i}{\varrho_i}}, \tag{D.1a}$$

where $\underline{X}_i$ is the molar fraction of the ith end-member (*e.g.* Ita and Stixrude, 1992); the underlying assumption here is ideal mixing, whereas the real density will be different, because differences in ion size lead to irregularities in lattice structure. The bulk density of a composite of several minerals is

$$\varrho_{\text{mc}} = \frac{1}{\sum\limits_{i=1}^{n} \frac{X_i}{\varrho_i}} = \sum_{i=1}^{n} \tilde{X}_i \varrho_i \tag{D.1b}$$

(Johnson and Olhoeft, 1984, p.7), with X_i and $\tilde{X}_i = X_i \varrho_{\text{mc}}/\varrho_i$ being the weight and volume fraction, respectively, in this case. The density depends in a complex way on the p–T conditions through the thermal expansivity α and the bulk modulus K_T resp. K_S (see sections D.1.3 and D.1.4). This relation can be expressed as

$$\varrho(p, T) = \varrho_0 \exp\left(\int_0^p \frac{\mathrm{d}p}{K_T} - \int_{T_0}^T \alpha\,\mathrm{d}T \right) \tag{D.2}$$

$$\approx \varrho_0 (1 - \alpha T) \left(\frac{K_{T_0} + K_T' p}{K_{T_0}} \right)^{\frac{1}{K_T'}},$$

the latter being a simple approximation for T-independent α and the p dependence of K_T (also see eq. II.5); for higher precision, one can use the exponential form for α, which is derived from eq. D.3, and an equation of state like eq. II.3c.

Partial melting induces substantial changes in the modal and chemical composition of mantle rock, thereby causing density variations which are a function of the melting degree. At least in the uppermost mantle, the residue has a lower density than the undepleted material, because the Fe components and the dense garnet phase enter the melt particularly easily, so that the remaining rock is up to 2.5 % less dense than the fertile; the subtraction of 20 mol% olivine basalt from a pyrolite *e.g.* produces a residue with a density reduced by 1.6 %. Similar relations are expected for spinel lherzolite (Jordan, 1979).

D.1.3 Thermal expansivity

The thermal expansion coefficient is defined as

$$\alpha := \frac{1}{V} \left(\frac{\partial V}{\partial T} \right)_p, \tag{D.3}$$

i.e. it describes the volume change with temperature at constant pressure. In their compilation of thermodynamic data, Saxena *et al.* (1993) give the following formula for fitting experimental data on the thermal expansivities of several chemically pure minerals at normal pressure:

$$\alpha = c_0 + c_1 T + \frac{c_2}{T} + \frac{c_3}{T^2}, \tag{D.4}$$

		$\varrho \left(10^3 \text{ kg}/\text{m}^3\right)$	m_{mol}	$V \left(10^{-6} \text{ m}^3/\text{mol}\right)$
forsterite	Mg_2SiO_4			
α		3.227	140.71	43.603
β		3.4729	140.71	40.515
γ		3.563	140.71	39.493
fayalite	Fe_2SiO_4			
α		4.402	203.78	46.29
γ		4.848	203.78	42.03
orthoenstatite	$Mg_2Si_2O_6$	3.204	200.792	62.676
diopside	$CaMgSi_2O_6$	3.279	216.56	66.039
orthoferrosilite	$Fe_2Si_2O_6$	4.002	263.862	65.941
clinohedenbergite	$CaFeSi_2O_6$	3.656	248.095	67.867
plagioclase				
anorthite	$CaAl_2Si_2O_8$	2.765	278.36	100.61
albite	$NaAlSi_3O_8$ (high)	2.61	262.23	100.452
spinel	$MgAl_2O_4$	3.578	142.27	39.762
hercynite	$FeAl_2O_4$	4.256	173.81	40.843
picrochromite	$MgCr_2O_4$	4.414	192.3	43.564
garnet				
pyrope	$Mg_3Al_2Si_3O_{12}$	3.565	403.15	113.08
almandine	$Fe_3Al_2Si_3O_{12}$	4.312	497.76	115.43
grossular	$Ca_3Al_2Si_3O_{12}$	3.6	450.45	125.12
majorite	$MgSiO_3$	3.513	100.4	28.581

Table D.2: Chemical compositions, densities, atomic weights and molar volumes of several mantle minerals at normal conditions (Smyth and McCormick, 1995, modified and corrected).

	$\varrho \left(10^3 \text{ kg}/\text{m}^3\right)$	reference
plagioclase lherzolite	3280	Cella and Rapolla (1997)
spinel lherzolite	3345	Oxburgh and Parmentier (1977)
	3370	Cella and Rapolla (1997)
garnet lherzolite	3360	Oxburgh and Parmentier (1977)
	3400	Cella and Rapolla (1997)
harzburgite/dunite	3295	Oxburgh and Parmentier (1977)

Table D.3: Average densities of different mantle rocks.

	$c_0/10^{-5}$	$c_1/10^{-9}$	$c_2/10^{-3}$	c_3
forsterite				
α	2.01	13.9	1.627	-0.338
β (wadsleyite)	2.319	9.04	-3.966	0.7496
γ (ringwoodite)	1.225	11.04	2.496	-0.511
	2.214^a	6.48^a	–	–
fayalite				
α	5.6728	1.6315	-2.5186	-1.61331
γ (ringwoodite)	8.897	2.803	-34.26	3.813
orthoenstatite	3.871	4.4633	0.34352	-1.7278
clinodiopside	2.51	9.821	5.191	-2.365
orthoferrosilite	4.95565	8.7617	-11.856	1.56019
clinohedenbergite	8.82146	3.7169	-38.431	5.55973
plagioclase				
anorthite	2.34041	12.1057	5.20652	-2.19267
albite	2.52098	8.3239	3.10442	-1.89468
spinel				
Mg spinel	6.96922	-1.08069	-30.7991	5.03946
hercynite	4.89524	8.62	-20.549	3.45353
picrochromite^b	1.43	11.191	–	-0.1063
garnet				
pyrope	0.991^c	11.65	10.624	-2.5
almandine	3.0836	6.659	-6.106	0.6453
grossular	2.529	6.97	2.724	-1.4794
majorite^a	2.45925	5.21	–	–

^{a}after Chopelas (2000)
^{b}Fei (1995)
^{c}corrected; original value apparently too high by factor 10

Table D.4: Thermal expansivity coefficients for eq. D.4 for several mantle minerals (Saxena *et al.*, 1993). The additional $\alpha_{\gamma\text{-ol}}$ value by Chopelas (2000) is higher than the fit from Saxena *et al.* (1993) and possibly rather applicable to higher T.

where T is in K; the values of the coefficients $c_{0...3}$ for several important upper-mantle minerals are listed in table D.4.

The mixing relation for solid solutions can be derived by plugging eq. D.1a into eq. D.3:

$$\alpha_{ss} = \frac{1}{\sum \underline{X}_i V_i} \left(\frac{\partial}{\partial T} \right)_p \left(\sum \underline{X}_i V_i \right) = \frac{\sum \underline{X}_i V_i \alpha_i}{V_{ss}}, \qquad (D.5a)$$

where the sums go over the end-members (see Ita and Stixrude, 1992). Similarly, for mineral composites follows from eqs. D.1b and D.3 that

$$\alpha_{mc}^R = \frac{1}{m_{mol} \sum X_i \frac{V_i}{m_{mol_i}}} \left(\frac{\partial}{\partial T} \right)_p \left(m_{mol} \sum X_i \frac{V_i}{m_{mol_i}} \right) = \varrho_{mc} \sum \frac{X_i}{\varrho_i} \alpha_i = \sum \tilde{X}_i \alpha_i \qquad (D.5b)$$

with the sum being over constituents; this is the Reuss approximation of the thermal expansivity (also see *e.g.* Stacey, 1998), which is based on the assumption of equal stresses on all composite components and is valid for complete intergranular elastic relaxation. In contrast, the Voigt limit is valid for the (physically unrealistic) case of equal strains and the corresponding existence of intergranular stresses:

$$\alpha_{mc}^{V} = \frac{1}{\varrho_{mc} \sum \frac{X_i}{\varrho_i \alpha_i}} = \frac{1}{\sum \frac{\tilde{X}_i}{\alpha_i}}. \tag{D.5c}$$

Different approaches for the calculation of thermodynamic variables of polycrystalline composites can be based on lattice dynamics using spatially averaged values of lattice quantities (*e.g.* Hama and Suito, 2001).

D.1.4   *Bulk modulus*

The isothermal bulk modulus K_T is the inverse of the isothermal compressibility

$$\beta := -\frac{1}{V}\left(\frac{\partial V}{\partial p}\right)_T, \tag{D.6}$$

and can be calculated from the fitting function

$$K_T = \frac{1}{c_0 + c_1 T + c_2 T^2 + c_3 T^3} \tag{D.7}$$

by Saxena *et al.* (1993). The adiabatic bulk modulus K_S can be related to K_T by

$$K_S = K_T\left(1 + K_T T V \frac{\alpha^2}{c_V}\right) = K_T\left(1 + \alpha\gamma T\right) = K_T\frac{c_p}{c_V}, \tag{D.8}$$

using eq. III.27. The data from table D.5 suggest that the pressure derivative of K_T is but weakly temperature-dependent in the materials of interest; while under normal conditions K_T' are quite similar to K_S', the difference is significant at high T (Pankov *et al.*, 1998).

The bulk modulus of a solid solution and a mineral assemblage can be determined with an approach similar to eqs. D.5 by using the definition of the compressibility eq. D.6 instead of eq. D.3. This yields for the bulk modulus of a solid solution

$$\beta_{Tss} = -\frac{1}{\sum \underline{X}_i V_i}\left(\frac{\partial}{\partial p}\right)_T\left(\sum \underline{X}_i V_i\right) = \frac{\sum \underline{X}_i V_i \beta_i}{V_{ss}} \Rightarrow K_{Tss} = \frac{V_{ss}}{\sum \frac{\underline{X}_i V_i}{K_{T_i}}}, \tag{D.9a}$$

(see Ita and Stixrude, 1992) and the Reuss approximation for a mineral composite

$$K_{mc}^{R} = \frac{1}{\varrho_{mc} \sum \frac{X_i}{\varrho_i K_i}} = \frac{1}{\sum \frac{\tilde{X}_i}{K_i}} \tag{D.9b}$$

(see Stacey, 1998; Stacey and Isaak, 2001), which is valid for the *per definitionem* thermally relaxed K_T or the thermally unrelaxed K_S. A totally or partially unrelaxed K_S appears in fast processes without full elastic and thermal equilibration such as the transition of seismic waves, whereas the thermally relaxed K_S is applicable to slow deformations and can be calculated from K_T with eq. D.8 (Stacey, 1998). The Voigt approximation of the composite modulus is given by

$$K_{mc}^{V} = \varrho_{mc} \sum \frac{X_i}{\varrho_i} K_i = \sum \tilde{X}_i K_i \tag{D.9c}$$

	$c_0/10^{-12}$	$c_1/10^{-15}$	$c_2/10^{-19}$	$c_3/10^{-22}$	K_T'	$\frac{\partial K_T'}{\partial T}$ $(10^{-4}/\mathrm{K})$	ref.[a]
forsterite							
α	7.427	1.24	0.69	1.702	5.2	2	1–3
β (wadsleyite)	5.51282	0.92017	0.8849	1.1529	4.3	3	3
γ (ringwoodite)	5.07778	1.3371	−2.9854	2.7822	5	6	2,3
fayalite							
α	6.63158	2.0846	−1.0135	1.1668	5.2	2.2	4,3
γ (ringwoodite)	3.86	3.978	−6.304	0.8275	4.1	10	4,3
orthoenstatite	8.892	1.3584	3.1613	1.14126	8.6	1.5	5,3
clinodiopside	9.082	−0.8384	11.65	0.2291	4.5	5	4,3
orthoferrosilite	9.27714	3.70379	−10.1	2.9097	4.2	0	3
clinohedenbergite	8.58338	1.13695	8.31098	0.26564	4.2	0	3
plagioclase							
anorthite	9.33561	2.17769	0.78758	2.5149	4	0	1,3
albite	14.0528	2.07138	2.4617	1.52582	4	0	1,3
spinel							
Mg spinel	4.80091	0.489213	5.9812	−0.37555	4.89	9	3
hercynite	4.59048	0.85524	−0.14274	0.93164	4	7.8	3
picrochromite[b]	5.26014	0.53601	6.5533	−0.41147	4.89	9	–
garnet							
pyrope	5.546	0.806	−0.46	1.567	4.8	0	3
almandine	4.8076	3.0405	0.1635	3.9961	5.6	0	3
grossular	5.972	−0.184	7.49	0.647	4.5	0	4,3
majorite	5.5	2.52	−11	5.58636	4.5	4	2,3

[a]Data sources: 1: from Knittle (1995), 2: Chopelas (2000), 3: Saxena *et al.* (1993), 4: from Ita and Stixrude (1992), 5: Angel and Jackson (2001)

[b]coefficients calculated from Mg spinel using $K_{\mathrm{pcr}} = K_{\mathrm{msp}}V_{\mathrm{msp}}/V_{\mathrm{pcr}}$ (Ita and Stixrude, 1992); derivatives assumed to be identical with those of Mg spinel

Table D.5: Isothermal bulk modulus coefficients for eq. D.7 (Saxena *et al.*, 1993) and derivatives after different authors.

(*e.g.* Stacey, 1998) and is valid for K_T and the thermally unrelaxed K_S; again, the thermally relaxed K_S can be determined by eq. D.8. However, Stacey (1998) came to the conclusion that thermal relaxation is of minor importance for the value of K_S, and for the task of calculating K_S for an assessment of seismic velocities, it will be sufficient to assume thermal, but not elastic relaxation and to use the Reuss–Voigt–Hill average:

$$K_{S,\mathrm{mc}}^{\mathrm{H}} = \frac{1}{2}\left(K_{S,\mathrm{mc}}^{\mathrm{R}} + K_{S,\mathrm{mc}}^{\mathrm{V}}\right). \tag{D.10}$$

In many cases, the Reuss–Voigt–Hill average lies within the Hashin–Shtrikman bounds for polycrystals, which are very tight unless crystal anisotropy becomes very large (Karki *et al.*, 2001).

D.1.5 Heat capacity

For the heat capacity at constant pressure, Saxena *et al.* (1993) put forth the fitting rule

$$c_p = c_0 + c_1 T + \frac{c_2}{T^2} + c_3 T^2 + \frac{c_4}{T^3} + \frac{c_5}{T} \tag{D.11}$$

with T in K; the original formula has an additional term with a $\sqrt{1/T}$-dependence which is not needed here. The heat capacity of a substance can be calculated approximately by adding the heat capacities of its components according to their molar fractions (Putnis, 1992; Ita and Stixrude, 1992).

	$c_0/10^2$	$c_1/10^{-2}$	$c_2/10^6$	$c_3/10^{-6}$	$c_4/10^8$	$c_5/10^4$
forsterite						
α	1.658	1.855	−3.971	0	2.861	−0.561
β (wadsleyite)	1.729	1.129	−1.0771	0	−2.1875	−1.3477
γ (ringwoodite)	1.5856	1.22	−12.297	0	14.841	0.79719
fayalite						
α	2.089	0.8064	5.894	0	−7.882	−3.466
γ (ringwoodite)	1.6786	2.8124	−5.6547	0	7.8238	−0.35644
orthoenstatite	1.4445	0.1882	−1.35	0	4.612	−1.938
clinodiopside	1.21665	0.814	−6.8095	0	9.384	0.012772
orthoferrosilite	1.096	1.388	−9.116	0	10.33	1.098
clinohedenbergite	1.5465	0.7025	15.395	0	−23.93	−4.661
plagioclase						
anorthite	2.909	2.76	−34.08	0	52.18	2.9625
albite	3.0974	1.527	−26.16	0	41.092	0.88361
spinel	2.0785	0.449	6.7109	0	−9.5984	−3.9681
hercynite	1.7572	2.22	8.1615	0	−17.335	−2.4517
garnet						
pyrope	4.768	3.167	−21.2	0	21.68	−0.117
almandine	7.13	−8.85	14.58	15.6	0	−14.96
grossular	5.426	1.294	−3.186	0	2.777	−5.602
majorite	1.275753	0.74603	−1.106	0	19.762	0

Table D.6: Coefficients of heat capacity at constant pressure for eq. D.11 for several mantle minerals (Saxena *et al.*, 1993); c_p is in J/mol K.

D.1.6 Phase transformations of the principal Al-bearing phase

Depending on the p–T conditions, most of the Al is bound in either plagioclase, spinel, or garnet, although pyroxenes can also absorb Al well under higher pressures (*e.g.* Longhi, 1995) and contain some smaller aluminous endmembers like jadeite (NaAlSi$_2$O$_6$) anyway. The respective stability fields are separated by rather sharp boundaries, which are also represented by thin linear transition intervals in the discrete model. The transition is not step-like, however, but does also extend over a certain interval in real systems as well, the width of the gap being determined by the composition: *e.g.*, for the plagioclase–spinel transition, the gap is about $0.5\,\text{GPa}$ wide in CMASN, but only $0.02\,\text{GPa}$ wide in CMASF (Presnall *et al.*, 2002), and is thus possibly dependent on the degree of depletion. For the plagioclase–spinel boundary, where reactions like

$$\underset{\text{anorthite}}{\text{CaAl}_2\text{Si}_2\text{O}_8} + \underset{\text{forsterite}}{2\,\text{Mg}_2\text{SiO}_4} \rightleftharpoons \underset{\text{spinel}}{\text{MgAl}_2\text{O}_4} + \underset{\text{diopside}}{\text{CaMgSi}_2\text{O}_6} + \underset{\text{enstatite}}{2\,\text{MgSiO}_3} \tag{D.12}$$

(Green and Falloon, 1998) take place, it seems to be justified to approximate the Clapeyron curve by a linear function; however, in the case of the spinel–garnet boundary, for which the reaction pair

$$\underset{\text{spinel}}{\text{MgAl}_2\text{O}_4} + \underset{\text{enstatite}}{n\,\text{Mg}_2\text{Si}_2\text{O}_6} \rightleftharpoons \underset{\text{forsterite}}{\text{Mg}_2\text{SiO}_4} + \underset{\text{aluminous enstatite}}{(n-1)\,\text{Mg}_2\text{Si}_2\text{O}_4 \cdot \text{MgAl}_2\text{SiO}_6} \tag{D.13a}$$

$$\underset{\text{spinel}}{\text{MgAl}_2\text{O}_4} + \underset{\text{enstatite}}{1.25\,\text{Mg}_2\text{Si}_2\text{O}_6} + \underset{\text{diopside}}{0.75\,\text{CaMgSi}_2\text{O}_6} \rightleftharpoons$$

$$\underset{\text{forsterite}}{\text{Mg}_2\text{SiO}_4} + \underset{\text{pyrope}}{2.25\,\text{Mg}_3\text{Al}_2\text{Si}_3\text{O}_{12}} + \underset{\text{grossular}}{0.75\,\text{Ca}_3\text{Al}_2\text{Si}_3\text{O}_{12}} \tag{D.13b}$$

(Green and Falloon, 1998; Johnson *et al.*, 1990), or, more comprehensively,

$$0.17\text{ol} + \text{gt} \rightleftharpoons 0.53\text{opx} + 0.47\text{cpx} + 0.17\text{sp} \tag{D.13c}$$

(Hauri and Hart, 1994), is characteristic, the curvature is stronger (Hall, 1996), so that a different parameterization should be more appropriate. Although a fitting procedure like the one used for the solidus and liquidus data (sections C.1–C.3.2) did yield an acceptable fit of data in the p–T range from $1.5\ldots2.3\,\text{GPa}$ and $850\ldots1350\,°\text{C}$ by O'Hara *et al.* (1971) for similar trial functions, the function

$$T_{\text{sp–gt}}(p) = 320\ln(p-1.5) + 10p + 1400 \tag{D.14}$$

with p in GPa and T in °C which was constructed by eye-fitting, was found to reproduce the curve drawn by these authors even better, while it does not have a much larger mean error; for $p = 1.6\ldots2.3\,\text{GPa}$ it yields values of $\text{d}p/\text{d}T = 0.312\ldots2.44\,\text{MPa/K}$ which coincides well with independent estimates (*e.g.* Schubert *et al.*, 2001; Iwamori *et al.*, 1995; Walter *et al.*, 2002). Apart from the p–T conditions, the modal and chemical composition resp. the melting degree is important: Robinson and Wood (1998), who fixed the transition at the solidus at about $(2.7\ldots2.8\,\text{GPa}, 1450\ldots1460\,°\text{C})$ for fertile MORB-parent lherzolite, assessed a shift towards higher pressures of $0.05\,\text{GPa/\%\,melt}$ for the sp–gt transition. On the other hand, the boundary for the CMAS system,

$$T_{\text{sp–gt}}(p) = -12.53 + 474.58\sqrt{4.86p - 1} \tag{D.14a}$$

(after Walter *et al.*, 2002), lies at lower p for a given T, *i.e.* at higher T at a given p, and crosses the solidus at $(2.51\,\text{GPa}, 1575\,°\text{C})$; this is almost exactly the same pressure at which the fit eq. D.14 would cross the solidus eq. II.25c by Hirschmann (2000), thus eqs. D.14 and D.14a are very similar in a *ca.* 80 K interval around the solidus if the latter is shifted by *ca.* 150 K downward. However, one general caveat should be borne in mind when considering all these high-p–T experiments: among others, Longhi (2002) and Walter *et al.* (2002) decry the large interlaboratory discrepancies in pressure determination, which are as high as several tenths of a gigapascal. – It is important to note that in a material other than peridotite such as pyroxenite, garnet is stable at pressures as low as 1.3–1.7 GPa (Hirschmann and Stolper, 1996).

One consequence of a phase transformation is a change in volume. In the most general case of a transition from n_i initial to n_f final phases, the volume change is

$$\Delta V = \sum_{k=1}^{n_i} V_{i_k} - \sum_{l=1}^{n_f} V_{f_l} = m \left(\sum_{k=1}^{n_i} \frac{\tilde{\varphi}_{i_k}}{\varrho_{i_k}} - \sum_{l=1}^{n_f} \frac{\tilde{\varphi}_{f_l}}{\varrho_{f_l}} \right), \qquad m = \sum_{k=1}^{n_i} m_{i_k} = \sum_{l=1}^{n_f} m_{f_l}; \quad \text{(D.15)}$$

for a monomineralic phase transition as the α, β, γ-olivine transitions, this simplifies to

$$\Delta V = m \frac{\varrho_f - \varrho_i}{\varrho_i \varrho_f}. \qquad \text{(D.15a)}$$

D.1.7   Thermal conductivity/diffusivity

The thermal conductivity $\varkappa$ is composed of a lattice term contributed by the phonons in the mineral crystal lattice and a radiative term caused by scattering of blackbody radiation. The lattice term dominates in the mantle; it increases with rising pressure and decreasing temperature, because at higher densities phonon frequencies and collision probabilities by which heat is transported are enhanced. In contrast, the radiative term is only temperature-dependent, but increases with temperature; compared with the lattice term it is of minor importance at upper-mantle temperatures, though (Ashcroft and Mermin, 1976; Hofmeister, 1999). To calculate the thermal conductivity under mantle conditions, Hofmeister (1999) derived a quantum-mechanical mechanical model of $\varkappa$ for the whole mantle; she proposed an average value of $\varkappa = 3\,\mathrm{W}/\mathrm{mK}$ for the upper mantle.

D.1.8   Seismic velocities and attenuation

The velocities of P- and S-waves are given by:

$$v_P = \sqrt{\frac{K_S + \frac{4}{3}\mu}{\varrho}}, \qquad v_S = \sqrt{\frac{\mu}{\varrho}}, \qquad \text{(D.16)}$$

whereby K_S, μ and ϱ are average properties in mineral aggregates like mantle rock. Hence, $v_{P,S}$ depend on composition; for instance, the seismic velocities of a garnet lherzolite become larger with increasing garnet content and decreasing Fe content. With respect to melting, it is particularly important that the subtraction of basalt from a fertile mantle rock leads to a net increase of v_P and v_S by 0.005 and 0.025 %/mol% ol.bas. (Jordan, 1979).

As already pointed out in appendices D.1.3 and D.1.4, the density and the elastic moduli of the rocks are temperature-dependent, so the same will hold for seismic velocities apart from the dependence on melt fraction discussed in section II.2.2. Two effects are important in this respect (Karato, 1993; Karato and Jung, 1998):

Anharmonicity. Anharmonic behaviour is frequency-independent and not dissipative; it is due to the deviation of a real material from the harmonic approximation made in the equation of motion for atoms in the crystal lattice.

Anelasticity. This property is frequency-dependent and causes attenuation and dispersion of
seismic waves, especially shear waves. It is significant at seismic frequencies ($\omega < 1\,\text{Hz}$)
and is likely to have a considerable effect at high temperatures. Appropriate inclusion
of anelasticity in the interpretation of seismic tomography with respect to temperature
is thus expected to result in lower T anomalies than would be achieved by taking
into account anharmonicity only. Apart from this, an increasing water content also
enhances anelastic damping of seismic waves and reduces their velocities.

For a weak dependence of type $Q^{-1} \sim \omega^{-\mathfrak{Q}}$ with $\mathfrak{Q} = 0.1 \ldots 0.3$ at seismic and sub-seismic ω,
which is of the same order of magnitude as the effect of melt measured by Gribb and Cooper
(2000) (see section II.2.2), the temperature dependence of the seismic velocity is

$$v_{\text{P,S}}(\omega, T) = v_{\text{P,S}_0}(T) \left[1 - \frac{1}{2} \cot \left(\frac{\pi \mathfrak{Q}}{2} \right) Q^{-1}(\omega, T) \right] \tag{D.17a}$$

$$\left(\frac{\partial \ln v_{\text{P,S}}}{\partial T} \right)_p = -\alpha \left(\frac{\partial \ln v_{\text{P,S}_0}}{\partial \ln \varrho} \right)_p - \frac{\pi \mathfrak{Q}}{2} \cot \left(\frac{\pi \mathfrak{Q}}{2} \right) \frac{Q^{-1}(\omega, T)}{\pi} \frac{\mathcal{E} + p \mathcal{V}}{RT^2}, \qquad (Q^{-1} \ll 1) \tag{D.17b}$$

(Karato, 1993). The first term in eq. D.17b is the anharmonic contribution to the temperature derivative; the most important parameter in the second, anelastic term is the activation
energy $\mathcal{E}$ (see section II.2.1). Reported values for Q_{S} in the mantle range from 80 in the
upper mantle low-velocity zone over 300 at the CMB to 600 in the lithosphere (Karki *et al.*,
2001). Combining the effects of temperature and water content on attenuation, Karato and
Jung (1998) propose a relation of the form

$$Q^{-1} = \left(\frac{B_{\text{dry}} + B_{\text{wet}} C_{\text{OH}}}{\omega} \right)^{\mathfrak{Q}} \exp \left(-\frac{\mathfrak{Q} c_Q T_{\text{s}}}{T} \right), \tag{D.18}$$

where B_{dry}, B_{wet}, and c_Q are constants; quantitative experimental constraints on the effect
are scarce, though.

It is of interest to consider the properties of the anharmonic term in eq. D.17b. By
differentiating eqs. D.16, one arrives at

$$\frac{\partial \ln v}{\partial \ln \varrho} = \frac{\varrho}{v} \frac{\partial v}{\partial \varrho} = -\frac{1}{2} \left(1 - \frac{\partial \ln \mathfrak{M}}{\partial \ln \varrho} \right), \tag{D.19a}$$

where $\mathfrak{M}$ is either $K_S + {}^{4\mu}/_3$ for the P-wave velocity or μ for the S-wave velocity. The RHS
can be transformed using the definition of α, eq. D.3, to yield the temperature derivative at
constant pressure:

$$\frac{\partial \ln \mathfrak{M}}{\partial \ln \varrho} = \frac{\varrho}{\mathfrak{M}} \frac{\partial \mathfrak{M}}{\partial \varrho} = \frac{\varrho}{\mathfrak{M}} \frac{\partial \mathfrak{M}}{\partial T} \frac{\partial T}{\partial \varrho} = -\frac{1}{\alpha \mathfrak{M}} \frac{\partial \mathfrak{M}}{\partial T}; \tag{D.19b}$$

For upper-mantle rock, $(\partial \ln v_{\text{P}} / \partial \ln \varrho)_p = 2.1$ and $(\partial \ln v_{\text{S}} / \partial \ln \varrho)_p = 2.5$ (Christensen, 1989). The
complete anharmonic term now reads

$$\left(\frac{\partial \ln v}{\partial T} \right)_{p,\text{anh}} = -\alpha \left(\frac{\partial \ln v}{\partial \ln \varrho} \right)_p = \frac{\alpha}{2} \left[1 + \frac{1}{\alpha \mathfrak{M}} \left(\frac{\partial \mathfrak{M}}{\partial T} \right)_p \right] = \frac{1}{2} \left[\alpha + \left(\frac{\partial \ln \mathfrak{M}}{\partial T} \right)_p \right] \tag{D.19c}$$

This expression highlights the important role of the thermal expansivity in this matter. As α
is smaller for the minerals in the deeper mantle under the corresponding p–T conditions (*e.g.*
Anderson, 1995), the variation of $v_{\text{P,S}}$ with T is also smaller, especially in the lower mantle.
Hence, it is to be expected that a thermal anomaly such as a plume is less well visible in
seismic tomography in the lower mantle than in the upper, in agreement with observations.

D.1.9 *Electrical conductivity*

The electrical conductivity of minerals, σ, depends on different thermally activated mechanisms and is also a function of pressure, although this relation is relatively weak in comparison with the T effect; it seems to rise from 10^{-2} to about $1\,\mathrm{S/m}$ throughout the upper $1000\,\mathrm{km}$ of the earth (Xu *et al.*, 2000). These authors found it convenient to express these relations by

$$\sigma(p, T) = \sigma_0 e^{-\frac{\mathcal{E}+p\mathcal{V}}{k_\mathrm{B} T}}. \tag{D.20}$$

	p (GPa)	T (°C)	σ_0 (S/m)	$\mathcal{E}$ (eV)	$\mathcal{V}$ (cm³/mol)
α-olivine (fo$_{90}$)	4–10	1000–1400	490	1.62	0.68
orthopyroxene (en$_{91}$)	5	1000–1400	5248	1.8	–
clinopyroxene	13	1000–1400	1778	1.87	–

Table D.7: Activation energies, activation volumes and reference conductivities for eq. D.20 (Xu *et al.*, 2000). Except for α-olivine, the pressure dependence is neglected.

D.2 MELT

D.2.1 *Density*

Lange and Carmichael (1990) have compiled molar volume data along with thermal expansivities and isothermal compressibilities for normal pressure and $T = 1673\,\mathrm{K}$ for several common oxides and proposed the formula

$$V(p, T, X_i) = \sum X_i \left[V_i(1673\,\mathrm{K}) + \frac{\partial V_i}{\partial T}(T - 1673\,\mathrm{K}) + \frac{\partial V_i}{\partial p}(p - 0.1\,\mathrm{MPa}) \right]$$

$$= \sum X_i V_i(1673\,\mathrm{K}) \left[1 + \alpha_i(T - 1673\,\mathrm{K}) + K_{T_i}(p - 0.1\,\mathrm{MPa}) \right] \tag{D.21}$$

to calculate the molar volume of a natural melt from its oxide components; here, the ferric–ferrous ratio of iron needs special treatment. For use with the melt-dynamical program, the assumption of a constant, given composition and direct use of the respective density measurements is though more convenient.

D.2.2 *Thermal expansivity*

The thermal expansivity of melts is larger than that of solids and seems to show some T dependence in some liquids, in particular near the glass transition temperature (Courtial *et al.*, 1997; Gottsmann and Dingwell, 2000, 2002). For instance, Toplis and Richet (2000) suggest the fitting formula

$$V = c_0 + c_1 \ln T \Rightarrow \alpha = \frac{c_1}{T(c_0 + c_1 \ln T)} \tag{D.22}$$

with T in K; Gottsmann and Dingwell (2002) measured $c_0 = 0.1151$ and $c_1 = 0.0364$ for diopside melt, $c_0 = 0.1893$ and $c_1 = 0.0267$ for an$_{42}$di$_{58}$ melt, and $c_0 = 0.2586$ and $c_1 = 0.0176$ for an$_{98}$di$_2$ melt (fits for T in °C). Both observed a decrease of T dependence with increasing

	ϱ $(10^3\,\text{kg}/\text{m}^3)$	m_{mol}	V $(10^{-6}\,\text{m}^3/\text{mol})$	ref.[a]
forsterite	2.786	140.71	50.506	1
fayalite	3.605	203.78	56.527	1
	3.536		57.63	2
enstatite	2.588	200.792	77.586	1
diopside	2.471	216.56	87.641	1
pyrope	2.488	403.15	162.038	1
almandine	3.032	497.76	164.169	1
basic liquid ($\text{di}_{35}\text{hd}_{15}\text{py}_{35}\text{al}_{15}$)	2.61	328.777	125.968	1
ultrabasic liquid				
($\text{di}_{16}\text{hd}_4\text{py}_{16}\text{al}_4\text{fo}_{36}\text{fa}_9\text{en}_{12}\text{fs}_3$)	2.795	229.995	82.288	1
PHN 1611 liquid	2.663			2

[a]Data sources: 1: Ohtani (1984), 2: after Courtial *et al.* (1997)

Table D.8: Density, atomic weight and molar volume of liquids of several mantle minerals at 1 atm and 2000 °C. The data of Ohtani (1984) were calculated from the 3rd-order Birch–Murnaghan equation of state. The fayalite and PHN 1611 values after Courtial *et al.* (1997) have been calculated from their linear fitting formulae $\varrho_{\text{fa}} = 3.9906 - 1.998 \cdot 10^{-4}T$ and $\varrho_{\text{PHN}} = 3.0508 - 1.706 \cdot 10^{-4}T$, with T in K.

anorthite content in these melts. Toplis and Richet (2000) remark that the T-dependent form of α of oxide components can be used straightforwardly in eq. D.21 to calculate the density of a melt. Given the relatively minor importance of thermal expansivity of the melt in the T range of the melting region compared with the compressibility effect, it is though acceptable to neglect T dependence and assume an average α.

	α $(^{10^{-5}}/\text{K})$	ref.[1]
forsterite	9.5	1
fayalite	8.7	1
	7.23[2]	2
	5.6[3]	3
enstatite	5.47	4
	6.4	1
diopside	7.29	4
	9.6	1
PHN 1611 liquid	6.3[3]	3

Table D.9: Thermal expansivity of liquids of several mantle minerals at normal pressure. The fayalite and PHN 1611 values after Courtial *et al.* (1997) have been calculated from their linear fitting formulae $\alpha_{\text{fa}} = 6.185 \cdot 10^{-5} - 3.1 \cdot 10^{-9}T$ and $\alpha_{\text{PHN}} = 7.098 \cdot 10^{-5} - 4 \cdot 10^{-9}T$, with α in $1/\text{K}$ and T in K.

D.2.3 Bulk modulus

Rigden *et al.* (1989) estimated K'_T for several silicate melts and arrive at $K'_T = 10.2$ for forsterite, which is much higher than the values listed in table D.10; for other melts, their estimates agree better with other values from the literature. A general problem with their shock-wave measurements, however, is that their K'_T actually integrate different mechanisms related to melt structure, but prevailing at largely different pressures, into one single value (Lange and Carmichael, 1990). It should also be noted that their synthetic melts do not contain iron.

	K_T (GPa)	K'_T	T (K)	ref.[4]
forsterite	58.815	3.75		1[5]
	61.2	3.3	2273	2[5]
fayalite	34.5	6.95	2273	2[5]
	24.4	10.1	1773	3
	19.8	10.1	1773	4
	13.1	21.5	1773	4
enstatite	21.3	6.4		5
	20.555	9.79		1[5]
	19.6	5.1	2273	2[5]
diopside	23.3	6.4		5
	24.218	5.31		1[5]
	13.8	6.6	2273	2[5]
	23.8	7.1	1773	6
anorthite	20.5	6.9	1923	6
albite	42.502	20.86		1[5]
pyrope	11.905	23.33		1[5]
basic liquid	13.8	6.6	2273	2[5]
basaltic liquid	22.3	7	1673	6
MORB liquid	19.3	4.4	1673	7
ultrabasic liquid	26.5	5.85	2273	2[5]
PHN 1611 liquid	32.45	4	2633	4[6]

Table D.10: Isothermal bulk moduli and pressure derivatives for liquids of several mantle minerals at normal pressure. The basic and the ultrabasic liquid are as in table D.8, the basaltic liquid is $di_{64}an_{36}$.

D.2.4   Heat capacity and melting enthalpy

	T_1 (K)	$c_p(T_f)$ ($^J/_{gK}$)	ΔL_m ($^J/_g$)	ΔS_m ($^J/_{gK}$)	ref.[a]
forsterite	2174	–	1009.17	0.464	1
	2163	1.905	810.18	0.375	2,3
	2163	–	730.58	0.338	4
fayalite (α)	1490	–	452.32	0.304	5
enstatite	1840	0.847	769.8	0.418	6
diopside	1665	1.537	639.55	0.383	6–8
plagioclase					
anorthite	1830	$1.44 + 7.272 \cdot 10^{-5}T$	481.39	0.263	7,6,2
albite	1373 (1393)	1.407^b	245.2	0.179	7,6
spinel	2408	–	752.09	0.312	1,5
garnet (pyrope)	1500	1.684	603	0.402	7,3
	1583		703.9	0.445	9
di-an system	1633^c	–	546.796	0.335	10
di-fo-an system	1543^c	1.96	506	0.328	11
		–	507.02	0.329	10
natural basalt	1533	–	676.17	0.441	10
peridotite (CMASNF)^d					
at 1.1 GPa	1523	1.68	590	0.387	12
at 3 GPa	1773	1.84	692	0.39	12
at 4 GPa	1923	1.51	807	0.42	12

[a]Data sources: 1: Richet *et al.* (1993), 2: Navrotsky (1994), 3: Navrotsky (1995), 4: Tangeman *et al.* (2001), 5: Saxena *et al.* (1993), 6: DeYoreo *et al.* (1995), 7: from Richet and Bottinga (1986), 8: Ziegler and Navrotsky (1986), 9: van Westrenen *et al.* (2001), 10: Fukuyama (1985), 11: Kojitani and Akaogi (1995), 12: Kojitani and Akaogi (1997)

[b]low albite

[c]eutectic temperature

[d]c_p values from experimental CMAS data

Table D.11: Heat capacities, melting enthalpies/entropies and melting points (at normal pressure, unless stated otherwise) for liquids of several mantle minerals. The forsterite data by Navrotsky (1994) are calculated, while those by Richet *et al.* (1993) are measured experimentally and according to Kojitani and Akaogi (1997) are to be preferred. The CMASNF peridotite values were calculated from 1 atm measurements on CMAS primary melt compositions for the respective pressures.

Determination of Thermoelastic Properties
of the Mantle

This appendix presents an application of the method for the determination of thermoelastic properties of the mantle developed in section III.4.4, which had also served for the definition of models VA3 and VA4 in section V.1.3.

The method can be used independently from any convection problem to construct depth profiles or maps of certain thermodynamical parameters in p–T space for a certain composition. As a tentative example, figure E.1 shows depth profiles of $v_{P,S}$, $K_{T,S}$ and ϱ for an estimated mantle adiabat with $\mathcal{T} = 1382\,°C$ (without a thermal boundary layer at the top), and figure E.2 shows such maps of α, $K_{T,S}$, c_p, γ, ϱ and v_P for a simplified, yet quite complex upper mantle composition, but without consideration of melting; therefore, the supersolidus values of these parameters in the maps cannot be considered reliable. The temperature range of the maps is restricted to values above $600\,°C$ because the current implementation forces the composition to be plagioclase lherzolite (with α-olivine) independent of p and would not give meaningful results in the high-p/low-T domain. The reference composition is a garnet lherzolite according to the assessment of Kostopoulos (1991) and Kostopoulos and James (1992), which seems to be a good average of several estimates from the literature (see appendix D.1.1, table D.1). All minerals were assumed to be ideally mixed solid solutions: olivine and its polymorphs of forsterite and fayalite (without a ficticious β-fayalite), orthopyroxene of orthoenstatite and orthoferrosilite, clinopyroxene of clinodiopside and clinohedenbergite, plagioclase of anorthite and albite, spinel of Mg spinel, hercynite and picrochromite, garnet of pyrope, almandine and grossular, and garnet majorite of aluminous garnet and $MgSiO_3$ garnet; solution of opx in cpx was not accounted for. No consideration was made of chemical species conservation, as is obvious *e.g.* from the fact that Na is only present in plagioclase. Further simplifications are a constant molar Mg:Fe ratio for ol, opx, cpx and sp of 9:1, a ratio of 7:1.5:1.5 of Mg:Fe:Ca in garnet, Cr#=0.091 for spinel, and a ratio of 9:1 of Ca:Na in plagioclase, which should correspond roughly to natural compositions. Phase changes are assumed to be linear, whereby the strongest simplification is made between the $410\,km$ and the $520\,km$ boundary: both the transitions from olivine to ringwoodite via wadsleyite and from the pyroxene–garnet assemblage to garnet majorite are assumed to be distributed over the whole interval, which is not the case in reality for the olivine transformations; the relatively small amount of Ca-perovskite $CaSiO_3$ existing below *ca.* $510\,km$ has also been neglected. The changes in modal composition in the pl–sp and sp–gt transitions are tuned to match the aforementioned compositions from table D.1; for the deeper phase changes it was assumed that the olivine system and the pyroxene–garnet system are independent, so that the modal fraction of olivine resp. its high-p polymorph is the same. Data used and further details of the phase changes are given in appendix D.1.

In figure E.1 the depth profiles calculated from the thermoelastic parameters are compared with the isotropic version of PREM (Dziewonski and Anderson, 1981; Masters and Shearer, 1995) and a parameterized version of IASP91 (Kennett and Engdahl, 1991; Kennett, 1995); the IASP91 values were calculated for the depths tabulated in table 1 of Masters and Shearer (1995). The agreement of the seismic velocities and the elastic moduli between calculations and earth models is acceptable for $p \gtrsim 4\,GPa$ resp. $z \gtrsim 120\,km$, whereby the

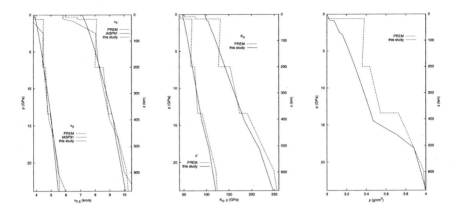

Figure E.1: Pressure resp. depth profiles of $v_{P,S}$ (in km/s), $K_{T,S}$ (in GPa) and ϱ (in g/cm³), calculated with the method described in section III.4.4 for an approximate mantle adiabat, and from the PREM and IASP91 models; the p and z ranges are based on the tabulated values for PREM. K_S is the Voigt–Reuss–Hill average applicable to seismic velocities, and μ was calculated from it using a constant Poisson ratio of 0.29.

two seismological models also differ notably in the uppermost mantle. At shallower depths, too low values are calculated, though, but this can probably be attributed to the fact that the mantle in the thermoelastic calculation was hot and oceanic, while the seismological models are expected to show a strong influence of cool, thick continental lithosphere, so that the upper 100 km are probably not directly comparable; interestingly, the average S-wave velocity observed in the upper 210 km beneath Iceland (Allen *et al.*, 2002a) is quite similar to the v_S profile calculated here. The low-velocity zone in PREM cannot be expected to be reproduced by the thermoelastic model, because melting or the influence of water, which are the most likely reasons for it, have not been taken into account; besides, it is also absent in IASP91. The differences in the wadsleyite stability zone are due to the simplifying treatment of the phase transition in the calculation. The choice of a constant Poisson number of 0.29 for calculating μ and v_S was apparently appropriate. – The agreement between the densities is less satisfying except for the deep upper mantle, and possibly due to compositional uncertainty; the PREM density profile is probably reasonably reliable, as it is a direct result of the joint inversion of real data.

The maps in figure E.2 show significant variations in several parameters which are commonly assumed constant in convection models. With respect to α, it confirms that the T dependence is strongest at low p and that the T effect would thus be strongest in or near the lithosphere, whereas on a larger vertical scale the pressure dependence is the most important aspect; note the significant gradient in α and the other parameters through the pressure range where the transformations to ringwoodite and garnet majorite take place. K_T and K_S show a strong increase (more than a factor 2) through the upper mantle which makes possible the observed increase of seismic velocities in spite of the increase in density. However, for $p \lesssim 4$ GPa, the density and the P-wave velocity are somewhat underestimated, which might be in part due to the several simplifications and assumptions made in the compositional model; by contrast, beyond this depth, ϱ and v_P look quite satisfying. The peak of c_p at higher temperatures in the wadsleyite depth range might be in part an artefact of the implementation of the phase transition, but its reason is unclear.

264 *Determination of thermoelastic properties of the mantle*

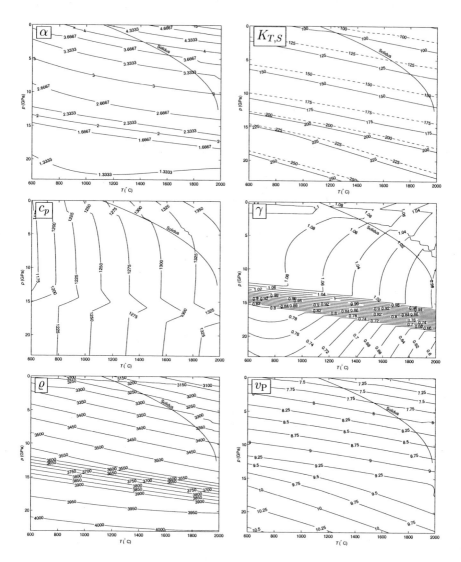

Figure E.2: p–T maps of α (in $10^{-5}\,1/\text{K}$), $K_{T,S}$ (in GPa), c_p (in $\text{J/kg\,K}$), γ, ϱ (in kg/m^3) and v_P (in km/s), calculated with the method described in section III.4.4. The solidus from Hirschmann (2000) is also shown, because melting, which was not considered here, would affect both the two-phase bulk properties and the matrix properties, thereby making the supersolidus part of the diagram increasingly unreliable with isobaric distance from the solidus. K_T in the top right map (solid lines) is the Reuss average, whereas K_S (dashed lines) is the Voigt–Reuss–Hill average applicable to seismic velocities.

LIST OF SYMBOLS

In general, the equations in the text are in non-dimensional form as usual in fluid dynamics. The units given here are for the convenience of the reader; in most cases, SI units are used, with exception of situations where usage of other units such as Ma leads to a more familiar magnitude of values. In many instances the subscripts s and f are used; they denote a property of the solid and the fluid (melt) phase, respectively.

$\mathbf{1}$	unity tensor	
α	volumetric coefficient of thermal expansion at constant pressure, see eq. D.3	$1/\text{K}$
$\beta_{T,S}$	isothermal/adiabatic compressibility, see eq. D.6	$1/\text{GPa}$
η	shear viscosity	Pa s
η_{b}	matrix bulk viscosity	Pa s
ε	strain	
ϕ_{r}	angle between surface and lithosphere base	$^\circ$ or rad
$\tilde{\varphi}$	melt fraction by weight	
φ	melt fraction by volume/porosity	
φ_{e}	eutectic melt fraction	
φ_{max}	threshold porosity	
Γ	melting/crystallization rate	$\text{kg}/\text{m}^3\,\text{s}$
γ_{sf}	surface tension between grain and melt	N/m
γ_{ss}	surface tension between grains	N/m
κ	thermal diffusivity	m^2/s
κ_C	chemical diffusivity	m^2/s
μ	shear modulus	GPa
ν	Poisson ratio	
ν_{ch}	number of channels per length unit (probability distribution)	
Ψ	streamfunction	m^2/s
ρ, ρ_{a}	(apparent) resistivity	$\Omega\,\text{m}$
ϱ	density	kg/m^3
ϱ_0	reference density	kg/m^3

$\boldsymbol{\sigma}$, σ_{ij}	stress tensor	Pa
σ	electrical conductivity	S/m
$\sigma_1, \sigma_2, \sigma_3$	maximum, intermediate, and minimum principal stress	Pa or MPa
Θ	dihedral/trihedral angle	$^\circ$
$\mathcal{A}$	pre-exponential factor for rheology law, see eq. I.11	m^3/Pa^{n+m}
A	area	m^2
a	grain size	m or rather mm
b	geometry factor	
b_{ch}	width of a channel/dike	m or rather mm
$\mathfrak{c}$	dislocation mobility increase factor	
C	concentration	
c_p	specific heat at constant pressure	J/gK
D	bulk partition coefficient	
D_i^j	partition coefficient of jth component with respect to the ith phase	
Di	$= \frac{z_m g \alpha}{c_p} = \left(\frac{\partial T}{\partial z}\right)_S \frac{z_m}{T}$, dissipation number	
$\mathcal{E}$	activation energy	J/mol
$\vec{e}_z$	unit vector in z direction	
e_0	elementary charge	$1.60219 \cdot 10^{-19}$ C
f	degree of melting	
f_{O_2}	oxygen fugacity	
$\mathcal{G}$	crustal growth rate	m/a
$\mathfrak{g}$	geoid	m
g	gravity acceleration	m/s^2
$\mathcal{H}$	Heaviside function	
H	internal specific heat production	J/kg
h	crustal thickness	m or km
$k_B = \frac{R}{N_A}$	Boltzmann's constant	$1.3807 \cdot 10^{-23}\, J/K = 8.6173 \cdot 10^{-5}\, eV/K$
k_φ	permeability, see eq. II.8	m^2
K_T'	pressure derivative of K_T	

$K_{T,S}$ isothermal/adiabatic bulk modulus   GPa

$\mathcal{K}_{e1-e2}^{x-liq}$ exchange coefficient for elements e1 and e2 between crystal x and liquid

$Le = \kappa/\kappa_C$ Lewis number

l_c compaction length   m

L_m latent heat of melting/freezing, see eq. I.3b J/g or J/mol

L_{X_i} latent heat of ith phase change, see eq. I.10 J/g or J/mol

$\mathcal{M}_i$ melting proportion of the ith phase

$\mathfrak{N}_{ch}$ cumulative distribution of channels

$\mathcal{P}_{SC}$ Saucier–Carron parameter

p pressure   Pa

Pe $= \frac{v z_m}{\kappa}$, Péclet number

Pr $= \frac{\eta}{\varrho \kappa}$, Prandtl number

$\mathfrak{q}$ buoyancy flux kg/s

R molar gas constant $8.314\,J/mol\,K$

Rc_{X_i} phase Rayleigh number for a certain phase transition

Ra $= \frac{\varrho_0 g \alpha \Delta T z_m^3}{\eta \kappa}$, Rayleigh number, see eq. I.2d

Rt $= \frac{b \eta_0 z^2}{\eta_0 a^2}$, melt retention number

ΔS_m entropy change due to melting (per unit mass) $J/g\,K$

ΔS_{X_i} entropy of ith phase change, see eq. I.10 $J/g\,K$ or $J/mol\,K$

S entropy (per unit mass) $J/g\,K$

$\mathcal{T}$ potential temperature, see eq. II.19   K or °C

$\mathfrak{t}$ topography   m

$\tilde{T}$ homologous temperature

$\underline{T}$ absolute temperature   K

T temperature   K or °C

t time   s, a or Ma

T_0 reference temperature   K or °C

T_l liquidus temperature   K or °C

T_s solidus temperature   K or °C

$\vec{u}_N$	normal unit vector	
$\vec{u}_B$	Burgers vector	m or nm
$\mathcal{V}$	activation volume	m^3/mol
$\vec{v}$	velocity	m/s or cm/a
V	molar volume	m^3
v_P	P-wave velocity	km/s
v_S	S-wave velocity	km/s
W	spectrum of weighting function	
w	weighting function	
X_i	fraction of a certain phase	
y_{cP}	crustal waist width	km
Δz_X	depression of the 410 km discontinuity	km
z_m	depth of layer (model)	km

COMMON ABBREVIATIONS

CFB	Courant–Friedrichs–Lewy (criterion)
CMAS	$CaO-MgO-Al_2O_3-SiO_2$
CMASF	$CaO-MgO-Al_2O_3-SiO_2-FeO$
CMASN	$CaO-MgO-Al_2O_3-SiO_2-Na_2O$
CMASNF	$CaO-MgO-Al_2O_3-SiO_2-Na_2O-FeO$
CMB	core–mantle boundary
cpx	clinopyroxene
di	diopside
EM-1	enriched member I (mantle reservoir)
en	enstatite
EVZ	Eastern Volcanic Zone
fo	forsterite
FOZO	focal zone (mantle reservoir)
gt	garnet
HIMU	high-μ (mantle reservoir)
LPO	lattice-preferred orientation
LVZ	low-velocity zone
MAS	$MgO-Al_2O_3-SiO_2$
MOR	mid-oceanic ridge
MORB	mid-oceanic ridge basalt
MPO	melt-preferred orientation
NVZ	Northern Volcanic Zone
OIB	oceanic island basalt
ol	olivine
opx	orthopyroxene
pl	plagioclase
py	pyrope
QMSL	quasi-monotone semi-Lagrangian
REE	rare earth element
sp	spinel
TFZ	Tjörnes Fracture Zone
WVZ	Western Volcanic Zone

BIBLIOGRAPHY

Abelson, M., Agnon, A. (2001): Hot spot activity and plume pulses recorded by geometry of spreading axes. *Earth Planet. Sci. Lett.* *189*(1–2), 31–47.

Adam, G., Gibbs, J. H. (1965): On the temperature dependence of cooperative relaxation properties in glass-forming liquids. *J. Chem. Phys.* *43*, 139–146.

Agee, C. B. (1992a): Isothermal compression of molten Fe_2SiO_4. *Geophys. Res. Lett.* *19*(11), 1169–1172.

Agee, C. B. (1992b): Thermal expansion of molten Fe_2SiO_4 at high pressure. *Geophys. Res. Lett.* *19*(11), 1173–1176.

Agee, C. B. (1998): Crystal-liquid density inversions in terrestrial and lunar magmas. *Phys. Earth Planet. Int.* *107*(1–3), 63–74.

Aharonov, E., Whitehead, J. A., Kelemen, P. B., Spiegelman, M. (1995): Channeling instability of upwelling melt in the mantle. *J. Geophys. Res.* *100*(B10), 20 433–20 450.

Albers, M., Christensen, U. R. (1996): The excess temperature of plumes rising from the core-mantle boundary. *Geophys. Res. Lett.* *23*(24), 3567–3570.

Albers, M., Christensen, U. R. (2001): Channeling of plume flow beneath mid-ocean ridges. *Earth Planet. Sci. Lett.* *187*, 207–220.

Allègre, C. J., Turcotte, D. L. (1986): Implications of a two-component marble-cake mantle. *Nature* *323*, 123–127.

Allen, R. M., Nolet, G., Morgan, W. J., Vogfjörd, K., Bergsson, B. H., Erlendsson, P., Foulger, G. R., Jakobsdóttir, S., Julian, B. R., Pritchard, M., Ragnarsson, S., Stefánsson, R. (1999): The thin hot plume beneath Iceland. *Geophys. J. Int.* *137*(1), 51–63.

Allen, R. M., Nolet, G., Morgan, W. J., Vogfjörd, K., Bergsson, B. H., Erlendsson, P., Foulger, G. R., Jakobsdóttir, S., Julian, B. R., Pritchard, M., Ragnarsson, S., Stefánsson, R. (2002a): Imaging the mantle beneath Iceland using integrated seismological techniques. *J. Geophys. Res.* *107*(B12), 2325, doi:10.1029/2001JB000595.

Allen, R. M., Nolet, G., Morgan, W. J., Vogfjörd, K., Nettles, M., Ekström, G., Bergsson, B. H., Erlendsson, P., Foulger, G. R., Jakobsdóttir, S., Julian, B. R., Pritchard, M., Ragnarsson, S., Stefánsson, R. (2002b): Plume driven plumbing and crustal formation in Iceland. *J. Geophys. Res.* *107*(B8), 2163, doi:10.1029/2001JB000584.

Anderson, D. L., Zhang, Y.-S., Tanimoto, T. (1992): Plume heads, continental lithosphere, flood basalt and tomography. In *Magmatism and the Causes of Continental Break-up*, edited by B. C. Storey, T. Alabaster, R. J. Pankhurst, number 68 in Geological Society Special Publication, pp. 99–124, The Geological Society, London.

Anderson, E., Bai, Z., Bischof, C., Blackford, S., Demmel, J., Dongarra, J., Du Croz, J., Greenbaum, A., Hammarling, S., McKenney, A., Sorensen, D. (1999): *LAPACK Users' Guide*. Society for Industrial and Applied Mathematics, Philadelphia, PA, 3rd edition. http://www.netlib.org/lapack/.

Anderson, O. L. (1995): *Equations of State of Solids for Geophysics and Ceramic Science*. Number 31 in Oxford Monographs on Geology and Geophysics, Oxford University Press.

Angel, R. J., Jackson, J. M. (2001): Elasticity and equation of state of orthoenstatite. In *LPI Contribution No. 1088*, p. Abstract #3268, Lunar and Planetary Institute, Houston. (CD-ROM).

Angenheister, G., Gebrande, H., Miller, H., RRISP Working Group (1980): Reykjanes Ridge Iceland Seismic Experiment (RRISP 77). *J. Geophys. 47*, 228–238.

Applegate, B., Shor, A. N. (1994): The northern mid-Atlantic and Reykjanes Ridges: spreading center morphology between 55°50'N and 63°00'N. *J. Geophys. Res. 99*(B9), 17 935–17 956.

Ashcroft, N. W., Mermin, N. D. (1976): *Solid State Physics*. Saunders College Publishing.

Asimow, P. D. (1999): A model that reconciles major- and trace-element data from abyssal peridotites. *Earth Planet. Sci. Lett. 169*(3–4), 303–319.

Asimow, P. D. (2002): Steady-state mantle-melt interactions in one dimension: II. Thermal interactions and irreversible terms. *J. Petrol. 43*(9), 1707–1724.

Asimow, P. D., Hirschmann, M. M., Ghiorso, M. S., O'Hara, M. J., Stolper, E. M. (1995): The effect of pressure-induced solid–solid phase transitions on decompression melting of the mantle. *Geochim. Cosmochim. Acta 59*(21), 4489–4506.

Asimow, P. D., Hirschmann, M. M., Stolper, E. M. (1997): An analysis of variations in isentropic melt productivity. *Phil. Trans. Roy. Soc. Lond. A 355*(1723), 255–281.

Asimow, P. D., Hirschmann, M. M., Stolper, E. M. (2001): Calculation of peridotite partial melting from thermodynamic models of minerals and melts. IV. Adiabatic decompression and the composition and mean properties of mid-ocean ridge basalts. *J. Petrol. 42*(5), 963–998.

Asimow, P. D., Langmuir, C. H. (2003): The importance of water to oceanic mantle melting regimes. *Nature 421*, 815–820.

Bagdassarov, N. (2000): Anelastic and viscoelastic behaviour of partially molten rocks and lavas. In *Physics and Chemistry of Partially Molten Rocks*, edited by N. Bagdassarov, D. Laporte, A. B. Thompson, pp. 29–66, Kluwer Academic Publishers.

Bagdassarov, N. S. (1988): Accumulation capacity of spinel lherzolite partial melts. *Geokhimiya 8*, 1168–1176. (english translation publ. by Scripta Technica, 1989).

Baker, M. B., Beckett, J. R. (1999): The origin of abyssal peridotites: a reinterpretation of constraints based on primary bulk compositions. *Earth Planet. Sci. Lett. 171*(1), 49–61.

Barnouin-Jha, K., Parmentier, E. M., Sparks, D. W. (1997): Buoyant mantle upwelling and crustal production at oceanic spreading centers: On-axis segmentation and off-axis melting. *J. Geophys. Res. 102*(B6), 11 979–11 990.

Batchelor, G. K. (1967): *An Introduction to Fluid Dynamics*. Cambridge University Press.

Beattie, P. (1993a): The generation of uranium series disequilibria by partial melting of spinel peridotite: constraints from partitioning studies. *Earth Planet. Sci. Lett. 117*, 379–391.

Beattie, P. (1993b): Uranium–thorium disquilibria and partitioning on melting of garnet peridotite. *Nature 363*, 63–65.

Beblo, M., Björnsson, A. (1980): A model of electrical resistivity beneath NE-Iceland, correlation with temperature. *J. Geophys. 47*, 184–190.

Becker, T. W., Braun, A. (1998): New program maps geoscience data sets interactively. *EOS Transactions, AGU 79*, 505.

Bell, D. R., Rossman, G. R. (1992): Water in Earth's mantle: The role of nominally anhydrous minerals. *Science 255*, 1391–1397.

Berber, J., Kacher, H., Langer, R. (eds.) (1986): *Physik in Formeln und Tabellen*. B.G.Teubner, Stuttgart, 3rd edition.

Bercovici, D. (1992): Wave dynamics in mantle plume heads and hot spot swells. *Geophys. Res. Lett. 19*(17), 1791–1794.

Bercovici, D., Lin, J. (1996): A gravity current model of cooling mantle plume heads with temperature-dependent buoyancy and viscosity. *J. Geophys. Res. 101*(B2), 3291–3309.

Bercovici, D., Ricard, Y., Schubert, G. (2001): A two-phase model for compaction and damage. 1. General Theory. *J. Geophys. Res. 106*(B5), 8887–8906.

Bermejo, R., Staniforth, A. (1992): The conversion of semi-Lagrangian advection schemes to quasi-monotone schemes. *Mon. Weather Rev. 120*(11), 2622–2632.

Berryman, J. G. (1995): Mixture theories for rock properties. In *Rock Physics & Phase Relations — A Handbook of Physical Constants*, edited by T. J. Ahrens, volume 3 of *AGU Reference Shelf*, pp. 205–228, AGU, Washington.

Bijwaard, H., Spakman, W. (1999): Tomographic evidence for a narrow whole mantle plume below Iceland. *Earth Planet. Sci. Lett. 166*, 121–126.

Bina, C. R., Helffrich, G. R. (1994): Phase transition Clapeyron slopes and transition zone seismic discontinuity topography. *J. Geophys. Res. 99*(B8), 15 853–15 860.

Bjarnason, I. Þ., Menke, W., Flóvenz, Ó. C., Caress, D. (1993): Tomographic image of the mid-Atlantic plate boundary in southwestern Iceland. *J. Geophys. Res. 98*, 6607–6622.

Bjarnason, I. Þ., Sacks, I. S. (2002): The plume, lithosphere and asthenosphere of Iceland. In *Geophys. Res. Abstr.*, EGS.

Bjarnason, I. Þ., Silver, P. G., Rümpker, G., Solomon, S. C. (2002): Shear-wave splitting across the Iceland hotspot: Results from the ICEMELT experiment. *J. Geophys. Res. 107*(B12), 2382, doi:10.1029/2001JB000916.

Bjarnason, I. Þ., Wolfe, C. J., Solomon, S. C., Guðmundsson, G. (1996): Initial results from the ICEMELT experiment: Body-wave delay times and shear-wave splitting across Iceland. *Geophys. Res. Lett. 23*(5), 459–462. Correction in *Geophys. Res. Lett.* 23(8), p.903 (1996).

Blackman, D. K., Kendall, J. M. (1997): Sensitivity of teleseismic body waves to mineral texture and melt in the mantle beneath a mid-ocean ridge. *Phil. Trans. Roy. Soc. Lond. A 355*(1723), 217–231.

Blankenbach, B., Busse, F., Christensen, U., Cserepes, L., Gunkel, D., Hansen, U., Harder, H., Jarvis, G., Koch, M., Marquart, G., Moore, D., Olson, P., Schmeling, H., Schnaubelt, T. (1989): A benchmark comparison for mantle convection codes. *Geophys. J. Int. 98*(1), 23–38.

Blundy, J. D., Falloon, T. J., Wood, B. J., Dalton, J. A. (1995): Sodium partitioning between clinopyroxene and silicate melts. *J. Geophys. Res. 100*(B8), 15 501–15 515.

Blundy, J. D., Robinson, J. A. C., Wood, B. J. (1998): Heavy REE are compatible in clinopyroxene on the spinel lherzolite solidus. *Earth Planet. Sci. Lett. 160*(3–4), 493–504.

Blundy, J. D., Wood, B. J. (1994): Prediction of crystal–melt partition coefficients from elastic moduli. *Nature 372*, 452–454.

Bottinga, Y. (1985): On the isothermal compressibility of silicate liquids at high pressure. *Earth Planet. Sci. Lett. 74*, 350–360.

Bottinga, Y., Richet, P. (1995): Silicate melts: The "anomalous" pressure dependence of the viscosity. *Geochim. Cosmochim. Acta 59*(13), 2725–2731.

Botz, R., Winckler, G., Bayer, R., Schmitt, M., *et al.* (1999): Origin of trace gases in submarine hydrothermal vents of the Kolbeinsey Ridge, north Iceland. *Earth Planet. Sci. Lett. 171*(1), 83–93.

Boudier, F., Nicolas, A. (1972): Fusion partielle gabbroïque dans la lherzolite de Lanzo (Alpes piémontaises). *Schweiz. Miner. Petrogr. Mitt. 52*, 39–56.

Brady, J. B. (1995): Diffusion data for silicate minerals, glasses, and liquids. In *Mineral Physics & Crystallography — A Handbook of Physical Constants*, edited by T. J. Ahrens, volume 2 of *AGU Reference Shelf*, pp. 269–290, AGU, Washington.

Brandsdóttir, B., Menke, W., Einarsson, P., White, R. S., Staples, R. K. (1997): Färoe-Iceland Ridge Experiment — 2. Crustal structure of the Krafla central volcano. *J. Geophys. Res. 102*(B4), 7867–7886.

Braun, M. G., Hirth, G., Parmentier, E. M. (2000): The effect of deep damp melting on mantle flow and melt generation beneath mid-ocean ridges. *Earth Planet. Sci. Lett. 176*, 339–356.

Braun, M. G., Kelemen, P. B. (2002): Dunite distribution in the Oman Ophiolite: Implications for melt flux through porous dunite conduits. *Geochem. Geophys. Geosyst. 3*(11), 8603, doi: 10.1029/2001GC000289.

Breddam, K. (2002): Kistufell: Primitive melt from the Iceland mantle plume. *J. Petrol. 43*(2), 345–373.

Breddam, K., Kurz, M. D., Storey, M. (2000): Mapping out the conduit of the Iceland mantle plume with helium isotopes. *Earth Planet. Sci. Lett. 176*, 45–55.

Bronstein, I. N., Semendjajew, K. A. (1991): *Taschenbuch der Mathematik.* B.G.Teubner/Verlag Nauka/Harri Deutsch, Leipzig/Moskau/Thun und Frankfurt a.M., 25th edition.

Buck, W. R. (1991): Modes of continental lithospheric extension. *J. Geophys. Res. 96*(B12), 20 161–20 178.

Buck, W. R. (1996): Shallow redistribution of crust produced by mantle plumes at mid-ocean ridges. In *Abstracts*, edited by B. Þorkelsson, p. 117, European Seismological Commission, Veðurstofa Íslands (Icelandic Meteorological Office), Ministry for Environment, University of Iceland, Reykjavík.

Buck, W. R., Su, W. (1989): Focused mantle upwelling below mid-ocean ridges due to feedback between viscosity and melting. *Geophys. Res. Lett. 16*(7), 641–644.

Bureau, H., Keppler, H. (1999): Complete miscibility between silicate melts and hydrous fluids in the upper mantle: experimental evidence and geochemical implications. *Earth Planet. Sci. Lett. 165*(2), 187–196.

Bussod, G. Y., Christie, J. M. (1991): Texture development and melt topology in spinel lherzolite experimentally deformed at hypersolidus conditions. *J. Petrol.* pp. 17–39. Special Volume: Orogenic lherzolites and mantle processes.

Buttkus, B. (1991): *Spektralanalyse und Filtertheorie in der angewandten Geophysik.* Springer-Verlag, Berlin-Heidelberg.

Bystricki, M., Kunze, K., Burlini, L., Burg, J.-P. (2000): High shear strain of olivine aggregates: rheological and seismic consequences. *Science 290*, 1564–1567.

Cella, F., Rapolla, A. (1997): Density changes in upwelling mantle. *Phys. Earth Planet. Int. 103*, 63–84.

Ceuleneer, G., Monnereau, M., Rabinowicz, M., Rosemberg, C. (1993): Thermal and petrological consequences of melt migration within mantle plumes. *Phil. Trans. Roy. Soc. Lond. A 342*, 53–64.

Ceuleneer, G., Rabinowicz, M. (1992): Mantle flow and melt migration beneath oceanic ridges: models derived from observations in ophiolites. In *Mantle flow and melt generation at mid-ocean ridges*, edited by J. Phipps Morgan, D. K. Blackman, J. M. Sinton, number 71 in Geophysical Monographs, pp. 123–154, AGU.

Chalmers, J. A., Larson, L. M., Pedersen, A. K. (1995): Widespread Palaeocene volcanism around the northern North Atlantic and Labrador Sea: evidence for a large, hot, early plume head. *J. Geol. Soc. Lond. 152*, 965–969.

Chauvel, C., Blichert-Toft, J. (2001): A hafnium isotope and trace element perspective on melting of the depleted mantle. *Earth Planet. Sci. Lett. 190*(3–4), 137–151.

Chauvel, C., Hémond, C. (2000): Melting of a complete section of recycled oceanic crust: Trace element and Pb isotope evidence from Iceland. *Geochem. Geophys. Geosyst. 1*, doi:1999GC000002.

Chen, G. Q., Ahrens, T. J., Stolper, E. M. (2002a): Shock-wave equation of state of molten and solid fayalite. *Phys. Earth Planet. Int. 134*(1–2), 35–52.

Chen, J., Inoue, T., Yurimoto, H., Weidner, D. (2002b): Effect of water on olivine–wadsleyite phase boundary in the $(Mg, Fe)_2SiO_4$ system. *Geophys. Res. Lett. 29*(18), 22, doi: 10.1029/2001GL014429.

Choblet, G., Parmentier, E. M. (2001): Mantle upwelling and melting beneath slow spreading centers: effects of variable rheology and melt productivity. *Earth Planet. Sci. Lett. 184*, 589–604.

Chopelas, A. (2000): Thermal expansivity of mantle relevant magnesium silicates derived from vibrational spectroscopy at high pressure. *Am. Mineral. 85*(2), 270–278.

Christensen, N. I. (1989): Seismic velocities. In *CRC Practical Handbook of Physical Properties of Rocks and Minerals*, edited by R. S. Carmichael, pp. 429–546, CRC Press, Boca Raton.

Christensen, U., Harder, H. (1991): 3-D convection with variable viscosity. *Geophys. J. Int. 104*, 213–226.

Christensen, U. R. (1995): Effects of phase transitions on mantle convection. *Ann. Rev. Earth Planet. Sci. 23*, 65–87.

Christensen, U. R., Yuen, D. A. (1985): Layered convection induced by phase transitions. *J. Geophys. Res. 90*, 10 291–10 300.

Chvátal, V. (1983): *Linear programming*. W.H.Freeman & Co., New York.

Clift, P. D., Turner, J., Ocean Drilling Program Leg 152 Scientific Party (1995): Dynamic support by the Icelandic plume and vertical tectonics of the northeast Atlantic continental margins. *J. Geophys. Res. 100*(B12), 24 473–24 486.

Cmíral, M., Fitz Gerald, J. D., Faul, U. H., Green, D. H. (1998): A close look at dihedral angles and melt geometry in olivine–basalt aggregates: a TEM study. *Contrib. Mineral. Petrol. 130*, 336–345.

Coffin, M. F., Eldholm, O. (1992): Volcanism and continental break-up: a global compilation of large igneous provinces. In *Magmatism and the Causes of Continental Break-up*, edited by B. C. Storey, T. Alabaster, R. J. Pankhurst, number 68 in Geological Society Special Publication, pp. 17–30, The Geological Society, London.

Cohen, R. E., Gülseren, O., Hemley, R. J. (2000): Accuracy of equation-of-state formulations. *Am. Mineral. 85*(2), 338–344.

Condie, K. C. (2001): *Mantle Plumes and Their Record in Earth History*. Cambridge University Press.

Connolly, J. A. D., Podlachikov, Y. Y. (1998): Compaction-driven fluid flow in viscoelastic rock. *Geodin. Acta 11*(2–3), 55–84.

Coogan, L. A., Saunders, A. D., Kempton, P. D., Norry, M. J. (2000): Evidence from oceanic gabbros for porous melt migration within a crystal mush beneath the Mid-Atlantic Ridge. *Geochem. Geophys. Geosyst. 1*, doi:1999GC000072.

Cordery, M. J., Phipps Morgan, J. (1993): Convection and melting at mid ocean ridges. *J. Geophys. Res. 98*(B11), 19 477–19 503.

Courtial, P., Ohtani, E., Dingwell, D. B. (1997): High-temperature densities of some mantle melts. *Geochim. Cosmochim. Acta 61*(15), 3111–3119.

Courtillot, V., Davaille, A., Besse, J., Stock, J. (2003): Three distinct types of hotspots in the Earth's mantle. *Earth Planet. Sci. Lett. 205*(3–4), 295–308.

Dagan, G. (1989): *Flow and Transport in Porous Formations.* Springer-Verlag.

Dahm, T. (2000): Numerical simulations of the propagation path and the arrest of fluid-filled fractures in the Earth. *Geophys. J. Int. 141*(3), 623–638.

Daines, M. J., Kohlstedt, D. L. (1993): A laboratory study of melt migration. *Phil. Trans. Roy. Soc. Lond. A 342*, 43–52.

Daines, M. J., Kohlstedt, D. L. (1994): The transition from porous to channelized flow due to melt/rock reaction during melt migration. *Geophys. Res. Lett. 21*(2), 145–148.

Daines, M. J., Kohlstedt, D. L. (1997): Influence of deformation on melt topology in peridotites. *J. Geophys. Res. 102*(B5), 10 257–10 271.

Darbyshire, F. A., Bjarnason, I. Þ., White, R. S., Flóvenz, Ó. G. (1998): Crustal structure above the Iceland mantle plume imaged by the ICEMELT refraction profile. *Geophys. J. Int. 135*(3), 1131–1149.

Darbyshire, F. A., Priestley, K. F., White, R. S., Stefánsson, R., Gudmundsson, G. B., Jakobsdóttir, S. S. (2000a): Crustal structure of central and northern Iceland from analysis of teleseismic receiver functions. *Geophys. J. Int. 143*(1), 163–184.

Darbyshire, F. A., White, R. S., Priestley, K. F. (2000b): Structure of the crust and uppermost mantle of Iceland from a combined seismic and gravity study. *Earth Planet. Sci. Lett. 181*, 409–428.

Davies, J. H., Bickle, M. J. (1991): A physical model for the volume and composition of melt produced by hydrous fluxing above subduction zones. *Phil. Trans. Roy. Soc. Lond. A 335*(1638), 355–364.

Davis, B. T. C., England, J. L. (1964): The melting of forsterite up to 50 kilobars. *J. Geophys. Res. 69*(6), 1113–1116.

DeYoreo, J. J., Lange, R. A., Navrotsky, A. (1995): Scanning calorimetric determinations of the heat contents of diopside-rich systems during melting and crystallization. *Geochim. Cosmochim. Acta 59*(13), 2701–2707.

Dick, H. J. B. (1977): Partial melting in the Josephine peridotite I, the effect on mineral composition and its consequence for geobarometry and geothermometry. *Am. J. Sci. 277*, 801–832.

Dick, H. J. B., Fisher, R. L., Bryan, W. B. (1984): Mineralogic variability of the uppermost mantle along mid-ocean ridges. *Earth Planet. Sci. Lett. 69*, 88–106.

Dixon, E. T., Honda, M., McDougall, I., Campbell, I. H., Sigurdsson, I. (2000): Preservation of near-solar neon isotopic ratios in Icelandic basalts. *Earth Planet. Sci. Lett. 180*(3–4), 309–324.

Donaldson, J. R., Tryon, P. V. (1990): STARPAC — The Standards Time Series And Regression Package, v.2.08. National Institute of Standards and Technology. http://www.scd.ucar.edu/softlib/STARPAC.html.

Du, Z., Foulger, G. R. (2001): Variation in the crustal structure across central Iceland. *Geophys. J. Int. 145*(1), 246–264.

Du, Z. J., Foulger, G. R. (1999): The crustal structure beneath the northwest fjords, Iceland, from receiver functions and surface waves. *Geophys. J. Int. 139*(2), 419–432.

Du, Z. J., Foulger, G. R., Julian, B. R., Allen, R. M., Nolet, G., Morgan, W. J., Bergsson, B. H., Erlendsson, P., Jakobsdóttir, S., Ragnarsson, S., Stefánsson, R., Vogfjörd, K. (2002): Crustal structure beneath western and eastern Iceland from surface waves and receiver functions. *Geophys. J. Int. 149*(2), 349–363.

Dubuffet, F., Yuen, D. A., Rabinowicz, M. (1999): Effects of a realistic mantle thermal conductivity on the patterns of 3-D convection. *Earth Planet. Sci. Lett. 171*(3), 401–409.

Duffy, T. S., Anderson, D. L. (1989): Seismic velocities in mantle minerals and the mineralogy of the upper mantle. *J. Geophys. Res. 94*(B2), 1895–1912.

Dziewonski, A. M., Anderson, D. L. (1981): Preliminary reference Earth model. *Phys. Earth Planet. Int. 25*, 297–356.

Ellam, R. M., Stuart, F. M. (2000): The sub-lithospheric source of north Atlantic basalts: evidence for, and significance of, a common end-member. *J. Petrol. 41*(7), 919–932.

Elliott, T. (1997): Fractionation of U and Th during mantle melting: a reprise. *Chem. Geol. 139*, 165–183.

ETOPO5 (1988): Digital Relief of the Surface of the Earth. *Data Announcement 88-MGG-02*, NOAA, National Geophysical Data Center, Boulder, Colorado.

Eysteinsson, H., Gunnarsson, K. (1995): Maps of gravity, bathymetry and magnetics for Iceland and surroundings. *Technical Report OS-95055/JHD-07*, Orkustofnun.

Eysteinsson, H., Hermance, J. F. (1985): Magnetotelluric measurements across the Eastern Neovolcanic Zone in south Iceland. *J. Geophys. Res. 90*(B12), 10 093–10 103.

Falloon, T. J., Green, D. H., Danyushevsky, L. V., Faul, U. H. (1999): Peridotite melting at 1.0 and 1.5 GPa: an experimental evaluation of techniques using diamond aggregates and mineral mixes for determination of near-solidus melts. *J. Petrol. 40*(9), 1343–1375.

Falloon, T. J., Green, D. H., Hatton, C. J., Harris, K. L. (1988): Anhydrous melting of a fertile and depleted peridotite from 2 to 30 kb and application to basalt petrogenesis. *J. Petrol. 29*(6), 1257–1282.

Farnetani, C. G. (1997): Excess temperature of mantle plumes: The role of chemical stratification across D''. *Geophys. Res. Lett. 24*(13), 1583–1586.

Farnetani, C. G., Legras, B., Tackley, P. J. (2002): Mixing and deformations in mantle plumes. *Earth Planet. Sci. Lett. 196*(1–2), 1–15.

Farnetani, C. G., Richards, M. A. (1995): Thermal entrainment and melting in mantle plumes. *Earth Planet. Sci. Lett. 136*, 251–267.

Farnetani, C. G., Richards, M. A., Ghiorso, M. S. (1996): Petrological models of magma evolution and deep crustal structure beneath hotspots and flood basalt provinces. *Earth Planet. Sci. Lett. 143*(1–4), 81–94.

Faul, U. H. (1997): The permeability of partially molten upper mantle rocks from experiments and percolation theory. *J. Geophys. Res. B102*, 10 299–10 311.

Faul, U. H. (2000): Constraints on the melt distribution in anisotropic polycrystalline aggregates undergoing grain growth. In *Physics and Chemistry of Partially Molten Rocks*, edited by N. Bagdassarov, D. Laporte, A. B. Thompson, pp. 67–92, Kluwer Academic Publishers.

Faul, U. H. (2001): Melt retention and segregation beneath mid-ocean ridges. *Nature 410*, 920–923.

Faul, U. H., Toomey, D., Waff, H. S. (1994): Intergranular basaltic melt is distributed in thin, elongated inclusions. *Geophys. Res. Lett. 21*, 29–32.

Fei, Y. (1995): Thermal expansion. In *Mineral Physics & Crystallography — A Handbook of Physical Constants*, edited by T. J. Ahrens, volume 2 of *AGU Reference Shelf*, pp. 283–291, AGU, Washington.

Fei, Y., Bertka, C. M. (1999): Phase transitions in the Earth's mantle and mantle mineralogy. In *Mantle Petrology: Field Observations and High Pressure Experimentation*, edited by Y. Fei, C. M. Bertka, B. O. Mysen, number 6 in Special Publications, pp. 189–207, The Geochemical Society.

Feigenson, M. D. (1986): Constraints on the origin of Hawaiian lavas. *J. Geophys. Res. 91*(B9), 9383–9393.

Feighner, M. A., Kellogg, L. H., Travis, B. J. (1995): Numerical modeling of chemically buoyant mantle plumes at spreading ridges. *Geophys. Res. Lett. 22*(6), 715–718.

Feighner, M. A., Richards, M. A. (1995): The fluid dynamics of plume-ridge and plume-plate interactions: An experimental investigation. *Earth Planet. Sci. Lett. 129*, 171–182.

Fitton, J. G., Saunders, A. D., Norry, M. J., Hardarson, B. S., Taylor, R. N. (1997): Thermal and chemical structure of the Iceland plume. *Earth Planet. Sci. Lett. 153*, 197–208.

Flóvenz, Ó. G. (1980): Seismic structure of the Icelandic crust above layer three and the relation between body wave velocity and the alteration of the basaltic crust. *J. Geophys. 47*, 211–220.

Flóvenz, Ó. G., Gunnarsson, K. (1991): Seismic crustal structure in Iceland and surrounding area. *Tectonophys. 189*, 1–17.

Forsyth, D. W. (1992): Geophysical constraints on mantle flow and melt generation beneath mid-ocean ridges. In *Mantle flow and melt generation at mid-ocean ridges*, edited by J. Phipps Morgan, D. K. Blackman, J. M. Sinton, number 71 in Geophysical Monographs, pp. 1–65, AGU.

Foulger, G. R., Pearson, D. G. (2001): Is Iceland underlain by a plume in the lower mantle? Seismology and helium isotopes. *Geophys. J. Int. 145*(3), F1–F5.

Foulger, G. R., Pritchard, M. J., Julian, J. R., Evans, B. R., Allen, R. M., Nolet, G., Morgan, W. J., Bergsson, B. H., Erlendsson, P., Jakobsdóttir, S., Ragnarsson, S., Stefánsson, R., Vogfjörd, K. (2001): Seismic tomography shows that upwelling beneath Iceland is confined to the upper mantle. *Geophys. J. Int. 146*(2), 504–530.

Fowler, A. C. (1990): A compaction model for melt transport in the Earth's asthenosphere. Part II: applications. In *Magma transport and storage*, edited by M. P. Ryan, pp. 15–32, Wiley, Chichester.

Fowler, A. C., Scott, D. R. (1996): Hydraulic crack propagation in a porous medium. *Geophys. J. Int. 127*, 595–604.

Fram, M. S., Lesher, C. E. (1993): Geochemical constraints on mantle melting during creation of the North-Atlantic basin. *Nature 363*, 712–715.

Frigo, M., Johnson, S. G. (1998): FFTW: an adaptive software architecture for the FFT. In *ICASSP Conference Proceedings*, volume 3, pp. 1381–1384. http://www.fftw.org/.

Fujii, N., Osamura, K., Takahashi, E. (1986): Effect of water saturation on the distribution of partial melt in the olivine–pyroxene–plagioclase system. *J. Geophys. Res. 91*(B9), 9253–9259.

Fukuyama, H. (1985): Heat of fusion of basaltic magma. *Earth Planet. Sci. Lett. 73*, 407–414.

Furman, T., Frey, F., Park, K.-H. (1995): The scale of source heterogeneity beneath the eastern neovolcanic zone, Iceland. *J. Geol. Soc. Lond. 152*, 997–1002.

Gable, C. W., O'Connell, R. J., Travis, B. J. (1991): Convection in three dimensions with surface plates: Generation of toroidal flow. *J. Geophys. Res. 96*(B5), 8391–8405.

Gaherty, J. B. (2001): Seismic evidence for hotspot-induced buoyant flow beneath the Reykjanes Ridge. *Science 293*, 1645–1647.

Gaherty, J. B., Wang, Y. B., Jordan, T. H., Weidner, D. J. (1999): Testing plausible upper-mantle compositions using fine-scale models of the 410-km discontinuity. *Geophys. Res. Lett. 26*(11), 1641–1644.

Gassmann, F. (1951): Über die Elastizität poröser Medien. *Vierteljahresschrift der Naturforschenden Gesellschaft Zürich 96*, 1–21.

Gebrande, H., Miller, H., Einarsson, P. (1980): Seismic structure of Iceland along RRISP-Profile I. *J. Geophys. 47*, 239–249.

Ghiorso, M. S., Sack, R. O. (1995): Chemical mass-transfer in magmatic processes. 4. A revised and internally consistent thermodynamic model for the interpolation and extrapolation of liquid–solid equilibria in magmatic systems at elevated temperatures and pressures. *Contrib. Mineral. Petrol.* *119*, 197–212.

Ghods, A., Arkani-Hamed, J. (2000): Melt migration beneath mid-ocean ridges. *Geophys. J. Int.* *140*(3), 687–697.

Ghods, A., Arkani-Hamed, J. (2002): Effects of melt migration on the dynamics and melt generation of diapirs ascending through asthenosphere. *J. Geophys. Res.* *107*(B1), 10.1029/2000JB000070.

Gill, R. C. O., Pedersen, A. K., Larsen, J. G. (1992): Tertiary picrites in West Greenland: melting at the periphery of a plume? In *Magmatism and the Causes of Continental Break-up*, edited by B. C. Storey, T. Alabaster, R. J. Pankhurst, number 68 in Geological Society Special Publication, pp. 335–348, The Geological Society, London.

Gottsmann, J., Dingwell, D. B. (2000): Supercooled diopside melt: confirmation of temperature-dependent expansivity using container-based dilatometry. *Contrib. Mineral. Petrol.* *139*, 127–135.

Gottsmann, J., Dingwell, D. B. (2002): Thermal expansivities of supercooled haplobasaltic liquids. *Geochim. Cosmochim. Acta* *66*(12), 2231–2238.

Grand, S. P. (1994): Mantle shear structure beneath the Americas and surrounding oceans. *J. Geophys. Res.* *99*(B6), 11 591–11 622.

Grand, S. P. (2002): Mantle shear-wave tomography and the fate of subducted slabs. *Phil. Trans. R. Soc. Lond.* *A360*(1800), 2475–2491.

Grand, S. P., van der Hilst, R. D., Widiyantoro, S. (1997): Global seismic tomography: a snapshot of convection in the Earth. *GSA Today* *7*(4), 1–7. Additional data and maps available from ftp.geo.utexas.edu.

Gravel, S., Staniforth, A. (1994): A mass-conserving semi-Lagrangian scheme for the shallow-water equations. *Mon. Weather Rev.* *122*(1), 243–248.

Green, D. H. (1973): Experimental melting studies on a model upper mantle composition at high pressure under water-saturated and water-undersaturated conditions. *Earth Planet. Sci. Lett.* *19*, 37–53.

Green, D. H., Falloon, T., Eggins, S. M., Yaxley, G. M. (2001): Primary magmas and mantle temperatures. *Eur. J. Mineral.* *13*(3), 437–451.

Green, D. H., Falloon, T. J. (1998): Pyrolite: A Ringwood concept and its current expression. In *The Earth's Mantle — Composition, Structure, and Evolution*, edited by I. Jackson, pp. 311–378, Cambridge University Press.

Gribb, T. T., Cooper, R. F. (2000): The effect of an equilibrated melt phase on the shear creep and attenuation behavior of polycrystalline olivine. *Geophys. Res. Lett.* *27*(15), 2341–2344.

Griffiths, R. W., Campbell, I. H. (1990): Stirring and structure in mantle starting plumes. *Earth Planet. Sci. Lett.* *99*, 66–78.

Grove, T. L., Kinzler, R. J., Bryan, W. B. (1992): Fractionation of mid-ocean ridge basalt (MORB). In *Mantle flow and melt generation at mid-ocean ridges*, edited by J. P. Morgan, D. K. Blackman, J. M. Sinton, volume 71 of *Geophys.Monogr.Ser.*, pp. 281–310, AGU, Washington.

Gudfinnsson, G. H., Presnall, D. C. (2000): Melting behaviour of model lherzolite in the system $CaO–MgO–Al_2O_3–SiO_2–FeO$ at 0.7–2.8 GPa. *J. Petrol.* *41*(8), 1241–1269.

Gudfinnsson, G. H., Presnall, D. C. (2001): A pressure-independent geothermometer for primitive mantle melts. *J. Geophys. Res.* *106*(B8), 16 205–16 211.

Guðmundsson, Ó. (2003): The dense root of the Icelandic crust. *Earth Planet. Sci. Lett.* *206*, 427–440.

Guðmundsson, Ó., Brandsdóttir, B., Menke, W., Sigvaldason, G. E. (1994): The crustal magma chamber of the Katla volcano in south Iceland revealed by 2-D seismic undershooting. *Geophys. J. Int. 119*, 277–296.

Haase, K. M. (1996): The relationship between the age of the lithosphere and the composition of oceanic magmas: Constraints on partial melting, mantle sources and the thermal structure of the plates. *Earth Planet. Sci. Lett. 144*, 75–92.

Hall, A. (1996): *Igneous Petrology.* Longman, 2nd edition.

Hall, C. E., Parmentier, E. M. (2000): Spontaneous melt localization in a deforming solid with viscosity variations due to water weakening. *Geophys. Res. Lett. 27*(1), 9–12.

Hama, J., Suito, K. (2001): Thermoelastic models of minerals and the composition of the Earth's lower mantle. *Phys. Earth Planet. Int. 125*(1–4), 147–166.

Hammond, W. C., Humphreys, E. D. (2000): Upper mantle seismic wave attenuation: Effects of realistic partial melt distribution. *J. Geophys. Res. 105*(B5), 10 987–10 999.

Hanan, B. B., Blichert-Toft, J., Kingsley, R., Schilling, J.-G. (2000): Depleted Iceland mantle plume geochemical signature: Artifact of multicomponent mixing? *Geochem. Geophys. Geosyst. 1*, doi:1999GC000009.

Hanan, B. B., Schilling, J.-G. (1997): The dynamic evolution of the Iceland mantle plume: the lead isotope perspective. *Earth Planet. Sci. Lett. 151*, 43–60.

Hansen, U., Yuen, D. A. (1994): Effects of depth-dependent thermal expansivity on the interaction of thermal–chemical plumes with a compositional boundary. *Phys. Earth Planet. Int. 86*, 205–221.

Hansen, U., Yuen, D. A., Kroening, S. E., Larsen, T. B. (1993): Dynamical consequences of depth-dependent thermal expansivity and viscosity on mantle circulations and thermal structure. *Phys. Earth Planet. Int. 77*, 201–223.

Hardarson, B. S., Fitton, J. G., Ellam, R. M., Pringle, M. S. (1997): Rift relocation — a geochemical and geochronological investigation of a palaeo-rift in northwest Iceland. *Earth Planet. Sci. Lett. 153*, 181–196.

Harris, P. G., Reay, A., White, I. G. (1967): Chemical composition of the upper mantle. *J. Geophys. Res. 72*, 6359–6369.

Harrison, W. (1981): Partitioning of REE between minerals and coexisting melts during partial melting of a garnet lherzolite. *Am. Mineral. 66*, 242–259.

Hart, S. R. (1993): Equilibration during mantle melting: a fractal tree model. *Proc.Nat.Acad.Sci.USA 90*(24), 11 914–11 918.

Hashin, Z., Shtrikman, S. (1963): A variational approach to the theory of the elastic behaviour of multiphase materials. *J. Mech. Phys. Solids 11*, 127–140.

Hauri, E. H., Hart, S. R. (1994): Constraints on melt migration from mantle plumes: A trace element study of peridotite xenoliths from Savai'i, Western Samoa. *J. Geophys. Res. 99*(B12), 24 301–24 321. Correction in *J. Geophys. Res.* 100, p.2003 (1995).

Hauri, E. H., Wagner, T. P., Grove, T. L. (1994a): Experimental and natural partitioning of Th, U, Pb and other trace elements between garnet, clinopyroxene and basaltic melts. *Chem. Geol. 117*, 149–166.

Hauri, E. H., Whitehead, J. A., Hart, S. R. (1994b): Fluid dynamic and geochemical aspects of entrainment in mantle plumes. *J. Geophys. Res. 99*(B12), 24 275–24 300.

Heinson, G., Constable, S., White, A. (2000): Episodic melt transport at mid-ocean ridges inferred from magnetotelluric sounding. *Geophys. Res. Lett. 27*(15), 2317–2320.

Hellebrand, E., Snow, J. E., Dick, H. J. B., Hofmann, A. W. (2001): Coupled major and trace elements as indicators of the extent of melting in mid-ocean-ridge peridotites. *Nature 410*, 677–681.

Hellebrand, E., Snow, J. E., Hoppe, P., Hofmann, A. W. (2002): Garnet-field melting and late-stage refertilization in 'residual' abyssal peridotites from the Central Indian Ridge. *J. Petrol. 43*(12), 2305–2338.

Heller, D.-A., Marquart, G. (2002): An admittance study of the Reykjanes Ridge and elevated plateaux between the Charlie-Gibbs and Senja fracture zones. *Geophys. J. Int. 148*(1), 65–76.

Helmberger, D. V., Wen, L., Ding, X. (1998): Seismic evidence that the source of the Iceland hotspot lies at the core–mantle boundary. *Nature 396*, 251–255.

Hémond, C., Arndt, N. T., Lichtenstein, U., Hofmann, A. W., Óskarsson, N., Steinthorsson, S. (1993): The heterogeneous Iceland plume: Nd-Sr-O isotopes and trace element constraints. *J. Geophys. Res. 98*, 15 833–15 850.

Hersir, G. P., Björnsson, A., Pedersen, L. B. (1984): Magnetotelluric survey across the active spreading zone in southwest Iceland. *J. Volc. Geoth. Res. 20*, 253–265.

Hertogen, J., Gijbels, R. (1976): Calculation of trace element fractionation during partial melting. *Geochim. Cosmochim. Acta 40*, 313–322.

Herzberg, C., Gasparik, T., Sawamoto, H. (1990): Origin of mantle peridotite: Constraints from melting experiments to 16.5 GPa. *J. Geophys. Res. 95*(B10), 15 779–15 803.

Herzberg, C., Raterron, P., Zhang, J. (2000): New experimental observations on the anhydrous solidus for peridotite KLB-1. *Geochem. Geophys. Geosyst. 1*, doi:2000GC000089.

Herzberg, C., Zhang, J. (1996): Melting experiments on anhydrous peridotite KLB-1: Compositions of magmas in the upper mantle and transition zone. *J. Geophys. Res. 101*, 8271–8295.

Hess, P. C. (1992): Phase equilibria constraints on the origin of ocean floor basalts. In *Mantle flow and melt generation at mid-ocean ridges*, edited by J. P. Morgan, D. K. Blackman, J. M. Sinton, volume 71 of *Geophys. Monogr. Ser.*, pp. 67–102, AGU, Washington, D.C.

Higo, Y., Inoue, T., Irifune, T., Yurimoto, H. (2001): Effect of water on the spinel–postspinel transformation in Mg_2SiO_4. *Geophys. Res. Lett. 28*(18), 3505–3508. Correction in *Geophys. Res. Lett.* 28(23), p.4415 (2001).

Hilton, D. R., Grönvold, K., Macpherson, C. G., Castillo, P. R. (1999): Extreme $^3He/^4He$ ratios in northwest Iceland: constraining the common component in mantle plumes. *Earth Planet. Sci. Lett. 173*, 53–60.

Hinz, K. (1981): A hypothesis on terrestrial catastrophes: Wedges of very thick oceanward dipping layers beneath passive continental margins. *Geol. J. E22*, 3–28.

Hirose, K. (2002): Phase transitions in pyrolitic mantle around 670-km depth: Implications for upwelling of plumes from the lower mantle. *J. Geophys. Res. 107*(B4), 10.1029/2001JB000 597.

Hirose, K., Kawamura, K. (1994): A new experimental approach for incremental batch melting of peridotite at 1.5 GPa. *Geophys. Res. Lett. 21*(19), 2139–2142.

Hirose, K., Kushiro, I. (1993): Partial melting of dry peridotites at high pressures: Determination of compositions of melts segregated from peridotite using aggregates of diamond. *Earth Planet. Sci. Lett. 114*, 477–489.

Hirschmann, M. M. (2000): Mantle solidus: Experimental constraints and the effects of peridotite composition. *Geochem. Geophys. Geosyst. 1*, doi:2000GC000070.

Hirschmann, M. M., Asimow, A. D., Ghiorso, M. S., Stolper, E. M. (1999): Calculation of peridotite partial melting from thermodynamic models of minerals and melts. III. Controls on isobaric melt production and the effect of water on melt production. *J. Petrol. 40*(5), 831–851.

Hirschmann, M. M., Ghiorso, M. S., Wasylenski, L. E., Asimow, P. D., Stolper, E. M. (1998): Calculation of peridotite partial melting from thermodynamic models of minerals and melts. I. Review of methods and comparison with experiments. *J. Petrol. 39*(6), 1091–1115.

Hirschmann, M. M., Stolper, E. M. (1996): A possible role for garnet pyroxenite in the origin of the "garnet signature" in MORB. *Contrib. Mineral. Petrol. 124*, 185–208.

Hirth, G., Kohlstedt, D. L. (1995): Experimental constraints on the dynamics of the partially molten upper mantle: 1. Deformation in the diffusion creep regime. *J. Geophys. Res. 100*(B2), 1981–2001.

Hirth, G., Kohlstedt, D. L. (1996): Water in the oceanic upper mantle: implications for rheology, melt extraction and evolution of the lithosphere. *Earth Planet. Sci. Lett. 144*, 93–108.

Hofmann, A. W., White, W. M. (1982): Mantle plumes from ancient oceanic crust. *Earth Planet. Sci. Lett. 57*, 421–436.

Hofmeister, A. M. (1999): Mantle values of thermal conductivity and the geotherm from phonon lifetimes. *Science 283*, 1699–1706.

Holbrook, W. S. Larsen, H. C., Korenaga, J., Dahl-Jensen, T., Reid, I. D., Kelemen, P. B., Hopper, J. R., Kent, G. M., Lizarralde, D., Bernstein, S., Detrick, R. S. (2001): Mantle thermal structure and active upwelling during continental breakup in the North Atlantic. *Earth Planet. Sci. Lett. 190*(3–4), 251–266.

Huber, P. J. (1981): *Robust Statistics.* J.Wiley, New York.

Huebner, J. S., Dillenburg, R. G. (1995): Impedance spectra of hot, dry silicate minerals and rock: Qualitative interpretation of spectra. *Am. Mineral. 80*, 46–64.

Hung, S.-H., Shen, Y., Chiao, L.-Y. (2003): Imaging the Iceland mantle plume by finite-frequency traveltime tomography. In *Geophys. Res. Abstr.*, volume 5, p. 11076.

Ita, J., King, S. D. (1994): Sensivity of convection with an endothermic phase change to the form of governing equations, initial conditions, boundary conditions, and equation of state. *J. Geophys. Res. 99*(B8), 15 919–15 938.

Ita, J., Stixrude, L. (1992): Petrology, elasticity, and composition of the mantle transition zone. *J. Geophys. Res. 97*(B5), 6849–6866.

Ito, G. (2001): Reykjanes 'V'-shaped ridges originating from a pulsing and dehydrating plume. *Nature 411*, 681–684.

Ito, G., Lin, J. (1995): Oceanic spreading center–hotspot interactions — constraints from along-isochron bathymetric and gravity anomalies. *Geology 23*(7), 657–660.

Ito, G., Lin, J., Gable, C. W. (1996): Dynamics of mantle flow and melting at a ridge-centered hotspot: Iceland and the mid-Atlantic ridge. *Earth Planet. Sci. Lett. 144*, 53–74.

Ito, G., Martel, S. J. (2002): Focusing of magma in the upper mantle through dike interaction. *J. Geophys. Res. 107*(B10), 2223, doi:10.1029/2001JB000251.

Ito, G., Shen, Y., Hirth, G., Wolfe, C. J. (1999): Mantle flow, melting and dehydration of the Iceland mantle plume. *Earth Planet. Sci. Lett. 165*, 81–96.

Iwamori, H. (1993a): Dynamic disequilibrium melting model with porous flow and diffusion-controlled chemical equilibration. *Earth Planet. Sci. Lett. 114*, 301–313.

Iwamori, H. (1993b): A model for disequilibrium mantle melting incorporating melt transport by porous and channel flows. *Nature 366*, 734–737.

Iwamori, H., McKenzie, D., Takahashi, E. (1995): Melt generation by isentropic mantle upwelling. *Earth Planet. Sci. Lett. 134*, 253–266.

Jackson, I., Fitz Gerald, J. D., Faul, U., Kokkonen, H., Carr, J., Tan, B. (2000): Seismic wave attenuation in polycrystalline olivine: frequency, temperature and grainsize sensitivity. In *Annual Report*, pp. 115–117, ANU, Canberra.

Jamtveit, B., Brooker, R., Brooks, K., Melchior Larsen, L., Pedersen, T. (2001): The water content of olivines from the North Atlantic Volcanic Province. *Earth Planet. Sci. Lett. 186*, 401–415.

Ji, S., Wang, Z., Wirth, R. (2001): Bulk flow strength of forsterite–enstatite composites as a function of forsterite content. *Tectonophys. 341*(1–4), 69–93.

Jin, Z.-M., Green, H. W., Zhou, Y. (1994): Melt topology in partially molten mantle peridotite during ductile deformation. *Nature 372*, 164–167.

Johnson, G. R., Olhoeft, G. R. (1984): Density of rocks and minerals. In *Handbook of Physical Properties of Rocks*, edited by R. S. Carmichael, volume III, pp. 1–38, CRC Press, Boca Raton, Florida.

Johnson, K. T. M. (1998): Experimental determination of partition coefficients for rare earth and high-field-strength elements between clinopyroxene, garnet, and basaltic melt at high pressures. *Contrib. Mineral. Petrol. 133*, 60–68.

Johnson, K. T. M., Dick, H. J. B., Shimizu, N. (1990): Melting in the oceanic upper mantle: An ion microprobe study of diopsides in abyssal peridotites. *J. Geophys. Res. 95*(B3), 2661–2678.

Jones, S. M., White, N., Maclennan, J. (2002): V-shaped ridges around Iceland: Implications for spatial and temporal patterns. *Geochem. Geophys. Geosyst. 3*(10), 1059, doi: 10.1029/2001GC000361.

Jordan, T. H. (1979): Mineralogies, densities and seismic velocities of garnet lherzolites and their geophysical implication. In *The Mantle Sample: Inclusions in Kimberlites and other Volcanics*, edited by F. R. Boyd, O. A. Meyer, pp. 1–14, American Geophysical Union, Washington.

Jull, M., Kelemen, P. B., Sims, K. (2002): Consequences of diffuse and channelled porous melt migration on uranium series disequilibria. *Geochim. Cosmochim. Acta 66*(23), 4133–4148.

Jull, M., McKenzie, D. (1996): The effect of deglaciation on mantle melting beneath Iceland. *J. Geophys. Res. 101*(B10), 21 815–21 828.

Jung, H., Waff, H. S. (1998): Olivine crystallographic control and anisotropic melt distribution in ultramafic partial melts. *Geophys. Res. Lett. 25*(15), 2901–2904.

Kaban, M. K., Flóvenz, O., Pálmason, G. (2002): Nature of the crust-mantle transition zone and the thermal state of the upper mantle beneath Iceland from gravity modeling. *Geophys. J. Int. 149*(2), 281–299.

Kamke, E. (1956): *Differentialgleichungen – Lösungsmethoden und Lösungen*, volume II. Akademische Verlagsgesellschaft Geest & Portig, Leipzig, 3rd edition.

Karato, S.-I. (1986): Does partial melting reduce the creep strength of the upper mantle? *Nature 319*, 309–310.

Karato, S.-I. (1990): The role of hydrogen in the electrical conductivity of the upper mantle. *Nature 347*, 272–273.

Karato, S.-I. (1993): Importance of anelasticity in the interpretation of seismic anisotropy. *Geophys. Res. Lett. 20*(15), 1623–1626.

Karato, S.-I., Jung, H. (1998): Water, partial melting and the origin of the seismic low velocity and high attenuation zone in the upper mantle. *Earth Planet. Sci. Lett. 157*, 193–207.

Karato, S.-I., Wu, P. (1993): Rheology of the upper mantle: a synthesis. *Science 260*, 771–778.

Karato, S.-I., Zhang, S., Zimmerman, M. E., Daines, M. J., Kohlstedt, D. L. (1998): Experimental studies of shear deformation of mantle materials: towards structural geology of the mantle. *Pure Appl. Geoph. 151*, 589–603.

Karki, B., Stixrude, L., Wentzcovitch, R. (2001): High-pressure elastic properties of major materials of Earth's mantle from first principles. *Rev. Geophys. 39*(4), 507–534.

Katsura, T., Ito, E. (1989): The system Mg_2SiO_4–Fe_2SiO_4 at high pressures and temperatures: Precise determination of stabilities of olivine, modified spinel, and spinel. *J. Geophys. Res. 94*(B11), 15 663–15 670.

Katz, R. F., Spiegelman, M., Langmuir, C. (2003): A new parameterization of hydrous mantle melting. *Geochem. Geophys. Geosyst. 4*(9), 1073, doi:10.1029/2002GC000433.

Keen, C. E., Boutilier, R. R. (2000): Interaction of rifting and hot horizontal plume sheets at volcanic margins. *J. Geophys. Res. 105*(B6), 13 375–13 387.

Kelemen, P. B., Braun, M., Hirth, G. (2000): Spatial distribution of melt conduits in the mantle beneath oceanic spreading ridges: Observations from the Ingalls and Oman ophiolites. *Geochem. Geophys. Geosyst.* 1, doi:1999GC000012.

Kelemen, P. B., Hirth, G., Shimizu, N., Spiegelman, M., Dick, H. J. B. (1997): A review of melt migration processes in the adiabatically upwelling mantle beneath oceanic spreading centers. *Phil. Trans. Roy. Soc. Lond. A 355*(1723), 283–318.

Keller, W. R., Anderson, D. L., Clayton, R. W. (2000): Resolution of tomographic models of the mantle beneath Iceland. *Geophys. Res. Lett. 27*(24), 3993–3996.

Kempton, P. D., Fitton, J. G., Saunders, A. D., Nowell, G. M., Taylor, R. N., Hardarson, B. S., Pearson, G. (2000): The Iceland plume in space and time: a Sr–Nd–Pb–Hf study of the North Atlantic rifted margin. *Earth Planet. Sci. Lett. 177*(3–4), 255–271.

Kennett, B. L. N. (1995): Seismic Traveltime Tables. In *Global Earth Physics — A Handbook of Physical Constants*, edited by T. J. Ahrens, volume 1 of *AGU Reference Shelf*, pp. 126–143, AGU, Washington.

Kennett, B. L. N., Engdahl, E. R. (1991): Traveltimes for global earthquake location and phase identification. *Geophys. J. Int.* 105, 429–465.

Kenyon, P. M. (1998): The effect of channel spacing during magma migration on trace element and isotopic ratios. *Geophys. Res. Lett. 25*(21), 3995–3998.

Kincaid, C., Schilling, J.-G., Gable, C. (1996): The dynamics of off-axis plume–ridge interaction in the uppermost mantle. *Earth Planet. Sci. Lett.* 137, 29–43.

Kinzler, R. J. (1997): Melting of mantle peridotite at pressures approaching the spinel to garnet transition: Application to mid-ocean ridge basalt petrogenesis. *J. Geophys. Res. 102*(B1), 853–874.

Knittle, E. (1995): Static compression measurements of equations of state. In *Mineral Physics and Crystallography — A Handbook of Physical Constants*, edited by T. J. Ahrens, volume 2 of *AGU Reference Shelf*, pp. 98–142, AGU, Washington.

Kogiso, T., Hirschmann, M. M. (2001): Experimental study of clinopyroxenite partial melting and the origin of ultra-calcic melt inclusions. *Contrib. Mineral. Petrol.* 142, 347–360.

Kohlstedt, D. L., Bai, Q., Wang, Z.-C., Mei, S. (2000): Rheology of partially molten rocks. In *Physics and Chemistry of Partially Molten Rocks*, edited by N. Bagdassarov, D. Laporte, A. B. Thompson, pp. 3–28, Kluwer Academic Publishers.

Kohlstedt, D. L., Keppler, H., Rubie, D. C. (1996): Solubility of water in the α, β and γ phases of $(Mg, Fe)_2SiO_4$. *Contrib. Mineral. Petrol. 123*(4), 345–357.

Kohlstedt, D. L., Zimmerman, M. E. (1996): Rheology of partially molten mantle rocks. *Ann. Rev. Earth Planet. Sci.* 24, 41–62.

Kojitani, H., Akaogi, M. (1995): Measurement of heat of fusion of model basalt in the system diopside–forsterite–anorthite. *Geophys. Res. Lett. 22*(17), 2329–2332.

Kojitani, H., Akaogi, M. (1997): Melting enthalpies of mantle peridotite: calorimetric determinations in the system CaO–MgO–Al$_2$O$_3$–SiO$_2$ and application to magma generation. *Earth Planet. Sci. Lett.* **153**, 209–222.

Kokfelt, T. F., Hoernle, K., Hauff, F. (2003): Upwelling and melting of the Iceland plume from radial variation of ^{238}U–^{230}Th disequilibria in postglacial volcanic rocks. *Earth Planet. Sci. Lett.* **214**(1–2), 167–186. Erratum in *Earth Planet. Sci. Lett.* 217 (1–2), 219–220.

Korenaga, J., Kelemen, P. B. (2000): Major element heterogeneity in the mantle source of the North Atlantic igneous province. *Earth Planet. Sci. Lett.* **184**(1), 251–268.

Kostopoulos, D. (1991): Melting of the shallow upper mantle: A new perspective. *J. Petrol.* **32**(4), 671–699.

Kostopoulos, D., James, S. D. (1992): Parameterization of the melting regime of the shallow upper mantle and the effects of variable lithospheric stretching on mantle modal stratification and trace-element concentrations in magmas. *J. Petrol.* **33**(3), 665–691.

Kreutzmann, A., Schmeling, H., Junge, A., Ruedas, T., Marquart, G., Bjarnason, I. (2004): Temperature and melting of a ridge-centered plume with application to Iceland, part II: Predictions for electromagnetic and seismic observables. *Geophys. J. Int.* Submitted.

Kumagai, I. (2002): On the anatomy of mantle plumes: effect of the viscosity ratio on entrainment and stirring. *Earth Planet. Sci. Lett.* **198**(1–2), 211–224.

Kushiro, I. (1986): Viscosity of partial melts in the upper mantle. *J. Geophys. Res.* **91**(B9), 9343–9350.

Kushiro, I. (2001): Partial melting experiments on peridotite and origin of mid-ocean ridge basalt. *Ann. Rev. Earth Planet. Sci.* **29**, 71–107.

Kushiro, I., Mysen, B. O. (2002): A possible effect of melt structure on the Mg–Fe^{2+} partitioning between olivine and melt. *Geochim. Cosmochim. Acta* **66**(12), 2267–2272.

Kushiro, I., Yoder, Jr., H. S., Mysen, B. O. (1976): Viscosities of basalt and andesite melts at high pressures. *J. Geophys. Res.* **81**(35), 6351–6356.

Landwehr, D., Blundy, J., Chamorro-Perez, E. M., Hill, E., Wood, B. (2001): U-series disequilibria generated by partial melting of spinel lherzolite. *Earth Planet. Sci. Lett.* **188**(3–4), 329–348.

Lange, R. A., Carmichael, I. S. E. (1990): Thermodynamic properties of silicate liquids with emphasis on density, thermal expansion and compressibility. In *Modern methods of igneous petrology: understanding magmatic processes*, edited by J. Nicholls, J. K. Russell, number 24 in Rev. Miner., pp. 25–64, MSA, Washington, D.C.

Langmuir, C. H., Klein, E. M., Plank, T. (1992): Petrological systematics of mid-ocean ridge basalts: Constraints on melt generation beneath ocean ridges. In *Mantle Flow and Melt Generation at Mid-Ocean Ridges*, edited by J. Phipps Morgan, D. K. Blackman, J. M. Sinton, number 71 in Geophysical Monographs, pp. 183–280, American Geophysical Union, Washington.

Laporte, D., Provost, A. (2000): The grain-scale distribution of silicate, carbonate and metallosulfide partial melts: a review of theory and experiments. In *Physics and Chemistry of Partially Molten Rocks*, edited by N. Bagdassarov, D. Laporte, A. B. Thompson, pp. 93–140, Kluwer Academic Publishers.

Laporte, D., Watson, E. B. (1995): Experimental and theoretical constraints on melt distribution in crustal sources: the effect of crystalline anisotropy on melt interconnectivity. *Chem. Geol.* **124**(3–4), 161–184.

Larsen, L. M., Pedersen, A. K., Pedersen, G. K., Piasecki, S. (1992): Timing and duration of Early Tertiary volcanism in the North Atlantic: new evidence from West Greenland. In *Magmatism and the Causes of Continental Break-up*, edited by B. C. Storey, T. Alabaster, R. J. Pankhurst, number 68 in Geological Society Special Publication, pp. 321–333, The Geological Society, London.

Larsen, T. B., Yuen, D. A., Storey, M. (1999): Ultrafast mantle plumes and implications for flood basalt volcanism in the Northern Atlantic Region. *Tectonophysics 311*(1–4), 31–43.

Lawver, L. A., Müller, R. D. (1994): Iceland hotspot track. *Geology 22*, 311–314.

le Roux, P. J., le Roex, A. P., Schilling, J.-G. (2002): MORB melting processes beneath the southern Mid-Atlantic Ridge (40–55°S): a role for mantle plume-derived pyroxenite. *Contrib. Mineral. Petrol. 144*(2), 206–229.

Leitch, A. M., Davies, G. F. (2001): Mantle plumes and flood basalts: Enhanced melting from plume ascent and an eclogite component. *J. Geophys. Res. 106*(B2), 2047–2059.

Levshin, A. L., Ritzwoller, M. H., Barmin, M. P., Villaseñor, A., Padgett, C. A. (2001): New constraints on the arctic crust and uppermost mantle: surface wave group velocities, P_n, and S_n. *Phys. Earth Planet. Int. 123*(2–4), 185–204.

Li, A., Detrick, R. S. (2001): Rayleigh wave constraints on shear wave velocities beneath Iceland. In *EOS Trans. AGU*, volume 82, pp. T32C–03.

Li, Y., Spohn, T. (1991): Vertical structure of the head of a cylindrical mantle plume from a model of two-phase flow with melting. *Geophys. J. Int. 107*, 103–115.

Longhi, J. (1995): Liquidus equilibria of some primary lunar and terrestrial melts in the garnet stability field. *Geochim. Cosmochim. Acta 59*(11), 2375–2386.

Longhi, J. (2002): Some phase equilibrium systematics of lherzolite melting: I. *Geochem. Geophys. Geosyst. 3*(3), doi:10.1029/2001GC000204.

Loper, D. E., Stacey, F. D. (1983): The dynamical and thermal structure of deep mantle plumes. *Phys. Earth Planet. Int. 33*, 304–317.

Lundstrom, C. (2000): Models of U-series disequilibria generation in MORB: the effects of two scales of melt porosity. *Phys. Earth Planet. Int. 121*, 189–204.

Lundstrom, C., Gill, J., Williams, Q. (2000): A geochemically consistent hypothesis for MORB generation. *Chem. Geol. 162*(2), 105–126.

Luth, R. W. (2002): Possible implications of modal mineralogy for melting in mantle lherzolites. *Geochim. Cosmochim. Acta 66*(12), 2091–2098.

Maaløe, S. (1982): Geochemical aspects of permeability controlled partial melting and fractional crystallization. *Geochim. Cosmochim. Acta 46*, 43–57.

Maaløe, S. (1998): Melt dynamics of a layered mantle plume source. *Contrib. Mineral. Petrol. 133*(1–2), 83–95.

Maaløe, S., Aoki, K. (1977): The major element composition of the upper mantle estimated from the composition of lherzolites. *Contrib. Mineral. Petrol. 63*, 161–173.

Mackwell, S. J., Kohlstedt, D. L. (1990): Diffusion of hydrogen in olivine: implications for water in the mantle. *J. Geophys. Res. 95*, 5079–5088.

Maclennan, J., McKenzie, D., Grönvold, K. (2001a): Plume-driven upwelling under central Iceland. *Earth Planet. Sci. Lett. 194*(1–2), 67–82.

Maclennan, J., McKenzie, D., Grönvold, K., Slater, L. (2001b): Crustal accretion under northern Iceland. *Earth Planet. Sci. Lett. 191*(3–4), 295–310.

Manglik, A., Christensen, U. R. (1997): Effect of mantle depletion buoyancy on plume flow and melting beneath a stationary plate. *J. Geophys. Res. 102*(B3), 5019–5028.

Marquart, G. (1991): Interpretation of geoid anomalies around the Iceland hotspot. *Geophys. J. Int. 106*, 149–160.

Marquart, G. (2001): On the geometry of mantle flow beneath drifting lithospheric plates. *Geophys. J. Int. 144*(2), 356–372.

Marquart, G., Schmeling, H. (2003): Modelling crustal accretion above the Iceland plume. In *8th European Workshop on Numerical Modeling of Mantle Convection and Lithospheric Dynamics*, Hrubá Skála.

Marquart, G., Schmeling, H., Ito, G., Schott, B. (2000): Conditions for mantle plumes to penetrate the mantle phase boundaries. *J. Geophys. Res. 105*(B3), 5679–5694.

Masters, T. G., Shearer, P. M. (1995): Seismic Models of the Earth: Elastic and Anelastic. In *Global Earth Physics — A Handbook of Physical Constants*, edited by T. J. Ahrens, volume 1 of *AGU Reference Shelf*, pp. 88–103, AGU, Washington.

McDonough, W. F., Rudnick, R. L. (1998): Mineralogy and composition of the upper mantle. In *Ultrahigh-pressure mineralogy*, edited by R. Hemley, D. Mao, number 37 in Rev. Mineral., pp. 139–164, MSA, Washington, D.C.

McKenzie, D. (1979): Finite deformation during fluid flow. *Geophys. J. R. astr. Soc. 58*, 689–715.

McKenzie, D. (1984): The generation and compaction of partially molten rock. *J. Petrol. 25*, 713–765.

McKenzie, D. (1985a): The extraction of magma from the crust and mantle. *Earth Planet. Sci. Lett. 74*, 81–91.

McKenzie, D. (1985b): ^{230}Th–^{238}U disequilibrium and the melting processes beneath ridge axes. *Earth Planet. Sci. Lett. 72*, 149–157.

McKenzie, D. (1987): The compaction of igneous and sedimentary rocks. *J. Geol. Soc. Lond. 144*, 299–307.

McKenzie, D. (2000): Constraints on melt generation and transport from U-series activity ratios. *Chem. Geol. 162*(2), 81–94.

McKenzie, D., Bickle, M. J. (1988): The volume and composition of melt generated by extension of the lithosphere. *J. Petrol. 29*(3), 625–679.

McKenzie, D., Bickle, M. J. (1990): A eutectic parameterization of mantle melting. *J. Phys. Earth 38*, 511–515.

McKenzie, D., O'Nions, R. K. (1991): Partial melt distribution from inversion of rare earth element concentrations. *J. Petrol. 32*, 1021–1091. Correction in *J. Petrol.*, 33(6), p.1453 (1992).

Mégnin, C., Romanowicz, B. (2000): The three-dimensional shear velocity structure of the mantle from the inversion of body, surface and higher-mode waveforms. *Geophys. J. Int. 143*(3), 709–728.

Mei, S., Bai, W., Hiraga, T., Kohlstedt, D. L. (2002): Influence of melt on the creep behavior of olivine–basalt aggregates under hydrous conditions. *Earth Planet. Sci. Lett. 201*(3–4), 491–507.

Mei, S., Kohlstedt, D. L. (2000a): Influence of water on plastic deformation of olivine aggregates 1. Diffusion creep regime. *J. Geophys. Res. 105*(B9), 21 457–21 469.

Mei, S., Kohlstedt, D. L. (2000b): Influence of water on plastic deformation of olivine aggregates 2. Dislocation creep regime. *J. Geophys. Res. 105*(B9), 21 471–21 481.

Menke, W. (1984): *Geophysical Data Analysis: Discrete Inverse Theory*. Academic Press, Orlando.

Menke, W. (1999): Crustal isostasy indicates anomalous densities beneath Iceland. *Geophys. Res. Lett. 26*(9), 1215–1218.

Menke, W., Brandsdóttir, B., Einarsson, P., Bjarnason, I. Þ. (1996): Reinterpretation of the RRISP-77 Iceland shear-wave profiles. *Geophys. J. Int. 126*, 166–172.

Menke, W., Levin, V. (1994): Cold crust in a hot spot. *Geophys. Res. Lett. 21*(18), 1967–1970.

Menke, W., Levin, V., Sethi, R. (1995): Seismic attenuation in the crust at the mid-Atlantic plate boundary in south-west Iceland. *Geophys. J. Int. 122*, 175–182.

Menke, W., Sparks, D. (1995): Crustal accretion model for Iceland predicts 'cold' crust. *Geophys. Res. Lett. 22*(13), 1673–1676.

Menke, W., West, M., Brandsdóttir, B., Sparks, D. (1998): Compressional and shear velocity structure of the lithosphere in northern Iceland. *Bull. Seis. Soc. Am. 88*(6), 1561–1571.

Menzies, M. (1973): Mineralogy and partial melt textures within a mafic–ultramafic body, Greece. *Contrib. Mineral. Petrol. 42*, 273–285.

Mertz, D. F., Devey, C. W., Todt, W., Stoffers, P., Hofmann, A. W. (1991): Sr-Nd-Pb isotope evidence against plume–asthenosphere mixing north of Iceland. *Earth Planet. Sci. Lett. 107*, 243–255.

Mertz, D. F., Haase, K. M. (1997): The radiogenic isotope composition of the high-latitude North-Atlantic mantle. *Geology 25*(5), 411–414.

Meyer, P. S., Sigurdsson, H., Schilling, J.-G. (1985): Petrological and geochemical variations along Iceland's Neovolcanic zones. *J. Geophys. Res. 90*(B12), 10043–10072.

Moore, W. B., Schubert, G., Tackley, P. J. (1999): The role of rheology in lithospheric thinning by mantle plumes. *Geophys. Res. Lett. 26*(8), 1073–1076.

Moreira, M., Breddam, K., Curtice, J., Kurz, M. D. (2001): Solar neon in the Icelandic mantle: new evidence for an undegassed lower mantle. *Earth Planet. Sci. Lett. 185*(1–2), 15–23.

Morgan, W. J. (1971): Convection plumes in the lower mantle. *Nature 230*, 42–43.

Muller, J. R., Ito, G., Martel, S. J. (2001): Effects of volcano loading on dike propagation in an elastic half-space. *J. Geophys. Res. 106*(B6), 11101–11113.

Murton, B. J., Taylor, R. N., Thirlwall, M. F. (2002): Plume-ridge interaction: a geochemical perspective from the Reykjanes Ridge. *J. Petrol. 43*(11), 1987–2012.

Nakano, T., Fujii, N. (1989): The multiphase grain control percolation: its implications for a partially molten rock. *J. Geophys. Res. 94*(B11), 15653–15661.

Navon, O., Stolper, E. (1987): Geochemical consequences of melt percolation: the upper mantle as a chromatographic column. *J. Geol. 95*(3), 285–307.

Navrotsky, A. (1994): *Physics and Chemistry of Earth Materials*. Cambridge University Press.

Navrotsky, A. (1995): Thermodynamic Properties of Minerals. In *Mineral Physics & Crystallography — A Handbook of Physical Constants*, edited by T. J. Ahrens, volume 2 of *AGU Reference Shelf*, pp. 18–28, AGU, Washington.

Nichols, A. R. L., Carroll, M. R., Höskuldsson, Á. (2002): Is the Iceland hot spot also wet? Evidence from the water contents of undegassed submarine and subglacial pillow basalts. *Earth Planet. Sci. Lett. 202*(1), 77–87.

Nicholson, H., Latin, D. (1992): Olivine tholeiites from Krafla, Iceland: Evidence for variations in melt fraction within a plume. *J. Petrol. 33*(5), 1105–1124.

Nicolas, A. (1986): A melt extraction model based on structural studies in mantle peridotites. *J. Petrol. 27*, 999–1022.

Nicolas, A. (1990): Melt extraction from mantle peridotites: hydrofracturing and porous flow, with consequences for oceanic ridge activity. In *Magma transport and storage*, edited by M. P. Ryan, pp. 159–173, Wiley, Chichester.

Nielsen, T. K., Larsen, H. C., Hopper, J. R. (2002): Contrasting rifted margin styles south of Greenland: implications for mantle plume dynamics. *Earth Planet. Sci. Lett. 200*(3–4), 271–286.

Niu, Y. (1997): Mantle melting and melt extraction processes beneath ocean ridges: Evidence from abyssal peridotites. *J. Petrol. 38*(8), 1047–1074. Also see: Walter, M.J.: Comments on 'Mantle melting and melt extraction processes beneath ocean ridges: Evidence from abyssal peridotites' by Yaoling Niu, *J. Petrol. 40*(7), pp.1187–1193 (1999) and Niu, Y.L.: Comments on some misconceptions in igneous and experimental petrology and methodology: a reply, *J. Petrol. 40*(7), pp.1195–1203 (1999).

Niu, Y., Batiza, R. (1991): In situ densities of MORB melts and residual mantle: implications for buoyancy forces beneath mid-ocean ridges. *J. Geol. 99*, 767–775.

Nixon, P. H., Boyd, F. R. (1973): Petrogenesis of the granular and sheared ultrabasic nodule suite in kimberlites. In *Lesotho Kimberlites*, edited by P. H. Nixon, pp. 48–56, Lesotho National Development Corporation.

Nolen-Hoeksema, R. C. (2000): Modulus-porosity relations, Gassmann's equations, and the low-frequency elastic-wave response to fluids. *Geophysics 65*(5), 1355–1363.

O'Connor, J. M., Stoffers, P., Wijbrans, J. R., Shannon, P. M., Morrissey, T. (2000): Evidence from episodic seamount volcanism for pulsing of the Iceland plume in the past 70 Myr. *Nature 408*, 954–958.

O'Hara, M. J., Richardson, S. W., Wilson, G. (1971): Garnet-peridotite stability and occurrence in crust and mantle. *Contr.Mineral.Petrol. 32*, 48–68.

Ohtani, E. (1984): Generation of komatiite magma and gravitational differentiation in the deep upper mantle. *Earth Planet. Sci. Lett. 67*, 261–272.

Ohtani, E., Kumazawa, M. (1981): Melting of forsterite Mg_2SiO_4 up to 15 GPa. *Phys. Earth Planet. Int. 27*, 32–38.

Oliveira, A., Fortunato, A. B. (2002): Toward an oscillation-free, mass conservative, Eulerian–Lagrangian transport model. *J. Comp. Phys. 183*(1), 142–164.

Olson, P. (1990): Hot spots, swells and mantle plumes. In *Magma transport and storage*, edited by M. P. Ryan, pp. 33–51, Wiley, Chichester.

Olson, P. (1994): Mechanics of flood basalt magmatism. In *Magmatic Systems*, edited by M. P. Ryan, pp. 1–18, Academic Press.

Ortoleva, P., Merino, E., C., M., Chadam, J. (1987): Geochemical self-organization, I, reaction-transport feedbacks. *Am.J.Sci. 287*, 979–1007.

Óskarsson, N., Steinþorsson, S., Sigvaldason, G. E. (1985): Iceland geochemical anomaly: origin, volcanotectonics, chemical fractionation and isotope evolution of the crust. *J. Geophys. Res. 90*(B12), 10 011–10 025.

Oxburgh, E. R., Parmentier, E. M. (1977): Compositional and density stratification in the oceanic lithosphere — causes and consequences. *J. Geol. Soc. Lond. 133*, 343–354.

Ozawa, K. (2001): Mass balance equations for open magmatic systems: Trace element behavior and its application to open system melting in the upper mantle. *J. Geophys. Res. 106*(B7), 13 407–13 434.

Pálmason, G. (1971): *Crustal Structure of Iceland from Explosion Seismology.* Soc. Sci. Iceland, Reykjavík, 1st edition.

Pálmason, G. (1973): Kinematics and heat flow in a Volcanic Rift Zone, with application to Iceland. *Geophys.J.Roy.astr.Soc. 33*, 451–481.

Pálmason, G. (1980): A continuum model of crustal generation in Iceland; Kinematic aspects. *J. Geophys. 47*, 7–18.

Pankov, V. L., Ullmann, W., Heinrich, R., Kracke, D. (1998): Thermodynamics of deep geophysical media. *Russ. J. Earth Sci. 1*(1), 11–49.

Parmentier, E. M. (1995): Spreading center dynamics and melt migration. *Rev. Geophys. 33S*, 385–400.

Paterson, M. S. (2001): A granular flow theory for the deformation of partially molten rock. *Tectonophys. 335*(1–2), 51–61.

Peate, D. W., Baker, J. A., Blichert-Toft, J., Hilton, D. R., Storey, M., Kent, A. J. R., Brooks, C. K., Hansen, H., Pedersen, A. K., Duncan, R. A. (2003): The Prinsen af Wales Bjerge formation lavas, East Greenland: the transition from tholeiitic to alkalic magmatism during palaeogene continental break-up. *J. Petrol. 44*(2), 279–304.

Peyret, R., Taylor, T. D. (1985): *Computational Methods for Fluid Flow*. Springer Series in Computational Physics, Springer.

Phipps Morgan, J. (1987): Melt migration beneath mid-ocean spreading centers. *Geophys. Res. Lett. 14*(12), 1238–1241.

Phipps Morgan, J. (1997): The generation of a compositional lithosphere by mid-ocean ridge melting and its effect on subsequent off-axis hotspot upwelling and melting. *Earth Planet. Sci. Lett. 146*, 213–232.

Phipps Morgan, J. (2001): Thermodynamics of pressure release melting of a veined plum pudding mantle. *Geochem. Geophys. Geosyst. 2*, doi:2000GC000049.

Phipps Morgan, J., Chen, Y. J. (1993): Dependence of ridge axis morphology on magma supply and spreading rate. *Nature 364*, 706–708.

Phipps Morgan, J., Morgan, W. J. (1999): Two-stage melting and the geochemical evolution of the mantle: a recipe for mantle plum-pudding. *Earth Planet. Sci. Lett. 170*(3), 215–239.

Phipps Morgan, J., Parmentier, E. M., Lin, J. (1987): Mechanisms for the origin of mid-ocean ridge axial topography: Implications for the thermal and mechanical structure of accreting plate boundaries. *J. Geophys. Res. 92*(B12), 12 823–12 836.

Pickering-Witter, J., Johnston, A. D. (2000): The effects of variable bulk composition on the melting systematics of fertile peridotitic assemblages. *Contrib. Mineral. Petrol. 140*, 190–211.

Poirier, J.-P., Tarantola, A. (1998): A logarithmic equation of state. *Phys. Earth Planet. Int. 109*, 1–8.

Pollitz, F. F., Sacks, I. S. (1996): Viscosity structure beneath northeast Iceland. *J. Geophys. Res. 101*(B8), 17 771–17 793.

Presnall, D. C. (1969): The geometrical analysis of partial fusion. *Am.J.Sci. 267*, 1178–1194.

Presnall, D. C. (1986): An algebraic method for determining equilibrium crystallization and fusion paths in multicomponent systems. *Am. Mineral. 71*, 1061–1070.

Presnall, D. C., Gudfinnsson, G. H., Walter, M. J. (2002): Generation of mid-ocean ridge basalts at pressures from 1 to 7 GPa. *Geochim. Cosmochim. Acta 66*(12), 2073–2090.

Presnall, D. C., Walter, M. J. (1993): Melting of forsterite, Mg_2SiO_4, from 9.7 to 16.5 GPa. *J. Geophys. Res. 98*(B11), 19 777–19 783.

Press, W. H., Teukolsky, S. A., Vetterling, W. T., Flannery, B. P. (1992): *Numerical Recipes in FORTRAN — The Art of Scientific Computing*. Cambridge University Press, 2nd edition.

Pritchard, M. J., Foulger, G. R., Julian, B. R., Fyen, J. (2000): Constraints on a plume in the mid-mantle beneath the Iceland region from seismic array data. *Geophys. J. Int. 143*(1), 119–128.

Putirka, K., Johnson, M., Kinzler, R., Longhi, J., Walker, D. (1996): Thermobarometry of mafic igneous rocks based on clinopyroxene–liquid equilibria, 0–30 kbar. *Contrib. Mineral. Petrol. 123*, 92–108.

Putnis, A. (1992): *Introduction to mineral sciences*. Cambridge University Press.

Rabinowicz, M., Briais, A. (2002): Temporal variations of the segmentation of slow to intermediate spreading mid-ocean ridges 2. A three-dimensional model in terms of lithosphere accretion and convection within the partially molten mantle beneath the ridge axis. *J. Geophys. Res. 107*(B6), doi:10.1029/2001JB000343.

Ribe, N. M. (1985a): The deformation and compaction of partial molten zones. *Geophys. J. R. astr. Soc. 83*, 487–501.

Ribe, N. M. (1985b): The generation and composition of partial melts in the earth's mantle. *Earth Planet. Sci. Lett. 73*, 361–376.

Ribe, N. M. (1986): Melt segregation by dynamic forcing. *Geophys. Res. Lett. 13*(13), 1462–1465.

Ribe, N. M. (1987): Theory of melt segregation — a review. *J. Volc. Geotherm. Res. 33*, 241–253.

Ribe, N. M. (1988): Dynamical geochemistry of the Hawaiian plume. *Earth Planet. Sci. Lett. 88*, 37–46.

Ribe, N. M. (1989): Seismic anisotropy and mantle flow. *J. Geophys. Res. 94*(B4), 4213–4223.

Ribe, N. M. (1996): The dynamics of plume-ridge interaction, 2: Off-ridge plumes. *J. Geophys. Res. 101*(B7), 16 195–16 204.

Ribe, N. M., Christensen, U. R. (1994): Three-dimensional modeling of plume-lithosphere interaction. *J. Geophys. Res. 99*(B1), 669–682.

Ribe, N. M., Christensen, U. R., Theißing, J. (1995): The dynamics of plume-ridge interaction, 1: Ridge-centered plumes. *Earth Planet. Sci. Lett. 134*, 155–168.

Ribe, N. M., Delattre, W. L. (1998): The dynamics of plume-ridge interaction — III. The effects of ridge migration. *Geophys. J. Int. 133*(3), 511–518.

Ribe, N. M., Smooke, M. D. (1987): A stagnation point flow model for melt extraction from a mantle plume. *J. Geophys. Res. 92*(B7), 6437–6443.

Richards, M. A., Duncan, R. A., Courtillot, V. E. (1989): Flood basalts and hot-spot tracks: Plume heads and tails. *Science 246*, 103–107.

Richardson, C. N. (1998): Melt flow in a variable viscosity matrix. *Geophys. Res. Lett. 25*(7), 1099–1102.

Richardson, C. N., Lister, J. R., McKenzie, D. (1996): Melt conduits in a viscous porous matrix. *J. Geophys. Res. 101*(B9), 20 423–20 432.

Richardson, K. R., Smallwood, J. R., White, R. S., Snyder, D. B., Maguire, P. K. H. (1998): Crustal structure beneath the Faroe Islands and the Faroe–Iceland Ridge. *Tectonophys. 300*, 159–180.

Richet, P. (1984): Viscosity and configurational entropy of silicate melts. *Geochim. Cosmochim. Acta 48*, 471–483.

Richet, P., Bottinga, Y. (1986): Thermochemical properties of silicate glasses and liquids: a review. *Rev. Geophys. 24*(1), 1–25.

Richet, P., Leclerc, F., Benoist, L. (1993): Melting of forsterite and spinel, with implications for the glass transition of Mg_2SiO_4 liquid. *Geophys. Res. Lett. 20*(16), 1675–1678.

Richter, F. M., McKenzie, D. (1984): Dynamical models for melt segregation from a deformable matrix. *J. Geol. 92*, 729–740.

Rigden, S. M., Ahrens, T. J., Stolper, E. M. (1984): Densities of liquid silicates at high pressures. *Science 226*, 1071–1074.

Rigden, S. M., Ahrens, T. J., Stolper, E. M. (1989): High-pressure equation of state of molten anorthite and diopside. *J. Geophys. Res. 94*(B7), 9508–9522.

Riisager, P., Abrahamsen, N. (1999): Magnetostratigraphy of Palaeocene basalts from the Vaigat Formation of West Greenland. *Geophys. J. Int. 137*(3), 774–782.

Ringwood, A. E. (1975): *Composition and Petrology of the Earth's Mantle.* McGraw-Hill Company, New York.

Ritsema, J., Allen, R. M. (2003): The elusive mantle plume. *Earth Planet. Sci. Lett. 207*(1–4), 1–12.

Ritsema, J., van Heijst, H. J., Woodhouse, J. H. (1999): Complex shear wave velocity structure imaged beneath Africa and Iceland. *Science 286*, 1925–1928.

Robinson, J. A. C., Wood, B. J. (1998): The depth of the spinel to garnet transition at the peridotite solidus. *Earth Planet. Sci. Lett. 164*(1–2), 277–284.

Rögnvaldsson, S. T., Guðmundsson, Á., Slunga, R. (1998): Seismotectonic analysis of the Tjörnes Fracture Zone, an active transform fault in north Iceland. *J. Geophys. Res. 103*(B12), 30 117–30 129.

Rubin, A. M. (1998): Dike ascent in partially molten rock. *J. Geophys. Res. 103*(B9), 20 901–20 919.

Ruedas, T., Schmeling, H., Marquart, G., Kreutzmann, A., Junge, A. (2004): Temperature and melting of a ridge-centered plume with application to Iceland, part I: Dynamics and crust production. *Geophys. J. Int. in press.*

Ryabchikov, I. D. (1998): Geochemical modelling of magma generation in passive and active mantle plumes. In *Challenges to chemical geology*, edited by M. Novak, J. Rosenbaum, pp. 103–120, Czech Geological Survey, Prague.

Saal, A. E., Hauri, E. H., Langmuir, C. H., Perfit, M. R. (2002): Vapour undersaturation in primitive mid-ocean-ridge basalt and the volatile content of Earth's upper mantle. *Nature 419*, 451–455.

Sæmundsson, K. (1979): Outline of the geology of Iceland. *Jökull 29*, 7–28.

Salters, V. J. M. (1996): The generation of mid-ocean ridge basalts from the Hf and Nd isotope perspective. *Earth Planet. Sci. Lett. 141*, 109–123.

Salters, V. J. M., Dick, H. J. B. (2002): Mineralogy of the mid-ocean ridge basalt source from neodymium isotopic composition of abyssal peridotites. *Nature 418*, 68–72.

Salters, V. J. M., Longhi, J. (1999): Trace element partitioning during the initial stages of melting beneath mid-ocean ridges. *Earth Planet. Sci. Lett. 166*, 15–30.

Salters, V. J. M., Longhi, J. E., Bizimis, M. (2002): Near mantle solidus trace element partitioning at pressures up to 3.4 GPa. *Geochem. Geophys. Geosyst. 3*(6), doi:10.1029/2001GC000148.

Sandwell, D. T., MacKenzie, K. R. (1989): Geoid height versus topography for oceanic plateaus and swells. *J. Geophys. Res. 94*(B6), 7403–7418.

Sandwell, D. T., Smith, W. H. F. (1997): Marine gravity anomaly from Geosat and ERS-1 satellite altimetry. *J. Geophys. Res. 102*(B5), 10 039–10 054.

Sato, H., Sacks, I. S., Murase, T., Muncill, G., Fukuyama, H. (1989): Qp-melting temperature relation in peridotite at high pressure and temperature: Attenuation mechanism and implications for the mechanical properties of the upper mantle. *J. Geophys. Res. 94*(B8), 10 647–10 661.

Saxena, S. K., Chatterjee, N., Fei, Y., Shen, G. (1993): *Thermodynamic Data on Oxides and Silicates.* Springer, Berlin/Heidelberg.

Scarrow, J. H., Curran, J. M., Kerr, A. C. (2000): Major element records of variable plume involvement in the north Atlantic province tertiary flood basalts. *J. Petrol. 41*(7), 1155–1176.

Schilling, J.-G. (1973): Iceland mantle plume: Geochemical study of Reykjanes ridge. *Nature 242*, 565–571.

Schilling, J.-G. (1991): Fluxes and excess temperatures of mantle plumes inferred from their interaction with migrating mid-ocean ridges. *Nature 352*, 397–403.

Schilling, J.-G., Bergeron, M. B., Evans, R. (1980): Halogens in the Mantle Beneath the North Atlantic. *Phil. Trans. Roy. Soc. Lond. A 297*(1431), 147–176.

Schilling, J.-G., Kingsley, R., Fontignie, D., Poreda, R., Xue, S. (1999): Dispersion of the Jan Mayen and Iceland mantle plumes in the Arctic: A He–Pb–Nd–Sr isotope tracer study of basalts from the Kolbeinsey, Mohns, and Knipovich Ridges. *J. Geophys. Res. 104*(B5), 10 543–10 569.

Schlindwein, V. (2001): On the use of teleseismic receiver functions for studying the crustal structure of Iceland. *Geophys. J. Int. submitted.*

Schmeling, H. (1985a): Numerical models on the influence of partial melt on elastic, anelastic, and electric properties of rocks. Part I: elasticity and anelasticity. *Phys. Earth Planet. Int. 41*, 34–57.

Schmeling, H. (1985b): Partial melt below Iceland: A combined interpretation of seismic and conductivity data. *J. Geophys. Res. 90*, 10 105–10 116.

Schmeling, H. (1986): Numerical models on the influence of partial melt on elastic, anelastic, and electric properties of rocks. Part II: electrical conductivity. *Phys. Earth Planet. Int. 43*, 123–136.

Schmeling, H. (2000): Partial melting and melt segregation in a convecting mantle. In *Physics and Chemistry of Partially Molten Rocks*, edited by N. Bagdassarov, D. Laporte, A. B. Thompson, pp. 141–178, Kluwer Academic Publishers.

Schmeling, H., Marquart, G., Ruedas, T. (2003): Pressure- and temperature-dependent thermal expansivity and the effect on mantle convection and surface observables. *Geophys. J. Int., 154*(1), 224–229.

Schubert, G., Sandwell, D. (1989): Crustal volumes of the continents and of oceanic and continental submarine plateaus. *Earth Planet. Sci. Lett. 92*, 234–246.

Schubert, G., Turcotte, D. L., Olson, P. (2001): *Mantle Convection in the Earth and Planets.* Cambridge University Press.

Schwab, B. E., Johnston, A. D. (2001): Melting systematics of modally variable, compositionally intermediate peridotites and the effects of mineral fertility. *J. Petrol. 42*(10), 1789–1811.

Schäfer, F. N., Foley, S. F. (2002): The effect of crystal orientation on the wetting behaviour of silicate melts on the surfaces of spinel peridotite minerals. *Contrib. Mineral. Petrol. 143*(2), 254–262.

Scott, D. R. (1988): The competition between percolation and circulation in a deformable porous medium. *J. Geophys. Res. 93*(B6), 6451–6462.

Scott, D. R. (1992): Small-scale convection and mantle melting beneath mid-ocean ridges. In *Mantle flow and melt generation at mid-ocean ridges*, edited by J. P. Morgan, D. K. Blackman, J. M. Sinton, volume 71 of *Geophys. Monogr. Ser.*, pp. 327–352, AGU, Washington, D.C.

Scott, D. R., Stevenson, D. J. (1984): Magma solitons. *Geophys. Res. Lett. 11*(11), 1161–1164.

Scott, D. R., Stevenson, D. J. (1986): Magma ascent by porous flow. *J. Geophys. Res. 91*(B9), 9283–9296.

Scott, D. R., Stevenson, D. J. (1989): A self-consistent model of melting, magma migration and buoyancy-driven circulation beneath mid-ocean ridges. *J. Geophys. Res. 94*(B3), 2973–2988.

Selby, N. D., Woodhouse, J. H. (2002): The Q structure of the upper mantle: Constraints from Rayleigh wave amplitudes. *J. Geophys. Res. 107*(B5), 2097, doi:10.1029/2001JB000257.

Seyler, M., Toplis, M. J., Lorand, J.-P., Luguet, A., Cannat, M. (2001): Clinopyroxene microtextures reveal incompletely extracted melts in abyssal peridotites. *Geology 29*(2), 155–158.

Shankland, T. J., Duba, A. G. (1990): Standard electrical conductivity of isotropic, homogeneous olivine in the temperature range 1200–1500 °C. *Geophys. J. Int. 103*(1), 25–31.

Shaw, D. M. (1970): Trace element fractionation during anatexis. *Geochim. Cosmochim. Acta 34*, 237–243.

Shearer, P. M., Flanagan, M. P. (1999): Seismic velocity and density jumps across the 410- and 660-kilometer discontinuities. *Science 285*(5433), 1545–1548.

Shen, Y., Forsyth, D. W. (1995): Geochemical constraints on the initial and final depths of melting beneath mid-ocean ridges. *J. Geophys. Res. 100*(B2), 2211–2237.

Shen, Y., Solomon, S. C., Bjarnason, I. Þ., Nolet, G., Morgan, W. J., Allen, R. M., Vogfjörd, K., Jakobsdóttir, S., Stefánsson, R., Julian, B. R., Foulger, G. R. (2002): Seismic evidence for a tilted mantle plume and north-south mantle flow beneath Iceland. *Earth Planet. Sci. Lett. 197*(3–4), 261–272.

Shen, Y., Solomon, S. C., Bjarnason, I. Þ., Wolfe, C. J. (1998): Seismic evidence for a lower mantle origin of the Iceland plume. *Nature 395*, 62–65.

Sigmundsson, F. (1991): Post-glacial rebound and asthenosphere viscosity in Iceland. *Geophys. Res. Lett. 18*(6), 1131–1134.

Sims, K. W. W., DePaolo, D. J., Murrell, M. T., Baldridge, W. S., Goldstein, S. J., Clague, D. A., Jull, M. (1999): Porosity of the melting zone and variations in the solid mantle upwelling rate beneath Hawaii: Inferences from ^{238}U–^{230}Th–^{226}Ra and ^{235}U–^{231}Pa disequilibria. *Geochim. Cosmochim. Acta 63*(23/24), 4119–4138.

Sims, K. W. W., Goldstein, S. J., Blichert-Toft, J., Perfit, M. R., Kelemen, P., Fornari, D. J., Michael, P., Murrell, M. T., Hart, S. R., DePaolo, D. J., Layne, G., Ball, L., Jull, M., Bender, J. (2002): Chemical and isotopic constraints on the generation and transport of magma beneath the East Pacific Rise. *Geochim. Cosmochim. Acta 66*(19), 3481–3504.

Sinha, M. C., Constable, S., Peirce, C., White, A., Heinson, G., MacGregor, L., Navin, D. A. (1998): Magmatic processes at slow spreading ridges: implications of the RAMESSES experiment at 57° 45'N on the Mid-Atlantic Ridge. *Geophys. J. Int. 135*(3), 731–745.

Skogseid, J., Pedersen, T., Eldholm, O., Larsen, B. T. (1992): Tectonism and magmatism during NE Atlantic continental break-up: the Vøring Margin. In *Magmatism and the Causes of Continental Break-up*, edited by B. C. Storey, T. Alabaster, R. J. Pankhurst, number 68 in Geological Society Special Publication, pp. 305–320, The Geological Society, London.

Skovgaard, A. C., Storey, M., Baker, J., Blusztajn, J., Hart, S. R. (2001): Osmium–oxygen isotopic evidence for a recycled and strongly depleted component in the Iceland mantle plume. *Earth Planet. Sci. Lett. 194*(1–2), 259–275.

Slater, L., McKenzie, D., Grönvold, K., Shimizu, N. (2001): Melt Generation and Movement beneath Theistareykir, NE Iceland. *J. Petrol. 42*(2), 321–354.

Sleep, N. H. (1984): Tapping of magmas from ubiquitous mantle heterogeneities: An alternative to mantle plumes? *J. Geophys. Res. 89*(B12), 10029–10041.

Sleep, N. H. (1988): Tapping of melt by veins and dikes. *J. Geophys. Res. 93*(B9), 10255–10272.

Sleep, N. H. (1990): Hotspots and mantle plumes: Some phenomenology. *J. Geophys. Res. 95*(B5), 6715–6736.

Sleep, N. H. (1996): Lateral flow of hot plume material ponded at sublithospheric depths. *J. Geophys. Res. 101*(B12), 28065–28083.

Small, C., Sandwell, D. (1989): An abrupt change in ridge-axis gravity with spreading rate. *J. Geophys. Res. 94*(B12), 17383–17392.

Smallwood, J. R., Staples, R. K., Richardson, K. R., White, R. S., the FIRE Working Group (1999): Crust generated above the Iceland mantle plume: From continental rift to oceanic spreading center. *J. Geophys. Res. 104*(B10), 22885–22902.

Smallwood, J. R., White, R. S., Minshull, T. A. (1995): Sea-floor spreading in the presence of the Iceland Plume: the structure of the Reykjanes Ridge at 61° 40' N. *J. Geol. Soc. Lond.* *152*, 1023–1029.

Smallwood, J. R., White, R. S., Staples, R. K. (1998): Deep crustal reflectors under Reydarfjörður, eastern Iceland: crustal accretion above the Iceland mantle plume. *Geophys. J. Int.* *134*(1), 277–290.

Smyth, J. R., Frost, D. J. (2002): The effect of water on the 410-km discontinuity: An experimental study. *Geophys. Res. Lett.* *29*(10), 10.1029/2001GL014 418.

Smyth, J. R., McCormick, T. C. (1995): Crystallographic Data for Minerals. In *Mineral Physics & Crystallography — A Handbook of Physical Constants*, edited by T. J. Ahrens, volume 2 of *AGU Reference Shelf*, pp. 1–17, AGU, Washington.

Sotin, C., Parmentier, E. M. (1989): Dynamical consequences of compositional and thermal density stratification beneath spreading centers. *Geophys. Res. Lett.* *16*(8), 835–838.

Sparks, D. W., Parmentier, E. M. (1991): Melt extraction from the mantle beneath spreading centers. *Earth Planet. Sci. Lett.* *105*, 368–377.

Sparks, D. W., Parmentier, E. M. (1994): The generation and migration of partial melt beneath ocean spreading centers. In *Magmatic Systems*, edited by M. P. Ryan, pp. 55–75, Academic Press.

Spiegelman, M. (1993a): Flow in deformable porous media. Part 1: Simple analysis. *J. Fluid Mech.* *247*, 17–38.

Spiegelman, M. (1993b): Physics of melt extraction: theory, implications and applications. *Phil. Trans. R. Soc. Lond. A* *342*, 23–41.

Spiegelman, M. (2000): Myths & Methods in Modeling. Lamont–Doherty Earth Observatory, Columbia University, New York.

Spiegelman, M., Elliott, T. (1993): Consequences of melt transport for uranium series disequilibrium. *Earth Planet. Sci. Lett.* *118*, 1–20.

Spiegelman, M., Kelemen, P. B., Aharonov, E. (2001): Causes and consequences of flow concentration during melt transport: The reaction infiltration instability in compactible media. *J. Geophys. Res.* *106*(B2), 2061–2077.

Spiegelman, M., McKenzie, D. (1987): Simple 2-D models for melt extraction at mid-ocean ridges and island arcs. *Earth Planet. Sci. Lett.* *83*, 137–152.

Stacey, F. D. (1998): Thermoelasticity of a mineral composite and a reconsideration of lower mantle properties. *Phys. Earth Planet. Int.* *106*, 219–236.

Stacey, F. D. (2000): The K-primed approach to high-pressure equations of state. *Geophys. J. Int.* *143*(3), 621–628.

Stacey, F. D., Isaak, D. G. (2001): Compositional constraints on the equations of state and thermal properties of the lower mantle. *Geophys. J. Int.* *146*(1), 143–154.

Staniforth, A., Côté, J. (1991): Semi-Lagrangian integration schemes for atmospheric models – a review. *Mon. Weather Rev.* *119*(9), 2206–2223.

Staples, R. K., White, R. S., Brandsdóttir, B., Menke, W., Maguire, P. K. H., McBride, J. H. (1997): Färoe-Iceland Ridge Experiment, 1, Crustal structure of northeastern Iceland. *J. Geophys. Res.* *102*(B4), 7849–7866.

Stecher, O., Carlson, R. W., Gunnarsson, B. (1999): Torfajökull: a radiogenic end-member of the Iceland Pb-isotopic array. *Earth Planet. Sci. Lett.* *165*(1), 117–127.

Stefánsson, R., Böðvarsson, P., Slunga, R., Einarsson, P., Jakobsdóttir, S., Bungum, H., Gregersen, S., Havskov, J., Hjelme, J., Korhonen, H. (1993): Earthquake prediction research in the South Iceland Seismic Zone and the SIL project. *Bull. Seis. Soc. Am.* *83*, 696–716.

Steinberger, B. (2000): Plumes in a convecting mantle: Models and observations for individual hotspots. *J. Geophys. Res. 105*(B5), 11 127–11 152.

Stevenson, D. J. (1989): Spontaneous small-scale melt segregation in partial melts undergoing deformation. *Geophys. Res. Lett. 16*(9), 1067–1070.

Stixrude, L. (1997): Structure and sharpness of phase transitions and mantle discontinuities. *J. Geophys. Res. 102*(B7), 14 835–14 852.

Stixrude, L., Bukowinski, M. S. T. (1990): Fundamental thermodynamic relations and silicate melting with implications for the constitution of D″. *J. Geophys. Res. 95*(B12), 19 311–19 325.

Stolper, E., Walker, D., Hager, B. H., Hays, J. F. (1981): Melt segregation from partially molten source regions: the importance of melt density and source region size. *J. Geophys. Res. 86*(B7), 6261–6271.

Stracke, A., Zindler, A., Salters, V. J. M., McKenzie, D., Blichert-Toft, J., Albarède, F., Grönvold, K. (2003): Theistareykir revisited. *Geochem. Geophys. Geosyst. 4*(2), 8507, doi: 10.1029/2001GC000201.

Su, W., Buck, W. R. (1993): Buoyancy effects on mantle flow under mid-ocean ridges. *J. Geophys. Res. 98*(B7), 12 191–12 205.

Suhr, G. (1999): Melt migration under oceanic ridges: Inferences from reactive transport modelling of upper mantle hosted dunites. *J. Petrol. 40*(4), 575–599.

Suhr, G., Hellebrand, E., Snow, J. E., Seck, H. A., Hofmann, A. W. (2003): Significance of large, refractory dunite bodies in the upper mantle of the Bay of Islands Ophiolite. *Geochem. Geophys. Geosyst. 4*(3), 8605, doi:10.1029/2001GC000277.

Takahashi, E. (1986): Melting of a dry peridotite KLB-1 up to 14 GPa: Implications on the origin of peridotitic upper mantle. *J. Geophys. Res. 91*(B9), 9367–9382.

Takahashi, E., Scarfe, C. M. (1985): Melting of peridotite to 14 GPa and the genesis of komatiite. *Nature 315*, 566–568.

Takahashi, E., Shimazaki, T., Tsuzaki, Y., Yoshida, H. (1993): Melting study of a peridotite KLB-1 to 6.5 GPa, and the origin of basaltic magmas. *Phil. Trans. Roy. Soc. Lond. A 342*, 105–120.

Takei, Y. (2002): Effect of pore geometry on V_P/V_S: From equilibrium geometry to crack. *J. Geophys. Res. 107*(B2), 10.1029/2001JB000 522.

Tan, B. H., Jackson, I., Fitz Gerald, J. D. (2001): High-temperature viscoelasticity of fine-grained polycrystalline olivine. *Phys. Chem. Minerals 28*, 641–664.

Tangeman, J. A., Phillips, B. L., Navrotsky, A., Weber, J. K. R., Hixson, A. D., Key, T. S. (2001): Vitreous forsterite (Mg_2SiO_4): Synthesis, structure, and thermochemistry. *Geophys. Res. Lett. 28*(13), 2517–2520.

Taniguchi, H. (1995): Universal viscosity-equation for silicate melts over wide temperature and pressure ranges. *J. Volc. Geotherm. Res. 66*(1–4), 1–8.

Tanimoto, T., Stevenson, D. J. (1994): Seismic constraints on a model of partial melts under ridge axes. *J. Geophys. Res. 99*(B3), 4549–4558.

Taylor, R. N., Thirlwall, M. F., Murton, B. J., Hilton, D. R., Gee, M. A. M. (1997): Isotopic constraints on the influence of the Icelandic plume. *Earth Planet. Sci. Lett. 148*, E1–E8. Express Letter.

Temperton, C., Staniforth, A. (1987): An efficient two-time-level semi-Lagrangian semi-implicit integration scheme. *Q. J. R. Meteorol. Soc. 113*, 1025–1039.

Thayer, R. E., Björnsson, A., Álvarez, L., Hermance, J. F. (1981): Magma genesis and crustal spreading in the northern neovolcanic zone of Iceland: telluric-magnetotelluric constraints. *Geophys. J. R. astr. Soc. 65*, 423–442.

Thirlwall, M. F. (1995): Generation of the Pb isotopic characteristics of the Iceland plume. *J. Geol. Soc. Lond. 152*, 991–996.

Thompson, A. B. (1992): Water in the Earth's upper mantle. *Nature 358*, 295–302.

Thompson, P. F., Tackley, P. J. (1998): Generation of mega-plumes from the core-mantle boundary in a compressible mantle with temperature-dependent viscosity. *Geophys. Res. Lett. 25*(11), 1999–2002.

Toplis, M. J., Richet, P. (2000): Equilibrium density and expansivity of silicate melts in the glass transition range. *Contrib. Mineral. Petrol. 139*, 672–683.

Toramaru, A., Fujii, N. (1986): Connectivity of the melt phase in partially molten peridotite. *J. Geophys. Res. 91*(B9), 9239–9252.

Torsvik, T. H., Mosar, J., Eide, E. A. (2001): Cretaceous–Tertiary geodynamics: a North Atlantic exercise. *Geophys. J. Int. 146*(3), 850–866.

Tryggvason, A., Rögnvaldsson, S. Þ., Flóvenz, Ó. G. (2002): Three-dimensional imaging of the *P*- and *S*-wave velocity structure and earthquake locations beneath Southwest Iceland. *Geophys. J. Int. 151*(3), 848–866.

Tryggvason, K., Husebye, E., Stefánsson, R. (1983): Seismic image of the hypothesized Icelandic hot spot. *Tectonophysics 100*, 97–118.

Turcotte, D. L., Phipps Morgan, J. (1992): The physics of magma migration and mantle flow beneath a mid-ocean ridge. In *Mantle flow and melt generation at mid-ocean ridges*, edited by J. Phipps Morgan, D. K. Blackman, J. M. Sinton, number 71 in Geophysical Monographs, pp. 155–182, AGU.

Turcotte, D. L., Schubert, G. (1982): *Geodynamics — Applications of Continuum Physics to Geological Problems.* John Wiley & Sons.

Tyburczy, J. A., Fisler, D. K. (1995): Electrical properties for minerals and melts. In *Mineral Physics & Crystallography — A Handbook of Physical Constants*, edited by T. J. Ahrens, volume 2 of *AGU Reference Shelf*, pp. 185–208, AGU, Washington.

van Keken, P. (1997): Evolution of starting mantle plumes: a comparison between numerical and laboratory models. *Earth Planet. Sci. Lett. 148*, 1–11.

van Keken, P. E., Gable, C. W. (1995): The interaction of a plume with a rheological boundary: A comparison between two- and three-dimensional models. *J. Geophys. Res. 100*(B10), 20 291–20 302.

van Westrenen, W., Wood, B. J., Blundy, J. D. (2001): A predictive thermodynamic model of garnet–melt trace element partitioning. *Contrib. Mineral. Petrol. 142*, 219–234.

Vasilyev, O. V., Podlachikov, Y. Y., Yuen, D. A. (1998): Modeling of compaction driven flow in poro-viscoelastic medium using adaptive wavelet collocation method. *Geophys. Res. Lett. 25*(17), 3239–3242.

Vasilyev, O. V., Podlachikov, Y. Y., Yuen, D. A. (2001): Modelling of plume–lithosphere interaction using the adaptive multilevel wavelet collocation algorithm. *Geophys. J. Int. 147*(3), 579–589.

Verhoogen, J. (1965): Phase changes and convection in the Earth's mantle. *Phil. Trans. Roy. Soc. Lond. A 258*(1088), 276–283.

Vinet, P., Ferrante, J., Rose, J. H., Smith, J. R. (1987): Compressibility of solids. *J. Geophys. Res. 92*, 9319–9325.

Vink, G. E. (1984): A hotspot model for Iceland and the Vøring plateau. *J. Geophys. Res. 89*(B12), 9949–9959.

Vogt, P. R. (1971): Asthenospheric motion recorded by the ocean floor south of Iceland. *Earth Planet. Sci. Lett. 13*, 155–160.

Vogt, P. R. (1974): The Iceland phenomenon: imprints of a hot spot on the ocean crust, and implications for flow below the plates. In *Geodynamics of Iceland and the North Atlantic area*, edited by L. Kristjánsson, volume 11 of *NATO Advanced Study Institutes Series C*, pp. 105–126, D. Reidel Publishing Company, Dordrecht/Boston.

Vogt, P. R. (1976): Plumes, sub-axial pipe flow, and topography along the mid-oceanic ridge. *Earth Planet. Sci. Lett. 29*, 309–325.

von Bargen, N., Waff, H. S. (1986): Permeabilities, interfacial areas and curvatures of partially molten systems: results of numerical computations of equilibrium microstructures. *J. Geophys. Res. 91*(B9), 9261–9276.

von Bargen, N., Waff, H. S. (1988): Wetting of enstatite by basaltic melt at 1350 °C and 1.0 to 2.5 GPa pressure. *J. Geophys. Res. 93*(B2), 1153–1158.

Waff, H. S., Bulau, J. R. (1979): Equilibrium fluid distribution in an ultramafic partial melt under hydrostatic stress conditions. *J. Geophys. Res. 84*, 6109–6114.

Waff, H. S., Faul, U. H. (1992): Effects of crystalline anisotropy on fluid distribution in ultramafic partial melts. *J. Geophys. Res. 97*, 9003–9014.

Walker, D., Agee, C. B., Zhang, Y. (1988): Fusion curve slope and crystal/liquid buoyancy. *J. Geophys. Res. 93*(B1), 313–323. Correction in *J. Geophys. Res.* 93(B6), p.6668.

Wallace, P. J. (1998): Water and partial melting in mantle plumes: Inferences from the dissolved H_2O concentrations of Hawaiian basaltic magmas. *Geophys. Res. Lett. 25*(19), 3639–3642.

Walter, M., Katsura, T., Kubo, A., Shinmei, T., Nishikawa, O., Ito, E., Lesher, C., Funakoshi, K. (2002): Spinel–garnet lherzolite transition in the system $CaO–MgO–Al_2O_3–SiO_2$ revisited: an in situ X-ray study. *Geochim. Cosmochim. Acta 66*(12), 2109–2121.

Walter, M. J. (1998): Melting of garnet peridotite and the origin of komatiite and depleted lithosphere. *J. Petrol. 39*(1), 29–60.

Walter, M. J., Presnall, D. C. (1994): Melting behaviour of simplified lherzolite in the system $CaO–MgO–Al_2O_3–SiO_2–Na_2O$ from 7 to 35 kbar. *J. Petrol. 35*(2), 329–359.

Wark, D. A., Williams, C. A., Watson, E. B., Price, J. D. (2003): Reassessment of pore shapes in microstructurally equilibrated rocks, with implications for permeability of the upper mantle. *J. Geophys. Res. 108*(B1), 2050, doi:10.1029/2001JB001575.

Weidner, D. J., Wang, Y. (2000): Phase Transformations: Implications for Mantle Structure. In *Earth's Deep Interior – Mineral Physics and Tomography. From the Atomic to the Global Scale*, edited by S.-I. Karato, A. Forte, R. Liebermann, G. Masters, L. Stixrude, volume 117 of *Geophys. Monogr. Ser.*, pp. 215–236, AGU, Washington, D.C.

Weir, N. R. W., White, R. S., Brandsdóttir, B., Einarsson, P., Shimamura, H., Shiobara, H., RISE Fieldwork Team (2001): Crustal structure of the northern Reykjanes Ridge and Reykjanes Peninsula, southwest Iceland. *J. Geophys. Res. 106*(B4), 6347–6368.

Wessel, P., Smith, W. H. F. (1998): New, improved version of the Generic Mapping Tools released. *EOS Transactions, AGU 79*, 579.

Wheeler, J. (1992): Importance of pressure solution and Coble creep in the deformation of polymineralic rocks. *J. Geophys. Res. 97*(B4), 4579–4586.

White, R. S. (1993): Melt production rates in mantle plumes. *Phil. Trans. Roy. Soc. Lond. A 342*, 137–153.

White, R. S. (1997): Rift-plume interactions in the North Atlantic. *Phil. Trans. Roy. Soc. Lond. A 355*(1723), 319–339.

White, R. S., Bown, J. W., Smallwood, J. R. (1995): The temperature of the Iceland plume and origin of outward-propagating V-shaped ridges. *J. Geol. Soc. Lond. 152*, 1039–1045.

White, R. S., McKenzie, D. P. (1989): Magmatism at rift zones: The generation of volcanic continental margins and flood basalts. *J. Geophys. Res.* *94*(B6), 7685–7729.

White, R. S., McKenzie, D. P. (1995): Mantle plumes and flood basalts. *J. Geophys. Res.* *100*(B9), 17 543–17 585.

White, R. S., McKenzie, D. P., O'Nions, R. K. (1992): Oceanic crustal thickness from seismic measurements and rare earth element inversions. *J. Geophys. Res.* *97*(B13), 19 683–19 715.

White, R. S., Minshull, T. A., Bickle, M. J., Robinson, C. J. (2001): Melt generation at very slow-spreading oceanic ridges: Constraints from geochemical and geophysical data. *J. Petrol.* *42*(6), 1171–1196.

Whitehead, J. A., Luther, D. S. (1975): Dynamics of laboratory diapir and plume models. *J. Geophys. Res.* *80*, 705–717.

Williams, T., Kelley, C., *et al.* (1986): gnuplot — An Interactive Plotting Program. http://www.gnuplot.info/.

Wilson, M. (1989): *Igneous Petrogenesis — A global tectonic approach.* Unwin & Hyman, Ltd., London.

Wolfe, C. J., Bjarnason, I. Þ., VanDecar, J., Solomon, S. (1997): Seismic structure of the Iceland mantle plume. *Nature* *385*(5727), 245–247.

Wolfe, C. J., Bjarnason, I. Þ., VanDecar, J., Solomon, S. (2002): Assessing the depth resolution of tomographic models of upper mantle structure beneath Iceland. *Geophys. Res. Lett.* *29*(2), 10.1029/2001GL013 657.

Wood, B. J. (1995): The effect of H_2O on the 410-kilometer seismic discontinuity. *Science* *268*, 75–76.

Wood, B. J., Blundy, J. D. (1997): A predictive model for rare earth element partitioning between clinopyroxene and anhydrous silicate melt. *Contrib. Mineral. Petrol.* *129*(2/3), 166–181.

Wood, B. J., Blundy, J. D., Robinson, J. A. C. (1999): The role of clinopyroxene in generating U-series disequilibrium during mantle melting. *Geochim. Cosmochim. Acta* *63*(10), 1613–1620.

Xu, Y., Shankland, T. J., Poe, B. T. (2000): Laboratory-based electrical conductivity in the Earth's mantle. *J. Geophys. Res.* *105*(B12), 27 865–27 875. Correction in *J. Geophys. Res.* 108(B6), 2314.

Yale, M. M., Phipps Morgan, J. (1998): Asthenosphere flow model of hotspot-ridge interactions: a comparison of Iceland and Kerguelen. *Earth Planet. Sci. Lett.* *161*, 45–56.

Yasuda, A., Fujii, T. (1998): Ascending subducted oceanic crust entrained within mantle plumes. *Geophys. Res. Lett.* *25*(10), 1561–1564.

Yaxley, G. M. (2000): Experimental study of the phase and melting relations of homogeneous basalt + peridotite mixtures and implications for the petrogenesis of flood basalts. *Geochim. Cosmochim. Acta* *139*(3), 326–338.

Yusa, H., Akaogi, M., Ito, E. (1993): Calorimetric study of $MgSiO_3$ garnet and pyroxene: Heat capacities, transition enthalpies, and equilibrium phase relations in $MgSiO_3$ at high pressures and temperatures. *J. Geophys. Res.* *98*(B4), 6453–6460.

Zhang, J., Herzberg, C. (1994): Melting experiments on anhydrous peridotite KLB-1 from 5.0 to 22.5 GPa. *J. Geophys. Res.* *99*(B9), 17 729–17 742.

Zhang, Y.-S., Tanimoto, T. (1993): High-resolution global upper mantle structure and plate tectonics. *J. Geophys. Res.* *98*(B6), 9793–9823.

Zhao, D. (2001): Seismic structure and origin of hotspots and mantle plumes. *Earth Planet. Sci. Lett.* *192*(3), 251–265.

Ziegler, D., Navrotsky, A. (1986): Direct measurement of the enthalpy of fusion of diopside. *Geochim. Cosmochim. Acta 50*, 2461–2466.

Zimmerman, M. E., Zhang, S., Kohlstedt, D. L., Karato, S.-I. (1999): Melt distribution in mantle rocks deformed in shear. *Geophys. Res. Lett. 26*(10), 1505–1508.

Zverev, S. M., Litvinenko, I. V., Pálmason, G., Yaroshevskaya, G. A., Osokin, N. N. (1980a): A seismic crustal study of the axial rift zone in Southwest Iceland. *J. Geophys. 47*, 202–210.

Zverev, S. M., Litvinenko, I. V., Pálmason, G., Yaroshevskaya, G. A., Osokin, N. N., Akhmetjev, M. A. (1980b): A seismic study of the rift zone in Northern Iceland. *J. Geophys. 47*, 191–201.

SOFTWARE

The convection/melt models of this study have been calculated with a code created by combining and further developping the 2D/3D convection program DECO3D by G. Marquart and several melting subroutines from the 2D convection/melting program FDCON by H. Schmeling. In the further development of the code, the linear algebra package LAPACK (Anderson *et al.*, 1999), some semi-Lagrange subroutines by Spiegelman (2000), and, for the computation of the error function and the determination of the roots of a function, the SLATEC/FNLIB and NAPACK libraries from Netlib (`www.netlib.org`) have been used; the original Fourier transform routines, which were limited to power-of-two array lengths, have been replaced by the FFTW routines (Frigo and Johnson, 1998). The model runs were performed on PCs running Linux. The code was compiled with the Intel® Fortran compiler (v.5 and 6).

All function plots (except figure III.4) were generated with the Gnuplot plotting software (Williams *et al.*, 1986), which has also been used to compute some data fits (see appendix C); additional fits were done with the STARPAC package (Donaldson and Tryon, 1990). The three-dimensional slice images, surface plots, and contour plots were generated with MATLAB®. The map plots of figures IV.1 and IV.2 have been made with the GMT mapping software (Wessel and Smith, 1998), making use of the frontend iGMT (Becker and Braun, 1998). Figures II.4 and IV.1 were created resp. edited with the vector drawing programs Xfig and Sketch. Figure V.46 was generated with the ray-tracing software POV-Ray. Some of the images of chapter IV and figure VI.2 were processed with The Gimp or xpdf.

Typesetting of the text was done with LaTeX and several additional macro packages.

Acknowledgments

> *If I have been able to see further, it*
> *was only because I stood on the*
> *shoulders of giants.*
> —Isaac Newton

I am very grateful to my advisers Harro Schmeling (University of Frankfurt am Main) and Gabriele Marquart (University of Frankfurt am Main/University of Utrecht) for giving me the opportunity to work in an interesting research project and finally get into a field of geophysics which has attracted me for quite a long time. For me as a newcomer in this field it was especially important and valuable to have two advisers who did not only give me much freedom in my work, but would also always take time to discuss problems and answer my questions with great patience. I am also grateful to G. Marquart for conducting the benchmark comparisons mentioned in sections III.1.2 and III.1.3, which would have been difficult with the code version used in this study, and for calling my attention to some critical points of the method in the latter.

I thank Wolfgang Jacoby (University of Mainz) for taking upon himself the burden of being the co-reviewer of this thesis, in spite of already being retired.

In several occasions, Nikolai Bagdassarov (University of Frankfurt/Main) has been helpful in explaining me rock-physical issues. Andreas Junge (University of Frankfurt/Main) took his time to discuss some statistical problems with me. The Iceland-specific aspects of this study profited from many discussions in the meetings of the Frankfurt–Mainz Iceland research group.

Advice from Gerhard Brey and Thomas Stachel (University of Frankfurt/Main) and from Dimitris Kostopoulos (University of Thessaloniki) has improved my understanding of the importance of rock chemistry and petrology for the melting history of peridotite.

Discussions with Ingi Bjarnason (Science Institute, Reykjavík) have been helpful in considering the tomographic results of ICEMELT and assessing the effect of partial melts on seismic velocities. He also has kindly shared results of his surface wave analysis during his stays in Frankfurt in May 2000 and May–August 2002.

Section V.2.1 is derived from an earlier draft of a now published paper by Ruedas *et al.* (2004); valuable comments by an anonymous reviewer of that paper stimulated the more detailed discussion of buoyancy effects, which has also been adopted for this study.

I am very grateful to Richard Katz (Lamont–Doherty Earth Observatory, Palisades) for giving me access to his paper on his melting parameterization at a very early stage. Ingi Þ. Bjarnason, Vera Schlindwein (University of Durham/AWI Bremerhaven), Dieter Mertz (University of Mainz) and Ruth Ziethe/Doris Breuer (University of Münster) also generously shared preprints of manuscripts.

Gábor Ország (University of Frankfurt/Main) kindly lent me a collection of papers on semi-Lagrangian methods, thereby sparing me a lot of time of searching for this material in libraries.

A large part of this work has been done using freely available open-source software like Linux, TeX, GNU software etc. I would like to express my gratitude to those many programmers who spent and are still spending lots of their time – in general their spare time – to write and maintain this software.

Several persons at the newsgroups `comp.text.tex`, `sci.math.num-analysis`, `comp.graphics.apps.gnuplot` and `comp.lang.fortran` gave helpful advice and pointers concerning technical problems.

The friendly and cooperative atmosphere established by my colleagues at the Institute of Geophysics in Frankfurt, and in particular in the geodynamics/rock physics group, have provided me with an excellent environment for completing this thesis. I also appreciate the good cooperation with Ralph Strobl in many computer system-related issues.

The image of the old map of the northern Atlantic area at the beginning of the book is a reproduction of the historic copy in Jónas Kristjánsson: *Icelandic Manuscripts: Sagas, History and Art* (Icelandic Literary Society, 1993) and was provided in digital form on the website of Eysteinn Björnsson (Reykjavík), who kindly also gave me the bibliographic information.

Last, but not least, I thank my parents for their continuous support which enabled me to study geophysics, to take my time looking for an interesting PhD opportunity afterwards, and to finish this work.

Most of this work was funded by the Deutsche Forschungsgemeinschaft (DFG) in the frame of the Iceland Plume Dynamics Project at the universities of Frankfurt am Main and Mainz, under grants Schm-672/1 and Schm-672/2. Additional financial support for participating in conferences from the sponsors of the 7th Workshop on Numerical Modeling of Mantle Convection and Lithospheric Dynamics in Aussois (France) in 2001, of the 8th European Workshop on Numerical Modeling of Mantle Convection and Lithospheric Dynamics in Hrubá Skála (Czech Republic) in 2003, from the Hermann Willkomm foundation, and from the Deutsche Geophysikalische Gesellschaft is also acknowledged.

NOTE ON COPYRIGHTED MATERIAL

THOMAS RUEDAS

2nd June, 1971	born in Hamburg, Germany, to Sophie-Luise and Gorgonio Ruedas
09/1977–07/1981	primary school in Ellerbek (district of Pinneberg)
09/1981–06/1990	Gymnasium am Bondenwald in Hamburg, Abitur
07/1990–09/1991	community service
10/1991–09/1997	studies of geophysics at the University of Hamburg Vordiplom: 12th October, 1993; Diplom: 26th September, 1997 Diploma thesis: "Entwicklung eines Programms zur zweidimensionalen seismischen Modellierung für transversalisotrope Medien mit einer Tschebyscheff–Fourier-Methode unter Einbeziehung der freien Oberfläche mit Topographie" ("Development of a program for two-dimensional seismic modelling for transversely isotropic media with a Chebyshev–Fourier method, incorporating the free surface with topography") (Adviser: Dr. Ekkehart Teßmer)
10/1997–03/1998	unemployed; research associated with the Institut für Geophysik of the University of Hamburg
04/1998–08/2003	research associate/PhD student in the geodynamics group (leader: Prof. Dr. Harro Schmeling) at the Institut für Meteorologie und Geophysik of the Johann Wolfgang Goethe University, Frankfurt am Main
since 09/2003	project researcher at the Danish Lithosphere Centre, Copenhagen, Denmark